Genetically Modified Diplomacy

Peter Andrée

Genetically Modified Diplomacy: The Global Politics of Agricultural Biotechnology and the Environment

UBCPress · Vancouver · Toronto

16 15 14 13 12 11 10 09 08 07   5 4 3 2 1

Printed in Canada on ancient-forest-free paper (100% post-consumer recycled) that is processed chlorine- and acid-free, with vegetable-based inks.

Library and Archives Canada Cataloguing in Publication

Andrée, Peter
Genetically modified diplomacy : the global politics of agricultural biotechnology and the environment / Peter Andrée.

Includes bibliographical references and index.
ISBN 978-0-7748-1268-9 (bound); 978-0-7748-1269-6 (pbk.)

1. Convention on Biological Diversity (1992). Protocols, etc., 2000 Jan. 29.
2. Agricultural biotechnology – Law and legislation. 3. Transgenic organisms.
4. Environmental law, International. 5. Precautionary principle.
I. Title.

S494.5.B563A53 2007   343'.0786606   C2007-901544-1

Canadä

UBC Press gratefully acknowledges the financial support for our publishing program of the Government of Canada through the Book Publishing Industry Development Program (BPIDP), and of the Canada Council for the Arts, and the British Columbia Arts Council.

This book has been published with the help of a grant from the Canadian Federation for the Humanities and Social Sciences, through the Aid to Scholarly Publications Programme, using funds provided by the Social Sciences and Humanities Research Council of Canada.

Printed and bound in Canada by Friesens
Set in Stone by Artegraphica Design Co. Ltd.
Copy editor: Judy Phillips
Proofreader: Jillian Shoichet
Indexer: David Luljak, Third Floor Research & Reference, Inc.

UBC Press
The University of British Columbia
2029 West Mall
Vancouver, BC V6T 1Z2
604-822-5959 / Fax: 604-822-6083
www.ubcpress.ca

Contents

Preface

I first became interested in the topic of genetically engineered crops in the mid-1990s when I was a graduate student, first at Trent University in Peterborough, Ontario, and then at York University in Toronto. It was a topic that came up from time to time while I was engaged in participatory research with local farmers on strategies they were adopting to make agriculture more sustainable. Some of the producers I spoke with saw genetic engineering as a way of achieving biological solutions to pest problems where agri-chemicals seemed to cause only more problems. Others were deeply suspicious of the new genetically engineered canola and soybeans, the first varieties of which were just being commercialized in Canada. As I learned about both the potential benefits and drawbacks of genetic engineering in agriculture, I found myself thinking less about the new organisms themselves and more about how society deals with this complex issue. I wanted to understand how we, collectively, make decisions about whether or not to use a new technology such as genetically engineered organisms in agriculture, given the complex array of facts and opinions out there about the consequences of those choices. I also wanted to know why the decision to pursue the use of genetically engineered crops in Canada, and to treat them as equivalent to non-engineered varieties (thereby evading more stringent regulatory scrutiny), appeared to have already been taken by the federal government long before most Canadians even started thinking much about them. This book is the product of eight years of research and critical thinking about genetic engineering in agriculture, about the North American policy direction on biotechnology, and about how the world community has responded to the biotech revolution in food and farming brought to it at the close of the twentieth century.

As an environmentalist, I sometimes feel that the last eight years of my research career were a major diversion. Genetically engineered organisms have not proven to be the environmental calamity that some were predicting. "Superweeds" with genetically engineered traits giving them resistance to multiple brands of herbicides are a problem in some Canadian Prairie communities, but they have not choked out native wild plants in the way that some exotic (and not genetically engineered) introductions have. However, these consequences may yet come to pass if strong regulatory approaches are not adopted in countries around the world. And, genetic engineering continues to raise a number of other major concerns, because its most widespread applications tend to exacerbate the worst features of the food system: corporate concentration, monocultures, chemical dependency, and short-term thinking about long-term problems. Genetic engineering may have a place in sustainable agriculture, but we have yet to see this potential realized on a large scale.

The main focus of this book, the Cartagena Protocol on Biosafety, is an international treaty that reflects a different way of thinking about genetic engineering than has been the norm in North America. This way of thinking is summed up in the idea of a "precautionary" approach. While precaution may not hold all the keys to a more considered treatment of genetically engineered organisms, its entrenchment in a treaty that even Canada has signed onto is a sign that we, collectively, may be learning to treat these organisms with more forethought and care in the twenty-first century than we did in the previous one.

Acknowledgments

I would like to express my sincere thanks to the following for helping to bring this project to fruition: those individuals who gave of their time and expertise when approached for interviews; the Canadian Institute for Environmental Law and Policy for access to its Biosafety Protocol files; Bradford Duplisea of the Canadian Health Coalition for documents he acquired under Canada's Access to Information Act; Cate Sandilands of York University for extensive critical feedback; David Holdsworth, Leesa Fawcett, Stewart Schoenfeld, Stephen Gill, Neil Evernden, Mora Campbell, Éric Darier, and Ray Bazowski for their comments on earlier versions of this work; three anonymous reviewers of this manuscript through UBC Press; Bob Paehlke, Jennifer Clapp, Stephen Bocking, John Wadland, Deb Mills, and the rest of my colleagues at Trent University; and especially Leslie McGrath for her incredible patience. This research was made possible through funding provided by the Social Sciences and Humanities Research Council of Canada. My thanks, as well, to the School of Geography and Environmental Sciences at Monash University, Melbourne, for support during the final stages of writing this book.

Parts of Chapter 3 on risks, hazards, and the regulation of novel plants; parts of Chapter 7 on the implications of precaution in Canada; and the description of Foucault's theoretical approach in Chapter 1 are substantially revised from a previously published paper, "The Biopolitics of Genetically Modified Organisms in Canada," *Journal of Canadian Studies* 37(3): 162-91. Sections of Chapter 2 and the description of Gramsci's theoretical approach in Chapter 1 were previously published in "The Genetic Engineering Revolution in Agriculture and Food: Strategies of the 'Biotech Bloc'" in *Business and Global Environmental Governance*, David Levy and Peter Newell, eds., Cambridge, MA: MIT Press, 2005, 135-66. A paper on the implications of a precautionary Biosafety Protocol (material presented in Chapter 7) was published in *Global Environmental Politics* 5(4): 25-46.

Frequently Used Abbreviations

AAFC	Agriculture and Agri-Food Canada
AIA	advance informed agreement
BCH	Biosafety Clearing-House
BIO	Biotechnology Industry Organization
BPAG	Biosafety Protocol Advisory Group (Canada)
BSE	Bovine Spongiform Encephalopathy
Bt	*Bacillus thuringiensis*
BSWG-*n*	*n*th meeting of the Open-Ended Ad Hoc Working Group on Biosafety
BWG	Biotechnology Working Group
CBD	Convention on Biological Diversity
CEE	Central and Eastern European
CEN	Canadian Environment Network
CFIA	Canadian Food Inspection Agency
COP-MOP-*n*	*n*th meeting of the Conference of the Parties to the CBD serving as the Meeting of the Parties to the Cartagena Protocol on Biosafety
COP-*n*	*n*th meeting of the Conference of the Parties to the CBD
CRG	Council for Responsible Genetics
CSO	civil society organization
DFAIT	Department of Foreign Affairs and International Trade (Canada)
EEC	European Economic Community
ERA	Environmental Risk Assessment
EU	European Union
ExCOP	Extraordinary Meeting of the Conference of the Parties to the CBD
FAO	Food and Agriculture Organization of the United Nations
FBCN	Food Biotechnology Communications Network
FDA	Food and Drug Administration (US)

G-77	Group of 77 nations of the developing world
GE	genetic engineering
GEF	global environment facility
GEMs	genetically engineered microorganisms
GEO	genetically engineered organism
GIC	Global Industry Coalition
GMO	genetically modified organism
GRAIN	Genetic Resources Action International
ICCP	Intergovernmental Committee for the Cartagena Protocol
INCCBD	Intergovernmental Negotiating Committee for a Convention on Biological Diversity
IPR	intellectual property right
ISAAA	International Service for the Acquisition of Agri-Biotech Applications
IUCN	International Union for the Conservation of Nature and Natural Resources
LMO	living modified organism
LMO-FFP	LMO intended for food, feed, or processing
MEA	multilateral environmental agreement
NGO	non-governmental organization
NIN	National Institute of Nutrition (Canada)
OECD	Organisation for Economic Co-operation and Development
PIC	prior informed consent
PNT	plant with novel traits
PrepCom	Preparatory Committee (for the United Nations Conference on Environment and Development)
PVP	Plant Variety Protection
RAFI	Rural Advancement Foundation International
rBGH	recombinant bovine growth hormone
SPS Agreement	Agreement on Sanitary and Phytosanitary Measures (of WTO)
TPS	Technology Protection System
TRIPS	Trade-Related Aspects of Intellectual Property Rights
TUA	technology use agreement
TWN	Third World Network
UNEP	United Nations Environment Programme
USDA	United States Department of Agriculture
WHO	World Health Organization
WRI	World Resources Institute
WTO	World Trade Organization

Introduction

In the mid-1990s, when genetically engineered (GE) seeds were first being planted commercially in North America, the biotechnology industry and its partners envisaged a world in which their crops would be widely accepted as the food of the future, providing a growing population with improved nutrition and farmers with more environmentally sustainable production options (Duvick 1995). To many environmentalists and others, however, the new seeds represented a troubling development. Some critics focused on the way that genetic engineering, along with new interpretations of intellectual property law, reinforced the trend towards oligopolies in the global food system while reducing the ability of farmers to use and save locally adapted seed varieties (Kloppenburg 1988; Shiva 1993). Others raised ethical concerns about the species crosses taking place in the engineering of plants and animals (Kneen 1999). The possibility of food safety and environmental hazards associated with GE organisms (GEOs) was also widely voiced by critics (Rissler and Mellon 1996).

The differences in perspective between the proponents of GEOs and their critics were part of a protracted struggle over the adoption of these organisms in the global agricultural and food system that began with the development of the first genetically engineered plants in the 1980s and continues to this day. This dispute has taken place in a host of fora, including farm meetings, protests against "Frankenfoods" at supermarkets, scientific meetings concerned with assessing the risks of GEOs, World Trade Organization (WTO) deliberations, and the stock exchanges of the world, where investors have had to decide whether or not to gamble on the agbiotech industry.

Although not obvious to observers, one of the issues at the heart of this wide-ranging dispute has been the very definition of the organisms in question. The terms "genetic engineering," "genetic modification," "transgenesis," and "biotechnology" have all featured centrally in the debate, and they are often assumed to carry the same meaning. However, these terms have not always been defined in the same way, as I explain in detail in later chapters. The use of one term over another has depended on various factors, including perceptions of how risky the new organisms actually are.

Because of the politicized nature of language in this field, it is important that I clarify my choice of terminology before going any farther. I use the terms "genetically engineered organisms" (or the *products* of genetic engineering) and "transgenic organisms" in the way that has become most widely accepted in recent years: these are organisms whose DNA has been purposefully altered in a way that does not occur naturally by mating or natural recombination. I also employ the term "genetically modified organism" (GMO) as a synonym for GEO, but restrict its use to discussions on policy deliberations from which it emerged as a regulatory category. I use "biotechnology" or simply "biotech" as a descriptor for the industry that employs genetic engineering, along with other tools, for agriculture (agbiotech) or other purposes, such as the development of GE pharmaceuticals.

Given the ambitions of the acolytes of genetic engineering for a revolution in global agriculture, their efforts appear to have resulted in mixed success to date. Between 1995 and 2005, the area planted in transgenic seeds grew from a small number of test plots to 90 million hectares (James 2005). This is an enormous achievement for the champions of GE seeds. However, in 2005, 94 percent of the area devoted to these crops was found in only five countries: the United States (55 percent), Argentina (19 percent), Brazil (10 percent), Canada (6 percent), and China (4 percent). Only sixteen other countries were growing GE crops on a commercial basis, with most growing on less than 100,000 hectares (James 2005).

Elsewhere in the world, the dawn of the twenty-first century brought with it a great deal of skepticism towards GEOs. An effective ban existed on the introduction of new GE crops and foods (other than those no longer containing novel DNA) in member states of the European Union (EU) from 1999 to 2004, and bans on specific GEOs remain in place in some EU countries (Pew Initiative on Food and Biotechnology 2005). In 2001, several countries, including Sri Lanka, Croatia, and Bolivia, considered banning engineered organisms altogether. A year later, three African states rejected food aid from the United States because it contains GEOs (Villar 2002; Vint 2002). One outcome of global skepticism towards GEOs is that a number of states, including Japan, Australia, and member states of the EU, developed laws that require the labelling and/or traceability of GEOs from field to plate. Another outcome is that many of the world's largest food manufacturers, including Coca-Cola and PepsiCo subsidiaries in China, have committed to producing foods without GE ingredients (Greenpeace 2005a). These are all signs that GE crops and foods have not received the global acceptance hoped for by the biotech industry but, instead, face major roadblocks; the revolution remains far from assured.

This book presents an analysis of the global politics of agricultural bio-technology through an in-depth examination of one particular forum that has been a central site of political struggle. This is a forum where the dispute over GEOs took a specific set of twists and turns in the late 1990s and early 2000s that continues to have major implications for politics of agbiotech and related fields today: the negotiation of the Cartagena Protocol on Biosafety.

The Cartagena Protocol on Biosafety

The Cartagena Protocol on Biosafety to the Convention on Biological Diversity (CBD) is an international treaty governing trade in living GEOs (known under the protocol as living modified organisms, or LMOs). This treaty is intended to protect the environments of importing countries from GEOs that could harm their environment. For example, a country with salt marshes supporting rare species of wildlife would want to know if a variety of wheat has been genetically engineered to be salt-tolerant, in order to protect against the possibility of the grain successfully invading the wildlife habitat. This protocol allows importing countries to demand such information and to block imports of the GEOs if they deem potential risks too severe or unmanageable, or if there is insufficient scientific knowledge to carry out a satisfactory risk assessment (CBD 2000a). The Cartagena Protocol, which has been ratified (or acceded to) by more than 130 countries, came into effect in September 2003 (CBD 2006a). At the time of this writing, there have been three Conferences of the Parties serving as the Meetings of the Parties to the Protocol (COP-MOPs) to work on its implementation. The first took place in Kuala Lampur, Malaysia, in 2004. The second meeting was in Montreal, Canada, in 2005, and the third took place in Curitiba, Brazil, in early 2006.

The negotiation of the protocol occurred between 1996 and 2000, although talks actually began as early as 1990, when, in the lead-up to the Rio Earth Summit of 1992, it was proposed that biotechnology be referred to in the CBD. After vigorous debate, the article was included in the final text of that treaty. Among other things, Article 19 of the CBD called on parties to the convention to consider "the need for and modalities of a protocol ... for the safe transfer, handling and use of any living modified organism resulting from biotechnology that may have adverse effect on the conservation and sustainable use of biological diversity" (UNEP 1992, Article 19.3). In 1996, following further debate, negotiations to create a protocol to meet these ends were set into motion under the CBD.

From the moment that it was first raised in international policy circles, the idea that the products of biotechnology should be given special attention in international law was strongly contested by countries such as the United States that were developing GEOs for export, and European countries such as Germany and the UK that saw international biosafety guidelines, rather than legally binding regulations, as the way forward. The call for a protocol was also strongly resisted by biotechnology industry organizations from North America, Europe, and Japan. Opposition from these quarters continued during the negotiation of the Cartagena Protocol. The name given to this protocol refers to the city in Colombia where negotiations were expected to be completed in February 1999. That final negotiations actually took place far from Cartagena, in Montreal almost a full year later, is a testimony to the challenges that faced the states and activist groups advocating strong, legally binding rules to govern trade in GEOs.

At the Cartagena meetings, a group of six nations named the Miami Group and chaired by Canada scuttled the talks at the last minute by stating that the emerging consensus around trade restrictions on LMOs was not in anyone's best interest (Chasek 1999). As late as one o'clock in the morning of 29 January 2000, this was still the position of the United States and Canada, in particular. However, for reasons that I explore in this book, just before five o'clock (CBD 2003, 13), after a night of intense negotiations, Canadian representatives, along with those of the United States and their other allies, gave in to a compromise protocol that they believed was weighted against their domestic biotechnology industry's interests.

The Cartagena Protocol breaks new ground in global environmental politics, and that it was completed and agreed to by international consensus surprised even many of those who participated in its negotiation. As one Canadian official put it,

> A large number of countries, like the small island states that never felt that the protocol would come into effect, but wanted it, have been stunned because it is there ... On the other hand, among the larger developed countries, and certainly among the Miami Group, you've got stunned disbelief that it ever came into effect; because they did all of their preparations on the premise that it would never happen. (Interview #1)[1]

The Cartagena Protocol is significant on many fronts. Particularly interesting with regards to environmental politics is that this multilateral environmental agreement (MEA) treats an entire class of new and unproven technologies in a *precautionary* manner.

Most multilateral environmental treaties are created in hindsight, after there is clear evidence that certain practices or industrial products, for example, are damaging the environment. By contrast, GEOs are relatively new and the extent to which they represent a serious safety hazard, either to the environment or human health, is still not clear. Furthermore, they have actually been promoted, as one representative of a public-interest civil society organization (CSO) active in the biosafety talks noted, as the "next economic basis for human existence on the planet" (Interview #2). (Instead of the term non-government organizations, or NGOs, I use the term "civil society organizations," or CSOs, to refer to all those voluntary organizations that exist between the individual and the state; see Chapter 1 for background on this conceptualization of civil society). That an international regulatory tool was negotiated before the widespread dispersal of genetically engineered organisms, and before clear evidence of the hazards they represent as a class of technologies, speaks to the unique political dynamics of this field.

The rules set out in the Cartagena Protocol for the import and export of these organisms are also unique when it comes to international norms for environmental regulation. The final text includes several clauses that empower states to make precautionary decisions on GEOs. For example, Article 10.6 of the protocol states:

> Lack of scientific certainty due to insufficient relevant scientific information and knowledge regarding the extent of the potential adverse effects of a living modified organism on the conservation and sustainable use of biological diversity ... shall not prevent that Party from taking a decision, as appropriate, with regard to the import of the living modified organism. (CBD 2000a)

That the Cartagena Protocol was developed with foresight and that it includes text supporting precautionary import decision making are two significant developments in the field of environmental politics. These developments are widely seen as part of an attempt, at the international level, to operationalize the precautionary principle in the field of genetic engineering.

The precautionary principle, as iterated in the Cartagena Protocol, has its legal origins in the German concept of *Vorsorgeprinzip*, which expresses the belief that society should seek to avoid environmental problems by careful forward-looking planning that blocks the flow of potentially harmful activities (Jordan and O'Riordan 1999, 4). However, precaution as a policy tool is not a German concept only. In the English-speaking world, concepts akin to the precautionary principle are found in such basic health care adages as "First, do no harm" and "An ounce of prevention is worth a pound of cure" (Raffensperger and Tickner 1999, 1). Before its evolution in Germany and then the EU in the 1990s, elements of a precautionary approach could be found in numerous North American laws, including the 1970 US National Environmental Policy Act, which institutionalized environmental impact assessment procedures for federally funded projects (Kleiss 2003).

That the final Cartagena Protocol text contains clauses that operationalize a precautionary approach towards GEOs at the international level – albeit without mentioning the phrase "precautionary principle" itself – was considered a surprising victory for the environmental CSOs that participated in the talks (Dawkins 2000a). It *was* a surprising victory, because the biotechnology industry, along with the US and Canadian governments and others, had worked through the 1980s and into the 1990s to establish a very different way of understanding GEOs and how they should be dealt with in regulatory policy. Even though these countries had earlier legal experience with precaution as an overarching approach to environmental issues, by the early 1990s they had chosen to frame most GEOs as minimally risky to the environment and human health, and as *substantially equivalent* to other crops and foods already in the international marketplace. An international protocol designed to allow precautionary decision making on LMO imports is clearly at odds with this equivalency framing.

The creation of an international legal instrument that allows for precautionary action on genetically engineered organisms raises important questions: Why do we have a protocol that singles out LMOs for special regulatory treatment? How were clauses that allow for precautionary decision making on LMOs, in advance of their import, embedded in the Cartagena Protocol on Biosafety? And what does the Biosafety Protocol, with its precautionary clauses, mean for the biotechnology revolution that was expected to change the face of modern agriculture? These questions are the central focus of this book. A secondary focus is on my home country's – Canada's – place in the global politics of biosafety.

Canada played an important role in the evolution of the Cartagena Protocol. On the one hand, from its inception in 1992, Canada had been seen as a key supporter of the Convention on Biological Diversity. As evidence of this relationship, Montreal was appointed as the home of the secretariat for the convention at the first Conference of the Parties to the CBD (COP-1) in 1995. On the other hand, in the late 1990s and early 2000s, Canada was third only to the United States and Argentina in the uptake of genetically engineered seeds in agricultural production (James 2001). Because of a perceived domestic economic interest in exporting GE crops grown in this country, combined with a grain trader's fears that even shipments with minimal GE content might be caught in new regulatory nets enabled by a strong protocol, Canada chaired the Miami Group. This negotiating group sought to minimize the potential impact of the protocol on the international grain trade. By taking this stance, many countries and activists felt that Canada had shifted from being a strong supporter of the CBD to being one of the principle opponents of a protocol negotiated under the auspices of that very convention. Such suspicions were apparently verified when Canada's position on biosafety, along with those of other Miami Group countries, was instrumental in precipitating the collapse of the talks in Cartagena in 1999.

Fortunately, the episode in Cartagena did not represent the end of the biosafety talks, and a similar collapse did not take place in January 2000 in Montreal. How did Canada justify its stances on biosafety in both Cartagena and Montreal, given its seemingly divided interests in the issue? Why was the compromise reached in Montreal acceptable to representatives of the Canadian government, along with other GEO-exporting countries, when a compromise had evidently not been possible a year earlier? And what, if anything, does the new precautionary treaty mean for the politics of GEO use in Canadian agriculture?

To address the two sets of questions raised above on the development and implications of a precautionary Cartagena Protocol in general and on the Canadian place in global GE politics, one needs to have a sense of where to begin looking for answers. This is where political theory offers direction. A theoretical framework helps illuminate certain features of the political landscape for scrutiny, although inevitably backgrounding others. The problem is that the adoption of one theoretical lens over another, or even a hybrid among several, involves a bit of a chicken-and-egg conundrum: in choosing what to focus on, one must have already made some educated guesses about which features of the political landscape could benefit from illumination.

Theory
My initial investigations into this field suggested that there were at least three factors at play in the politics of GEO regulation and trade. First, economic interests, such as Canada's interests as a wheat and canola exporter, looked to be important in shaping a national position on GEO trade. The position of most other countries of the world as primarily GEO importers in the mid-1990s, which entailed a different set of economic interests, appeared to be equally relevant to their domestic policies on GEO trade. However, economic interests on their own did not appear sufficient to explain GEO policies. If GEOs were inherently risky for human health and the environment, as some activists were suggesting, the Canadian government would have a hard time justifying commercial production, let alone export, of these products. Conversely, if there were no grounds upon which the safety of GEOs could be called into question, any movement to limit their trade would never have reached the proportions it had. So scientific knowledge, and its interpretation by the public and politicians, appeared to be a second factor influencing GEO policies. A third factor also came to the fore: cultural values, and their definition in institutions and norms of regulation, appeared to play an important role in the politics of GEO trade. This factor surfaced in the differences emerging between Europe and North America in the late 1990s on how to regulate the exact same GE foods.

Before beginning this research, then, I knew that policies on GEOs were likely influenced by at least these three factors: scientific knowledge, economic interests, and norms of governance. To understand the internationalization of agricultural biotechnology and its regulation, I would want to draw on a theoretical tradition, or traditions, that would help me examine, at a minimum, how these three factors interacted at the global level.

I first looked to the international relations literature. The dominant theoretical tradition in this field for examining environmental issues, neo-institutionalist regime theory, focuses on the international dynamics of establishing global regulatory rules and norms. Regime theory did prove useful, for example, in interpreting the influence of already established norms of governance on subsequent international institutions. However, as I explain in Chapter 1, I found this approach wanting in many other ways. Regime theory was particularly weak when it came to theorizing the relationships between knowledge and interests in establishing regulatory norms in the first place. In contrast, I found those scholars who had the most to offer on the relationships between interests, knowledge, and norms drew either on the theoretical insights of Antonio Gramsci (1891-1937) or Michel Foucault (1926-84) in their work.

One example of scholarship that draws on the work of Gramsci is Purdue's study of biotechnology and the global institutionalization of new norms of intellectual property rights to living materials through the Trade-Related Aspects of Intellectual Property Rights (TRIPS) agreement of the WTO (Purdue 1995). A second example is the work of Levy and Egan (2003), who develop Gramscian concepts for the purpose of examining the way that corporations and other political actors interacted in the formation of the international climate change regime. Of the many political scientists who work with Foucault's ideas, Wright and Litfin's studies were particularly inspiring. Wright (1994) looks at the relations of power that led to the creation of American and British regulatory policies for genetic engineering in the 1970s and early 1980s, while Litfin (1994) examines the way that the Montreal Protocol on Substances That Deplete the Ozone Layer came into existence amid divergent understandings of the ozone-depletion problem, competing commercial interests, and a range of possible regulatory responses. These studies and others provided guidance as I developed my own hybrid theoretical approach to the politics of biosafety, rooted in the theories of Gramsci and Foucault.

Both Gramsci and Foucault offer political philosophies that appear daunting to the uninitiated. However, the reward of being able to illuminate contemporary problems through concepts they introduced makes up for the challenges presented by their texts. This is why in Chapter 1 I go into some depth in explaining, as accessibly as I can, the contributions of both theorists to understanding environmental politics. In general terms, Gramsci and Foucault are useful to a study of the global politics of agricultural bio-technology and the environment because both of these thinkers are highly attuned to the power of ideas in politics, and clearly it was ideas – whether ideas about the economic importance (or threat) of GEOs, ideas about the nature and risks of GEOs, or ideas about how to govern such risks – that I had first set out to examine in order to shed light on the politics of biosafety. Equally important, both theorists are interested in the way that ideas, whether conceptualized as ideologies or discourses, are embedded *within* social relations. To put it simply, both argue that ideas don't just appear out of thin air; they arise from, are acted upon in, and help give shape to spe-cific material, social, and political arrangements. While there are signifi-cant differences between Gramsci and Foucault, together these two thinkers offer a variety of tools for interpreting the way that knowledge, interests, and norms (along with other factors) helped shape the Cartagena Protocol, and how this protocol, in turn, has affected the wider politics of genetic engineering and the environment.

Gramsci's specific contributions to this work include his notions of "he-gemony" and "historical bloc." While each of these concepts is distinctive, their commonality is that they were formulated to describe relations of power that are reinforced across different areas of social and political engagement, particularly the realms of material capabilities (including economic forces), organizations (including institutions), and ideas. I use these two concepts to help elucidate the formidable relationships among the biotechnology indus-try, governments, and civil society that formed around the biotechnology revolution in its North American heartland, and to illustrate the various levels on which these relationships had to take hold in order to be successful.

Foucault's contributions to this work include his theorization of power and resistance in "discourse," as well as the political dynamics of "govern-mentality" and "biopower." Most simply, "discourses" refer to ingrained (yet always still contested) patterns of thought and action, and the power that accompanies these normalized patterns for defining possible futures. "Governmentality" indicates a specific set of discourses regarding the prac-tice of modern governance, while "biopower" refers to the particular forms that governance of people and other living beings have taken in the indus-trialized West over the past two centuries or so.

I develop each of these three concepts in the early chapters of this book to help me dig more deeply into the assumptions and practices that under-lie debates over the risks of genetic engineering than mainstream approaches to international relations would allow. Notably, the works of both Gramsci and Foucault exemplify the importance of grounding political theory in historical, real-world struggles and in day-to-day relations of power. In ret-rospect, it is clear that this was the approach I had adopted in this research.

Synopsis

The argument I develop through the six empirical chapters of this book is that the Cartagena Protocol on Biosafety was enabled by, and further embedded, an emergent discourse of precaution in the field of agricultural biotechnology at the international level in the late 1990s and early 2000s. This precautionary discourse is gradually supplanting the discourse of "risk" in international policy fora, the discourse that had been promoted by the proponents of genetically engineered organisms since the early 1980s. The risk discourse holds that the potential harms of GEOs could be easily characterized and calculated, and that any hazards associated with GEOs could be easily managed. Because the precautionary discourse presupposes that the risks of GEOs are not necessarily easily understood and managed, the institutionalization of this alternate framing of GEOs at the international level occurred in the face of stiff resistance from countries such as Canada and the United States that had become, through the 1980s and 1990s, deeply enmeshed in the biotech revolution led by the agri-chemical industry. That these countries assented to the Cartagena Protocol in 2000 represented a major concession by proponents of genetic engineering towards their critics, a concession that continues to have impacts on the biotech revolution in agriculture and in other areas of environmental governance.

At the same time, it is important to recognize that a precautionary approach to GEO governance, in terms of day-to-day regulatory practice, is not yet universally accepted as the norm. The positions of the US, Canadian, and Argentine governments in their 2003 WTO trade dispute demonstrates the continued existence of the risk framing of GEOs (although one that has subtly shifted from the position those countries might have taken three years earlier, as I explain in Chapter 7) (WTO 2006). It is also important to recognize that a precautionary protocol was not simply a victory for the critics of GEOs; the institutionalization of "precaution" in the Cartagena Protocol came at a price.

Following Foucault's characterization of discourse, the precautionary discourse can be understood as a dispersed set of practices and truth claims about genetic engineering, and about how GEOs should be dealt with in governmental regulation and by industry, that is grounded in a particular social and political context. As a discourse, precaution does reflect the precautionary principle in that it reinforces practices designed to anticipate, assess, and prevent problems that may be caused by GEOs in the environment, even when full scientific evidence of those harms may be lacking. But the way precaution is framed in the Cartagena Protocol has other productive effects, not all of which sit as well with critics of GEOs. Consider the following three examples. First, given its origins in the earlier risk discourse, many of the actors operating within the precautionary discourse still assume that eventually all environmental risks will be predictable and manageable. Although a large number of GEO critics who argued for a precautionary response to GEOs in the protocol negotiations would challenge these assumptions, the precautionary elements of the Cartagena Protocol text do not clearly back one side or the other on this issue. Second, the precautionary discourse institutionalized in the Cartagena Protocol does not appear to accept the notion that the full range of potential ethical, environmental, human health, social, and economic implications of GEOs ought to be part of precautionary decision making, even though many of the initial advocates of precaution saw these as fundamentally interrelated risks of genetic engineering. Third, because of the protocol's relationship to international trade agreements, its iteration of precaution must be seen to uphold the view that proactive actions taken to prevent harm in the face of uncertainty must still represent the "least trade restrictive" option, along with conforming to other standards of trade law. Among its repercussions, this framing rules out blanket bans on GEOs on the basis of precaution such as that enacted in Upper Austria in 2003 (Anonymous 2005b).

Emerging from earlier debates about the hazards, risks, and uncertainties of genetic engineering, this specific precautionary response to the genetic engineering issue crystallized during the late 1990s as a compromise among divergent ways of framing the genetic engineering issue. The negotiation of the Cartagena Protocol was the focal point of this process. As a result, I term this particular regulatory narrative the "Cartagena" discourse of precaution.

What is interesting about the negotiation of the Cartagena Protocol is that here the precautionary response was not simply a force coming from outside the negotiations to shape perceptions of the issue, as Litfin (1994) sees precautionary discourse in relation to the international negotiations on ozone depletion. In this case, precaution entered the debate as a way of framing the issue on the part of certain key actors and, eventually, the widespread adoption of this discourse resulted in a critical shift in the negotiation dynamic by late 1999 and early 2000. At the same time, however, many of the elements of the Cartagena discourse of precaution – a discourse that by now has had wide-ranging effects even outside the realm of GEOs – were actually shaped through the negotiation process among representatives from nation-states and civil society. While it is true that by Foucault's definition discourse is necessarily the product of political struggle and negotiation through the medium of language, it is rare that one can actually trace the micro-politics of discourse formation through the positions and counter-positions taken in a specific conversation, and that one can relate these shifting positions to domestic political struggles as well as to the dynamics internal to government and intergovernmental relations.

Chapters 1 through 6 are the result of a fine-grained analysis of biosafety protocol negotiation documents and other policy documents, as well as interviews and first-hand written accounts of the negotiations. I also draw on previous studies of the politics of agricultural biotechnology. I am particularly indebted to Charles (2001) for his book *Lords of the Harvest*, to Krimsky (1991) for *Biotechnics and Society*, to Wright (1994) for her monograph *Molecular Politics*, and to Gupta's (1999, 2000a, 2000b, and 2000c) research on the Cartagena Protocol.

I undertook twenty-six interviews for this book, most between one and two hours long. I also sifted through more than sixty first-hand accounts of the Cartagena Protocol's negotiation. The majority of these can be found in an excellent anthology that features chapters written by participants in the biosafety negotiations edited by Bail, Falkner, and Marquard (2002). I have treated these written accounts in much the same way as my interviews, since they represent first-hand (as well as politicized and often contradictory) memoirs of the negotiation process. Most of the interviews, along with the chapters in the Bail, Falkner, and Marquard volume, were written or recorded within the first two-and-a-half years after the Cartagena Protocol's negotiation, rather than at the time of the negotiation itself.

Political theorists recognize that discursive politics can be fully appreciated only at the micro-analytical level (Darier 1999). At the same time, Gramsci and Foucault would argue that the researcher must also be aware of the larger historical webs within which these details are situated. In an attempt to achieve both of these ends, I trace the micro-politics of biosafety in the context of macro-trends in the field of biotechnology and environmental governance.

Chapter 1 lays out the theoretical framework adopted in this study. I begin by considering the strengths and limitations of mainstream approaches to international relations. I then explain how I employ Gramscian and Foucauldian concepts in this study of the politics of agbiotech.

Chapters 2 and 3 present an overview of the global politics of agricultural biotechnology in the 1980s and 1990s and are organized by the Gramscian conceptualization of politics as involving three sets of relations of force: the material, organizational, and ideational. Chapter 2 highlights activities in the material and organizational arenas. Drawing largely on Canadian examples, it documents the emergence of what I call the "biotech bloc" within the agriculture and food system. This historical bloc, which first coalesced in the late 1970s and early 1980s, involves a set of alliances among molecular biologists, agri-chemical corporations, and the US, Canadian, and Argentine governments. These actors had all set their sights on the genetic engineering revolution in agriculture and worked together, over the ensuing decades, to make this global revolution a reality.

Chapter 3 presents a detailed examination of the ideational politics of biotechnology and its regulation. In this chapter, I trace the way that four particular sets of ideas organize the politics of agricultural biotechnology.

These are ideas of the gene, the environment, liberalism, and risk. I begin by examining each idea independently and then turn to consider where they intersect, with attention to the ways that they shaped, through the late 1980s and early 1990s, specific debates about the regulation of the hazards and risks associated with the products of genetic engineering. In Canada, the risk discourse was mobilized in limited regulatory overviews that allowed authorities to approve the commercial introduction of plants with novel traits to the environment while also ensuring the public that these plants had undergone comprehensive assessments of their risks. In the EU and several other jurisdictions, however, more complex understandings of the possible risks of genetically modified organisms led to slightly different regulatory systems that would eventually create the space for an alternate discourse of regulation rooted in precaution.

Chapters 4 through 6 trace the development of the biosafety issue and the subsequent negotiation of a protocol on biosafety under the CBD. These chapters tell the story of the gradual emergence, mobilization, and entrenchment of the discourse of precaution in the field of genetic engineering through this international regulatory instrument, and the material, organizational, and ideational dynamics that made this possible. I argue that the institutionalization of this emergent discourse took place over four distinctive discursive shifts. These shifts were actively spearheaded by the leadership of specific GEO critics from both the North and the South but depended on the confluence of various forces to be realized.

The first discursive moment took place during the meetings of the Intergovernmental Negotiating Committee for a Convention on Biological Diversity (INCCBD) in the late 1980s and early 1990s. During this period, biosafety came to be defined as an issue worthy of specific attention by the CBD. A second moment occurred between late 1994 and early 1995, when it came to be accepted that a biosafety instrument would focus narrowly on GEOs, and not on a wider class of "novel" organisms that could include those produced through traditional breeding. These developments were soon followed by the third moment: the acceptance that the field of biosafety required a legally binding international regulatory framework, rather than a voluntary one. The final discursive moment occurred between late 1998 and early 2000, towards the end of the protocol negotiations. This is when the operational language of the Cartagena Protocol came to reflect the specific precautionary framing of biosafety outlined above. The first three discursive moments are discussed in Chapter 4, while the final one is the subject of Chapters 5 and 6.

Chapters 5 and 6 discuss those aspects of the Cartagena Protocol negotiations that are most relevant to demonstrating the evolution and impact of the discourse of precaution through these talks. These include the debates about the definition of "living modified organisms" and the role of advance informed agreement, risk assessment, and scientific uncertainty, as well as socio-economic considerations in LMO import decisions. I also discuss the creation of separate import procedures and documentation requirements for bulk shipments of living GEOs for food, feed, or processing, as well as the debates over the relationship between the protocol and other international agreements. (These chapters do not cover all aspects of the protocol's negotiation. For example, compliance issues and debates over liability and redress, as important as they are to the future of the protocol, are not addressed. Readers seeking a negotiating history of the entire Cartagena Protocol should consult Mackenzie et al. 2003 and CBD 2003.)

Chapter 5 begins with an examination of initial positions on the central issues related to the emergence of a precautionary protocol held by key negotiating groups, demonstrating how those positions emerged in the context of material, institutional, and ideational relations of force. The latter half of the chapter focuses on the development of the Canadian position, showing how it was shaped by domestic factors and by Canada's chairing of the Miami Group from 1998 to 2000. Chapter 6 follows the key debates through the final stages of negotiation, from Cartagena in February 1999 to Montreal in January 2000, showing how each of these issues was concluded in the final text.

Chapter 7 examines the implications of the Cartagena Protocol, with its precautionary framing, for the politics of agricultural biotechnology, and environmental politics more broadly, from 2000 to 2006. I pay attention to both international and Canadian implications. This chapter also examines how the protocol has evolved in the first three years since its implementation, including in particular the evolution of the debate over documentation requirements for bulk shipments of living GEOs. The book concludes with observations on the implications of this empirical study for the theoretical framework employed.

A great deal has been written on the benefits and drawbacks, real and potential, of genetically engineered crops for both the North and the South. While this literature informs my work, the intention here is not to summarize what others have written. Rather, this book examines where genetically engineered organisms have come from, how they came to assume a major position in the global agricultural system, and how political institutions, first in the United States and Canada, and eventually internationally, came to respond to these new technologies in ways that may be as unique as the modified organisms themselves.

1
Theorizing International Environmental Diplomacy

Regime Theory and Its Limitations

How does one begin to assess an international agreement such as the Cartagena Protocol on Biosafety? Most political scientists turn to neo-institutionalist regime theory to help make sense of multilateral environmental treaty making and its implications. Neo-institutionalists are interested in investigating the forms of cooperation that arise among nation-states in an increasingly interdependent global political and economic order. In the context of transborder environmental issues, this cooperation is evident in the development of new environmental regimes. Some scholars define regimes narrowly as a set of norms or rules specified by a multilateral legal instrument (Porter and Brown 1996, 20). However, it is more common to see the concept defined broadly to encompass, as Levy and his colleagues (1995, 274) put it, a set of "principles, norms, rules, decision-making procedures and programs that govern the interaction of actors in specific issue areas." Significantly, this broader definition sees a regime as more than an institution: it is a form of governance as well as a common set of understandings of a problem and how it is best solved. This definition also does not restrict the development and administration of environmental regimes to the actions of nation-states. As Levy and Newell (2002, 85) point out, regimes are theorized to comprise "networks of actors, routines, principles, and rules, simultaneously constituting and disciplining their subjects, constraining and enabling patterns of behavior." Neo-institutionalist research tends to focus on the ways in which regimes take shape in the context of interstate politics, how they evolve, and how they influence one another and the actions of nation-states (e.g., O. Young 1997, 2002; Oberthür 2001). There is also a growing literature attempting to assess the effectiveness of international regimes.

Regimes are not restricted to norms and routines that have been formally institutionalized at the international level. In theory, then, regime analysts should be interested in the norms of environmental governance that emerged in industrial and state practice in various parts of the world before these become codified in international law. In practice, however, researchers in this field tend to focus on formal multilateral environmental agreements (MEAs), possibly because these MEAs provide tidy case studies of truly global environmental regimes (e.g., Haas, Keohane, and Levy 1993; O. Young 2001; Mitchell 2002; and Hovi, Sprinz, and Underdal 2003).

Regime theory provides a useful starting point for an analysis of the Cartagena Protocol on Biosafety. This approach draws our attention to the role of international institutions, such as the United Nations Environment Programme (UNEP), the Convention on Biological Diversity, and their advisory bodies, in structuring the international GE regulatory agenda around the question of biosafety. Another important thread in neo-institutional writings on environmental MEAs examines the role of science in regime formation; this thread is also clearly relevant here. In the case of the Cartagena Protocol, scientific assessments of the GE issue were important in the generation of shared perspectives on this issue, and regime theorists such as Haas (1992) would encourage inquiry into the scientific debates over GEOs during the Cartagena Protocol negotiations to see whether and how shared understandings emerged. Despite these insights, however, neo-institutionalist writings on regimes have several limitations which are by now well recognized in the literature (for a review of some of the key limitations of regime theory, see Newell 2005). These limitations suggest the need for a more complex theoretical framework.

One key limitation of regime theory is the tendency of researchers in this field to focus narrowly on nation-states as the primary actors in global politics. This approach can lead one to lose sight of the political dynamics at the domestic and sectoral levels, where many of the issues and ideas that come to be contested internationally are first defined and shaped. In terms of the politics of agbiotech, in which consumer boycotts, scientific controversies, and industry product strategies can each be seen to be as important as state actions in defining the political terrain, a state-centred political theory is clearly misleading. In response to this issue, one current of thought in the regime literature has focused its attention on the role of non-governmental organizations (NGOs) in environmental politics. NGOs are variously defined to include a range of organizations, from the grassroots level to the international, engaged in advocacy on social and environmental issues (for an overview, see Betsill and Corell 2001). In keeping with the general orientation of regime theory, however, much of this literature remains trained on how NGOs contribute to international environmental policy-making processes. Betsill and Corell, for example, have developed a method for systematically measuring NGO "influence" in MEA negotiations (Betsill and Corell 2001; Corell and Betsill 2001). Although such methods offer useful tools, they help reinforce the assumption that interstate relations (and the influence of other actors on them) are at the centre of global environmental politics.

Some analysts have tried to break with this assumption by focusing their attention on the theatre of "world civic politics" (a term introduced by Lipshutz) or "global civil society" (Wapner's preferred term) (Lipshutz 1996; Wapner 1996). For example, Wapner (1996) argues that transnational environmental activist groups work to bring about change at the level of civil society (in terms of building alternative international institutions, consciousness raising, and impacts on consumer behaviour) as much as at the level of international policy making, and that there needs to be more attention given to the effects of these activities on the production of international environmental norms and practices. Jasanoff (1997, 579) has made a similar case about the work of NGOs at multiple levels in global environmental politics, from the provision of local knowledge to policy formulation, implementation, and technology transfer.

In general, the growing attention focused on environmental advocacy groups is important. The insights of Corell, Betsill, Jasanoff, and Wapner, among others, are particularly helpful when trying to understand the role of organizations such as Greenpeace, for example, in international biotech politics both inside and outside the negotiation of MEAs. Furthermore, the terminology employed by Wapner, in particular, suggests an important conceptual shift that I believe needs to be further developed.

As noted, the term NGO (which has its origins in UN parlance) is vague and used to refer to a wide range of groups active in environmental politics. Unfortunately, it is clear that many researchers who employ this term in the environmental politics literature are actually referring to a fairly narrow class of international environmental advocacy organizations – implicitly excluding many other organizations that are not environmentally oriented (such as faith-based organizations) or that have other agendas (e.g., educational institutions, community service clubs, trade organizations). Because these other organizations also play a strong role in determining the norms and values of a society, whether actively involved in shaping environmental policy or not, it is important to make room for them in any broadening of the conceptual terrain of global environmental politics. For these reasons (in addition to the fit with a Gramscian understanding of politics), instead of the term "NGO," I adopt the descriptor of "civil society organization" (CSO) to refer to any of a wide range of organizations, rooted in voluntary participation, that function between the individual and the state. At the same time, I frequently qualify this term with the addition of an adjective, as in "industry" CSO or "environmentalist" CSO, to give a clearer sense of where specific CSOs are located in material and/or ideological terms.

Wapner's work, along with that of other researchers interested in the global impacts of environmental CSOs, offers an important contribution to a more comprehensive framing of environmental politics. In general, however, this approach to conceptualizing politics beyond the state has taken only one step where there remain others to be taken. In his move to document the activities of organizations such as Greenpeace, for example, Wapner sidesteps the disquieting fact that such advocacy groups may be particularly active on fronts other than international policy making because the official fora of international politics are all too often hijacked by business interests. In my view, Wapner is not really taking the bull by the horns. To do so would be to see that what is ultimately required is a wider, more historicist framing of global political dynamics.

In the environmental politics literature, those scholars most likely to attempt to contextualize interstate politics within broader social, economic, and political relationships tend to work within the historical materialist tradition of international political economy, and many draw, in particular, on the theoretical works of Gramsci (e.g., Purdue 1995; D. Humphreys 1996; Levy and Egan 2003). Gramscian historical materialism has its origins in Marxism and, like Marxism, sees social and political relationships as historically situated within class struggles over production. This approach differs from other strands of Marxism (which Gramsci terms "economism") in that it begins with Gramsci's insight that economic or technological circumstances do not themselves produce change, "they simply create a terrain more favorable to the dissemination of certain modes of thought, and certain ways of posing and resolving questions involving the entire subsequent development of national life" (Gramsci 1971, 184).[1] The actual path taken from technological innovation to political transformation is far from linear and predictable. Yet, Gramsci proposes, this path can be analyzed with the appropriate theoretical tools.

In the context of a study of genetic engineering politics, the Gramscian approach would suggest that while the advent of the techniques of genetic engineering provided new possibilities for agriculture, the biotechnology revolution that has been taking place in agriculture over the last twenty years is necessarily dependent on a host of supportive shifts across civil society, states, and the global order in order to become widespread and accepted as the new norm. Like that of Wapner, this approach emphasizes the centrality of civil society in politics, but Gramscians go beyond activist CSOs to consider a broader range of civil society actors. Like neo-institutionalists, Gramscians see the organization of state and interstate institutions as a critical form of political activity. However, this activity is conceptualized as being only one of three sets of "relations of force," the other two being the material and the ideational (Gramsci 1971, 181-84).

A second key limitation of regime theory is the epistemological stance found in neo-institutionalist writings. Like economistic Marxists, most regime theorists are objectivist: a state's (or any other actor's) interests are assumed to be rooted in its material capabilities and/or position in the international system. This assumption is particularly problematic in the field of environmental politics, where actors develop their positions in relation to how they understand the issues at hand.

Haas' work on the role of science in environmental politics has contributed to a rethinking of objectivism in the neo-institutionalist literature. His studies of the Mediterranean Action Plan and the politics of ozone depletion and climate change all suggest that the international political consensus that develops around certain environmental regimes can be directly correlated to the development of scientific consensus on the issue among technical experts from the negotiating states – the "epistemic community" (Haas 1989, 1992). Although this approach is an important contribution, its weakness, as identified by Litfin (1994, 4), is that it still views scientific knowledge as somehow outside politics: knowledge remains divorced from power. A more comprehensive approach would recognize the importance of subjective perceptions of environmental issues in the formation of regimes while also acknowledging that these perceptions, and the scientific knowledge upon which they may be based, are themselves produced within power relations and reflect these origins in one way or another.

On this issue, the literature suggests at least two possible avenues for exploration. The Gramscian approach proposes that ways of understanding the world are actively produced within class relations in order to build and maintain consent for those relationships (or to challenge them). From this perspective, knowledge – including scientific knowledge – is both produced and productive within relationships that have material, ideational, and institutional dimensions. Another possibility is to pursue a Foucauldian analysis of knowledge as discourse, that is, as a system of interwoven truth claims embedded in social relations and material practices. This discursive approach is suggested by Litfin in her critique of Haas. Litfin (1994, 4) illustrates the way that an understanding of discursive power is applicable to controversies involving the interface between policy and the natural sciences. She notes that in such controversies, scientific knowledge is never simply a body of concrete and objective facts; it is deeply implicated in questions of framing and interpretation, and these are shaped in relation to perceived interests.

Scientific discourses are enormously powerful in the contemporary global order because it is within these webs of knowledge that most of us define ourselves and our world. With the politics of GEOs, we can see that "facts" that originated within the scientific disciplines of molecular biology and ecology, in particular, have played a central role in political struggles over agbiotech. A Foucauldian approach would draw our attention to this phenomenon, focusing on the productive effects of discourses of genetic engineering and the forms of resistance that emerge in this context.

In sum, while regime theory offers a useful starting point, it does not pay enough attention to the complex relationships among the full cast of actors, material forces, and ideas that together shape an MEA such as the Cartagena Protocol. However, this discussion suggests that Gramsci and Foucault have a great deal to offer analyses of global environmental politics.

Gramsci's Relations of Force

At the centre of Gramsci's approach are his theories of hegemony, historical blocs, relations of force, and organic intellectuals. Mainstream international relations scholars look at hegemony in the international arena in terms of dominant states that guarantee regimes through economic or military power over others, or through the provision of public goods. For theorists who draw on Gramsci's work, such as Gill and Law (1989), global hegemony is never seen as having been achieved solely by a nation-state, narrowly construed, because hegemony must be achieved, first and foremost, in the sphere of civil society. Drawing on Hegel, Gramsci sees civil society as the arena of social engagement that exists above the individual and below the state, in and through which individuals form political identities. Murphy (1994, 31) calls civil society the realm where "I" becomes "we." It is in civil society that a leading class, or class fraction, initially constructs the values, programs, and ideologies that represent its own interests as the interests and values of society as a whole. (While many Gramscians see classes as being formed in relation to production, like structural Marxism, others emphasize the subjective dimensions of class formation in relation to ideas [Hall 1988]. I favour this latter interpretation, which allows for recognition of other types of identity-based social actors, such as social movements. These are not necessarily reducible to their position in the relations of production.) These ideologies allow for the building of alliances with other classes, political parties, social movements, important arms of governments, and so on. When the interests of these different actors and institutions converge on a strategic and coherent set of ideas, Gramsci uses the descriptor "historical bloc" (Gill and Law 1989). It is an historical bloc, in this view, and not simply the nation-state, that attempts to consolidate global hegemony.

An historical bloc is made possible when there is a convergence of the three sets of relations of force: first, the material forces of production; second, the relations of political forces through which the interests of one class fraction come to be accepted across that class, and eventually as the common interest of society in general; and third, the relations of military forces and other coercive actions taken by governmental institutions

(Gramsci 1971, 181-84). Those who want to change the status quo, as well as those who wish to preserve it, must develop strategies in each of these arenas of political activity as they engage in "wars of position" designed to gain influence across civil and political society (which, taken together, Gramsci characterizes as the "extended state"). This emphasis on three distinct sets of relations of force is tied with Gramsci's observation that hegemony is dependent on both coercion and consent. Coercion may be exercised by the state or other institutions, while consent takes place in the realm of ideas.

Gramsci (1971, 350) stresses that the construction of consent involves an educational effort:

> Every relationship of "hegemony" is necessarily an educational relation-ship and occurs not only within the nation, between the various forces of which the nation is composed, but in the international and world-wide field, between complexes of national and continental civilizations.

He terms "organic intellectuals" those individuals and groups, emerging from an historical bloc, who educate society on the need for change. The work of these agents involves framing transformations in a way that makes sense to the public at large. This work is necessary because people are rarely simply blind followers; as Rupert (1995, 26) puts it, they actively constitute their own internal relations with society and nature. The goal of hegemony formation is "to transcend a particular form of common sense and to create another which [is] closer in conception of the world of the leading group" (Gramsci 1971, 423). This effort to reshape common sense involves the introduction of a more homogenous, coherent, and systematic philosophy or "ideology," with ideology defined not as a necessary product of material structures but as a set of ideas that serves to cement and unify activities across law, economic activity, art, and so on (Gramsci 1971, 348, 420, 331).

Contemporary political analysts have worked with Gramsci's ideas in at least two distinct ways. Cox (1996), on the one hand, argues that global transformations in the political economy can be seen to depend on congruence among material capabilities, ideas, and institutions across the levels of civil society, state, and world order. Neo-liberal globalization, as such a global transformation, can be seen to be associated with the ascendancy of a new historical bloc since the early 1970s (Gill 1998). This "transnational historical bloc" is centred in countries belonging to the Organisation for Economic Co-operation and Development (OECD), and is led by the transnational capitalist class. At its material foundations, this class fraction is associated with the growing importance of the high technology and service sectors of the global economy, as well as with the increased power of internationally mobile financial capital. In the ideological and institutional realms, the work of international institutions such as the OECD and the World Bank, and of planning groups such as the Trilateral Commission and the World Economic Forum, have been key to the universalization and consensual adoption of neo-liberal ideology across many sectors of society. Meanwhile, the WTO has emerged as the key coercive vehicle for transnational capital and its most willing state partner, the United States.

Levy and Egan (2003), on the other hand, show how Gramsci's ideas are useful for the study of corporate behaviour in a particular sector and issue area, such as the activities of US oil and automobile industries in response to international climate change negotiations. They liken efforts to establish hegemony to field stabilization within institutional theory. In an interpretation of Gramsci's relations of force suited to this scale, they note that hegemony within a given field requires supportive economic systems of production (including product and industrial strategies), organizational capacity (including links with other sectors, governments, and civil society), and discursive structures (to guide behaviour and lend legitimacy to activities). This analysis of the micro-politics of hegemony formation underscores the notion that effective hegemony necessarily involves compromise and accommodation on the part of leaders in order to generate consent from other participants in the bloc and in civil society more broadly. The resultant historical blocs are contingent and unstable, prone to change both from outside and from within.

Both the macro-level approach of Cox and Gill and the field-level analysis developed by Levy and Egan are consistent with Gramsci's framework. The differences in scale found in these two interpretations may be because of ambiguities in Gramsci's own writings on historical blocs. As Levy and Newell (2002) point out, Gramsci uses this term in two ways. It is sometimes used to refer to the alliances that come to be formed among social forces in each of the three political arenas in order to move a particular agenda of change forward. In other places, "historical bloc" describes the ultimate objective of these efforts: the complete alignment of material, organizational, and discursive formations that stabilize and reproduce relations of production and meaning.

Gramsci also uses his concept of hegemony in two distinct ways. In some cases, hegemony refers to the ability of a class fraction, through the active building of consent, to "gain the upper hand, to propagate itself throughout society" at the level of political forces, which he refers to as the level of the "ethico-political" (Gramsci 1971, 181). Elsewhere, he stresses that hegemony cannot exist simply as a relation of cultural or ideological influence: "It must also be economic, must necessarily be based on the decisive function exercised by the leading group in the decisive nucleus of economic activity" (Gramsci 1971, 211-12). These apparent contradictions in Gramsci's concepts do not take away from their theoretical value. Rather, they emphasize the close interrelationships that exist among all three sets of relations of force in order for a change to become widespread. Hegemony is necessarily ethico-political, but it cannot be only that, especially not if a bloc that is working to establish its new conception of the good as the common good in a particular field hopes to entrench this view among all the important arms of the economy, civil society, state, and international order. And, only when this goal is reached will we see the emergence of a truly historical bloc that can reproduce and extend its own capacities.

International political economists working within the Gramscian theoretical framework emphasize the necessity of supportive ideas in the formation of hegemony at the international level. The focus in such work is on the way that ideologies are established to structure a political field in line with the interests of a group of actors, on how they are mobilized as resources by actors in global politics, and on how they act as a mooring around which otherwise disparate classes and groups achieve consensus on a common purpose. Ideologies, according to this Gramscian framing, are *embedded* in social relations, and strategically *employed* to generate consensus around a particular group's interests. These insights of Gramsci and his interlocutors regarding the function of ideologies are important, and the case of international biotech politics presented in subsequent chapters illustrates how ideas do indeed function in these ways. Foucault adds further depth to Gramsci's theorization of the formation and impact of ideas, by elaborating on the *discursive* power of ideas.

Foucault's Discourse
Like Gramsci, Foucault recognizes that interests and predispositions of networks of actors influence the accounts they construct of the world. Also like Gramsci, Foucault rejects the economistic view that sees ideology as a simple product of material interests. Instead, Foucault would be particularly interested in Wynne's studies in the field of environmental policy formation, for example, which demonstrate how a wide variety of tacit social commitments (not just material interests but also disciplinary biases) can and do influence purportedly objective knowledge claims (Wynne 1994). Foucault identifies two main ways that ideas or, rather, "discourses" are embedded in social relations. First, he takes the poststructuralist linguistic turn by emphasizing that discourses are shaped through the shared medium of language. By adopting particular terms, metaphors, and modes of reasoning, networks of actors draw on the cultural narratives that give these linguistic forms meaning. Second, he emphasizes that embeddedness means that influence goes both ways. Discourses are tied to, and reinforced by, particular social and institutional practices that make them visceral in daily life (Foucault 1978, 97).

In addition to his insights on the social relations of discourse formation, Foucault also offers important observations on the effects of discourses. These effects are associated with the way that the shaping of discourse is a truth-claiming activity: discourses purport to describe reality in an authoritative and objective way. Because of their status as truth claims, discourses have a kind of agency, or normalizing power, in politics. This agency has two faces, one disciplinary, the other productive.

As disciplines, discourses restrict the boundaries of what makes sense, marginalizing actors who do not have the tools to participate in the debate on the terms that the discourse defines as acceptable and restricting avenues of credible resistance to the status quo. This disciplinary effect of scientific discourse has long been observed by analysts of environmental controversies (e.g., Wynne 1989; Shrader-Frechette 1991). In debates over genetic engineering, for example, those without technical expertise have limited power in both policy fora and the wider public sphere, even though they may have strong opinions on the matter and strong interests in the outcomes of deliberation.

Discursive truth claims are also productive, in that they define what *does* make sense, thereby *producing* new possibilities. In our example of scientific debates over GEOs, discourses empower certain actors, such as those with technical expertise, or those who are able to mobilize such expertise and arguments, even if these actors have little material or institutional power. (Although not all analysts frame such actions as discursive resistance, the ability to mobilize scientific knowledge claims has been widely recognized as central to the power of citizen's organizations in a broad range of environmental risk controversies. See Beck 1992; Richardson, Sherman, and Gismondi 1993; Irwin 1995; and Jasanoff 1997.) This characteristic of discourse means that actors routinely define themselves, their interests, and their perceptions of an issue in relation to dominant discursive frames. It also means that, rather than being silenced, certain forms of resistance are actually enabled within a discursive field. Examples in Chapters 2 and 3 demonstrate how the adoption of the scientific language of a permissive GE policy discourse organized around the notion of risk assessment, followed by a reinterpretation of this notion in a broader reading of GE risks, has enabled groups critical of the introduction of GEOs in farming to change the policy outcomes in at least two specific cases: the introduction of crops with pesticidal properties ("Bt" potatoes, engineered to produce a protein toxic to potato beetles), and the introduction of recombinant bovine growth hormone (rBGH) in Canada. Given these productive dynamics of discursive politics, Foucault stresses the inherent unpredictability of outcomes.

Foucault's keen insights into the nature of discursive power, and the action of this kind of dispersed "power/knowledge" in society, are best revealed through his observations on two particular (and interrelated) sets of discourses and practices that are enormously relevant to the study of the emerging international genetic engineering regime: governmentality and biopower/biopolitics. (Foucault [1978] appears to use the terms "biopolitics" and "biopower" interchangeably.) Foucault uses the term "governmentality" to refer to the particular historical configuration of modern government that evolved in the eighteenth century. It was the product of two often conflicting tendencies. The first was a growing emphasis on the role of the state in governing human social life rather than simply governing territory (Rutherford 1999, 48). This *raison d'état* assumed that the state should develop a total knowledge of its resources and populations in order to maximize their productivity, thus bringing prosperity to the nation as a whole. At the same time, liberalism was emerging as a critique of state reason. Liberalism is typically understood as stressing the limits of governmental rationality and the rights of the individual to freedom from state control (49). While the ideology of liberalism does promote these values, Foucault argues that the discourse of liberalism did not actually undermine state reason. Rather, liberalism shifted its goals, centres of authority, and modes of application in a process that Foucault refers to as the "governmentalization of the state" (Foucault 1991).

In terms of ends, governmentality is intent on directing human conduct towards maximizing the public good. Because of this seemingly objective goal – framed as an object of truth that begs expert analysis and advice – governmentality involves a new relationship between knowledge and government. Institutions and expert bodies that are actively engaged in the study of society, political economy, public health, and ecology, among other areas – whether inside or outside the formal state apparatus – have taken on an important role in shaping the discourses of public life and population management within governmentality. Modes of administration are also diffuse and multi-layered. While the state maintains an active role, regulation is no longer its exclusive domain. There is also a critical role for what Rose and Miller (1992), drawing on Foucault, describe as "government at a distance": the apparatuses, procedures, and tactics of government that are carried out by institutions that are formally at arm's-length from the state, and the internalization of governmental norms by individuals in their daily activities. Because of its openness and complexity, governmentality is a highly unstable politico-epistemic configuration prone to change from within (Gordon 1991, 16).

I interpret Foucault's concept of biopower as the manifestation of governmentality in the discursive field of human life. This term refers to the forms of power/knowledge concerned with fostering and administering human beings that first emerged in the eighteenth and nineteenth centuries. In terms of social relations, biopower is bound up with the aspirations of industrial capital: "The investment of the body, its valorization, and the distributive management of its forces were at the time indispensable" (Foucault 1978, 141).

Foucault (1978, 140) refers to biopower as the "entry of life into history," the arrival of *life* as a distinct object of concern, with a wide range of accompanying knowledges and practices. Biopower involves administration of human beings by state and by non-state bodies, expertise, and "practices of the self," all intended to further the production and supervision of life across "two poles of development linked together by a whole intermediary cluster of relations" (140). At the one pole, the disciplines of medicine and nutrition, among others, focused on shaping the body of the individual – the production of a particular kind of human subject – to increase its utility and docility (139). At the opposite pole, the emergence of studies of population, agricultural productivity, living conditions, and biology focused on the supervision of the human being as object. Sex and reproduction sit at the pivot of the biopolitics axis. The biopower of sex is directed at organizing and regulating sexuality into a "concerted economic and political behaviour" (26).

Foucault's ideas, especially those about the power of discourse, have been employed by numerous researchers in studies of environmental policy (e.g., Litfin 1994; Wright 1994). However, this move often comes only after these scholars make an effort to clarify their position on what is perceived to be an absence of agency in Foucault's writings, the notion that, for Foucault, "power is everywhere and so ultimately nowhere" (Harstock 1990, 169-70). Litfin (1995, 253), for her part, argues that Foucault's analysis of power is actually incoherent without a conception of social agency, and so takes as a point of departure the interrelationship between discourse and agents. She argues that without agents promoting them, identifying with them, and struggling over them, discourses would not exist.

Litfin's is an important clarification, but it seems to take away from Foucault's observations on the strength of discursive power and the unpredictability of discursive politics. Wright's way of addressing this issue fits more closely with how I work with Foucault's insights here. She argues that policy analysts can take seriously Foucault's observations about the dispersal of power through discursive practices while still maintaining a voluntarist concept of agency. As Wright (1994, 14) puts it,

> There is no reason not to assume that power may be expressed at the center as well as at the periphery, by agents as well as through institutions and discursive practices. The historical problem of revealing the processes through which power is diffused – from the center to the periphery and possibly vice versa – should not be foreclosed.

An Integrated Theoretical Approach

Both Gramsci and Foucault offer a great deal to an examination of the politics of genetic engineering and the Cartagena Protocol. However, there are significant differences between these two theorists that cannot be glossed over. For example, while both consider the central role of ideas in power relations, for Gramsci, the material dimensions of power relations remain the starting point for analysis. For Foucault, power has no "possessor," and his works were written, in part, as a reaction to historical materialism and its fairly rigid notions of class-based power relations. Still, there are many ways that these theorists complement one another. There are even overlaps between their approaches and that of regime theory, such as on conceptualizations of governance.

Each of these three approaches would recognize that MEAs are about more than nation-states collectively determining how to govern the international environment. For its part, regime theory begins with the understanding that while regimes usually have institutional dimensions, they also have normative and cognitive dimensions, and that these are as likely to have been shaped by the participation of non-state actors as they are to have involved nation-states. Gramsci also emphasizes this kind of governance through his theory of the "extended state," which includes both political and civil society, with the latter fundamentally important to the authority and stability of the former. A similar theory of governance, broadly conceived, is developed in Foucault's concept of governmentality. Foucault stresses the role of non-governmental bodies, expertise, and disciplinary *savoirs* (originating both inside and outside the state) in establishing the norms of modern governmentality.

These shared observations on governance point to the need to examine the emergence of an MEA such as the Cartagena Protocol on Biosafety historically and contextually. This treaty needs to be studied as one element of a wider regime – the body of discourses, practices, and institutions that have come to constitute the field of genetic engineering governance across civil and political society – that is in the process of being constructed. How has this context shaped the Cartagena Protocol, and what are the protocol's effects within this regime as a whole?

When neo-institutionalists study an emergent regime, they tend to focus on the mechanics of cooperation, and the development of shared understandings. While there certainly are elements of cooperation and consensus-formation in the field of biosafety that need to be documented, this field has also been a terrain of conflict and struggle among actors who hold widely divergent perspectives on the nature of the genetic engineering "problem" and how it might be solved. To gain a grasp of this field of both conflict and consensus, the Gramscian theorization of politics as a war of position engaged in by groups trying to normalize their own perspectives as hegemonic, and needing to make accommodations in the process, offers a strong overarching conceptual frame.

This approach means trying to understand the politics of genetic engineering in terms of the three sets of relations of force, or arenas of political activity. Following Cox and Levy and Newell's interpretations of Gramsci, I refer to these three realms as material capabilities, organizations, and ideas. Each is a site of action, of alliances, and of struggle, through which actors with an interest in GE issues (including individuals, states, businesses, and environmental organizations) engage to build structures supportive of their interests. While conceptualized independently, these three sets of relations of force are closely interrelated, with the balance of one set of forces being, at any given time, potentially both a product of and productive in activities and struggles in the other two fields. Figure 1 depicts the three sets of relations of force as intersecting circles, with the realm of ideas (understood to function ideologically and/or discursively) larger than the other two circles. This size differential conveys the notion that for both Gramsci and Foucault, ideas represent the most important arena of modern political activity. Furthermore, for both of these theorists, it would be impossible to speak of material capabilities or organizations with any real political impact (power) that are not intrinsically linked to, made sense of within, and defended through ideational structures.

What does it mean to examine each of these three sets of relations of force, and how do the theoretical approaches I draw upon inform such an

Figure 1

Relations of force

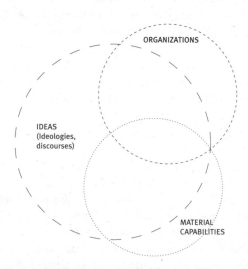

examination? Material capabilities, for Gramscians, remain the foundation of national and global politics. To take material forces seriously as an arena of political engagement means to pay attention to the dynamics of finance, production, consumption, and the material dimensions of technological change. In this case, looking at material capabilities requires an analysis of the major productive actors in the field of agbiotech, from the academic researchers that developed the techniques of genetic engineering to the seed companies, farmers, food processors, distributors, and consumers who buy and sell GEOs. It also means examining the actions of those actors who actively reject GEOs and define themselves accordingly. The ways that these groups conduct themselves, in terms of the products they grow and market and how they interact with one another, create the material conditions around which struggles over the ideas and the organization of genetic engineering governance take place.

The second set of relations of force, of organization and institution building, is a central preoccupation of both regime theorists and Gramscian analysts. Regime theorists point to the need to study the way that states, under the influence of domestic pressures and international obligations, are able to negotiate shared norms on the question of biosafety. Their approach urges a focus on the development of scientific consensus, the formation of shared interests, and the role of international institutions in facilitating cooperation.

While the dynamics of interstate cooperation are important to study, Gramscians would argue for the need to look at these relationships in the context of social relations in the extended states that make up the international community. It is at the levels of civil society that alliances are first formed, where organic intellectuals initially exercise leadership, where concessions towards potentially divergent interests are made, and where consensus is generated. These alliances then help shape political society (the state) and eventually international treaties. By the time a country such as Canada enters international negotiations on an issue such as biosafety, its stance is not simply a national position; it is rooted in the social relations of biotechnology within Canada's state–civil society complex and has also already been influenced by transnational state and civil society networks.

When it comes to organizational relations of force, there is one role that Gramscians do recognize as being reserved for the nation-state, narrowly defined: coercion. Consensus may be developed in formal institutional negotiating fora, but states also use military or, as is more likely in the environmental field, trade pressures to bring other states into conformity with their positions on the best definition of an international regime. The role of coercion is often ignored by neo-institutionalists focused on the *negotiation* of consent, but it must be a part of the analysis of biosafety politics undertaken here.

Ideas are the third set of relations of force and arguably the most important. As with the other forms of political activity in this integrated framework, ideas need to be studied as sites of struggle that crystallize within the
context of social and material relations, rather than being brought to politics from somewhere on the outside. This model theorizes ideational struggles in two ways in order to draw attention to two distinct aspects of the
power of ideas. Following Gramsci, it is useful to think of some ideas as ideologies. While designed to represent a common interest, an ideology can be
traced back to the interests (whether material or otherwise) of specific classes
or groups in society, and to the ideological apparatus and organic intellectuals who gave it shape. At the same time, because they act as resources in
the process of alliance formation and accommodation of otherwise disparate social groups, ideologies can genuinely come to reflect a plurality of
interests.

This framework also adopts Foucault's insights on the discursive function
of ideas. The distinctions and overlaps between the Gramscian concept of
ideology and the Foucauldian notion of discourse are outlined in Table 1,
which presents simplified definitions, along with key concepts related to
the formation and effects of these two types of ideational forces.

It is clear that there is considerable overlap between the theoretical concepts of discourse and ideology, yet there are also important distinctions.
Among the most important concepts that a Foucauldian analysis adds to a
researcher's analytical toolbox is greater attention to the way discourses are
embedded in language and conventions of truth (whether scientific or otherwise). Gramscians may recognize that ideas can have productive effects by
legitimating certain courses of action, for example, but a discursive study of
the normalizing effects of ideas as truth would look for more far-reaching
impacts. It is also important to consider that among the impacts of a discourse is the way that interests, and the very self-perception of agents, are
constructed and reinterpreted in a discursive context. This suggests a more
complex relationship between knowledge, interests, and agency than the
Gramscian reading of ideology. Furthermore, not only do powerful ideas
require negotiation and consensus building, they offer loci of resistance.
This relationship between power and resistance within discourse is important to explore and, once again, it is a direction that a Gramscian reading of
ideology does not suggest on its own. Finally, because of the way debates
over truth empower those groups able to mobilize technical expertise, and
because of the internal logics of these truths, ideas are not as easily manipulated by specific groups in society as the Gramscian reading of ideology
suggests, or as those who engage with them might hope.

Viewing the politics of ideas as discourses in the realm of agricultural
biotechnology and the environment through a Foucauldian lens clearly adds
directions of inquiry to a study of the power of ideologies in this field.

Table 1

A comparison of Gramscian ideology and Foucauldian discourse

	Ideology	Discourse
Simplified definition	A coherent, systematic philosophy designed to be the new "common sense"	An ingrained pattern of talking, thinking, and doing held as truth
Formation	• Driven by specific agents • A manifestation of interests (material or otherwise) but not necessarily a product of material structures • The outcome of struggle and compromise • Eventually reflects a plurality of interests	• Discourse/agency relationship not linear • Often defined by experts • Rooted in language, history, and culture • A product of contestation • Mutually constitutive with social and institutional practices
Effects	• Guides behaviour • Legitimates certain activities • Unifies law, economic activity, and art • Enables consent and the formation of alliances	• Produces specific outcomes by defining what makes sense • Disciplines actors and outcomes by marginalizing other ways of knowing and doing • Creates opportunities for resistance

Methodologically, this combined approach requires an analysis of the positions taken in the formation of the biosafety regime – in the Cartagena Protocol negotiations and outside it – and the social and material relations to which they are tied, the layers of meaning on which they depend to make sense, and the full range of their productive and disciplinary effects.

The term "historical bloc," as I use it here, refers to the functional units of allied social forces that form around a set of ideas in a specific field. An historical bloc is the product of hegemony construction on the part of a class or other social actor, rooted first and foremost in civil society, and necessarily allied with arms of the state. Once formed at the level of extended states, an historical bloc can be expected to continue working to build and maintain hegemony through both coercion and consent, with the latter involving the formation of a new "common sense." The centrality of civil society in an historical bloc, including norms, values, and expertise established at this level, as well as civil society's ideal relation to the state and the global order, are depicted in Figure 2. While the three levels are shaded, there are no firm boundaries between them. This is meant to indicate the

Figure 2

Historical bloc

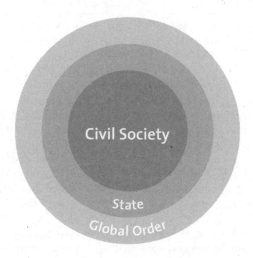

close relationships that need to exist between the three levels for an historical bloc to be successfully established, or hegemonic. To achieve its
hegemonic status, a great deal of consensus building and accommodation
must also have taken place. As a result, the historical bloc that achieves
global hegemony is never the same bloc, with exactly the same agenda, as
that which began the process at the heart of civil society.

Another way of conceptualizing an historical bloc and its hegemonic
ambitions, one that demonstrates more clearly the importance of the three
sets of relations of force described above in its creation and maintenance, is
presented in Figure 3. This diagram builds on Figure 1. The area at the intersection of the three concentric circles of relations of force in these two figures can be thought of as the material-ideational-institutional convergence
that is necessary for an historical bloc to gain a foothold. In consolidating
hegemony, whether at national or international levels, an historical bloc
works to solidify and extend its alliances through strategic actions, eventually (if successful) expanding the area of overlap among the three circles, as
Figure 3 illustrates. By definition, however, a hegemonic constellation, even
if temporarily stable, is never fully consolidated and always remains contested from outside as well as from within. This is illustrated by the spaces
that remain outside the hegemonic formation but within the political constellation as a whole.

Figure 3

Hegemonic relations of force

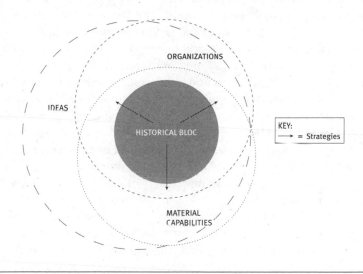

The Gramscian view of temporary, contested hegemony is not unlike what Foucault describes as a discursive constellation of power/knowledge. Both perspectives invoke pervasive structures that involve ways of thinking, talking, and doing that together define the world, objectively and subjectively, at a particular point in time. Foucault's description of governmentality and of biopower, I would suggest, can each be seen as descriptions of constellations that became hegemonic within their respective fields. Both Gramsci and Foucault point towards the need to conceptualize the field of biosafety, with its own internal political dynamics, as emergent within the context of already extant hegemonic/discursive constellations.

Gramscian scholarship would suggest that the genetic engineering industry emerged within an already hegemonic constellation centred on transnational capital and the ideology of neo-liberalism. This wider constellation can be expected to have influenced genetic engineering through material relations as well as through its particular conceptions of property rights, norms of acceptable (and unacceptable) government intervention, and assumptions about the inherent value of growth and liberalized trade. The institutions of neo-liberalism, such as the WTO and the OECD, need to be studied to see what roles they may have played in the development of agbiotech. Being hegemonic, the neo-liberal order is also necessarily contested and productive of its own resistance. One source of resistance is the environmental movement. This resistance raises the question, how have the political dynamics of environmentalism, shaped in the context of neo-liberalism, affected politics in the field of biotech?

A Foucauldian analysis points towards the structures of biopower as a locus of dispersed power active in the genetic engineering field. This constellation is centred on discourses and practices of the governance of life. It brings together industrial capital, military interests, science, and practices of governmental administration and regulation. Biopower appears to have shaped the field of biotech on multiple fronts, from the engineering drive of molecular biology to the practices of environmental regulation that are employed in this field. Foucauldian scholarship also stresses the points of resistance within biopower related to practices of environmental management and to the quest of individuals to regain control over their own food and health, for example. How have each of these factors shaped the biopolitics of agbiotech?

Insights gained from the study of these two hegemonic/discursive constellations are enormously useful in contextualizing the politics of biosafety and the Cartagena Protocol, and they are unpacked in Chapters 2 and 3. While I don't wish to obscure differences by suggesting that conceptualizations of biopower and the transnational historical bloc are inherently complementary in all respects, they do both point to related sets of ideas and practices that have likely been influential in shaping the politics of GE. Specifically, they point to the importance of liberalism (as an ideology and/ or as a discourse), contemporary state-scientific-industrial relations in the West, and the politics of environmental resistance as key elements of the hegemonic/discursive order within which the biotech industry, and the politics of biosafety, emerged.

The integrated theoretical framework presented here is ambitious. It sets out to interpret the Cartagena Protocol on Biosafety as an MEA shaped within an emerging global regime of genetic engineering governance. This framework suggests looking at biotech governance and, indeed, at the agbiotech project itself, in the context of activities and debates that have taken place across three sets of relations of force. That these relations of force are not narrowly bounded by the field of biotechnology itself points to the need to examine these struggles in the context of the larger hegemonic/discursive constellations of the transnational historical bloc and biopower.

Figure 4 presents an overall conceptual map for this study. The object at the centre of our inquiry is the project of agbiotech and the actors, ideas, and institutions that came together to initially support this revolution in agriculture. The state and civil society complex (or nascent historical bloc) identified in this figure is what I describe in Chapter 2 as the biotech bloc. It is dependent on overlapping institutional, material, and ideational support structures for its existence, so it is located in the centre of the diagram where

Figure 4

Integrated conceptual map

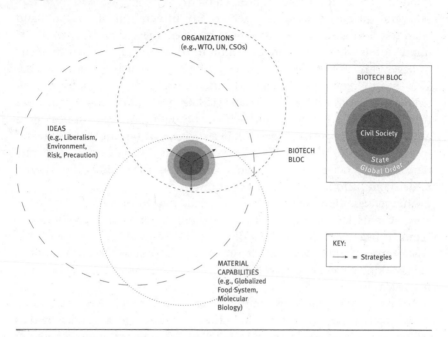

such overlaps occur. At the same time, surrounding the biotech bloc is a wider array of institutions, material capabilities, and ideas, some of which support the bloc's ambitions and some of which do not. The relationships between wider relations of force and the biotech bloc (at the level of civil society, the state, and global order), and the strategies the biotech bloc developed to build support in each of these arenas of political activity, are worked out in the next two chapters. The larger question that really drives our interest here, that of the relationship between the three sets of relations of force, the biotech bloc, and the Cartagena Protocol itself, are developed in Chapters 4 through 7. One way of phrasing the query that this conceptual framework can help us answer is: What is the relationship between the regime that has enabled agbiotech to come into being, and the institution, norms, and discourses of the Cartagena Protocol on Biosafety?

Because of its hybrid nature, my theoretical framework raises important questions. First, is the distinction between the function of discourses and ideologies truly helpful in theorizing ideational politics, or am I simply splitting hairs in drawing out these distinctions? This is important given that some Gramscian scholars, such as Levy and Newell (2002), speak of ideologies and discourses as if these concepts are synonymous. Second, can this framework really help make sense of a political event, such as the emergence of the Cartagena Protocol, in a way that any of the three theoretical traditions I draw from could not on its own?

Several questions are also raised by my integration of Foucault and Gramsci's ideas, since the differences between these theorists may be as important as what they share in common. For example, consider the issue of hegemony and change. Foucault's analysis of discursive politics emphasizes the inherent unpredictability of outcomes and thus the fluidity of politics. Gramsci acknowledges that change to hegemonic formations often arises from within a hegemonic formation, but his definitions of historical blocs and hegemony appear to suggest rigid structures with static leadership and overarching material goals. If we want to take the possibility of incremental (and unpredictable) change seriously, how would this affect the way we understand historical blocs and hegemony?

Another question concerns the conceptualization of resistance politics. Many Gramscians theorize political struggles in terms of hegemonic and "counter-hegemonic" forces (e.g., Hisano 2005). Foucault, on the other hand, characterizes normalization and resistance as inextricably linked. If Foucault's observations have merit, how might this affect an understanding of counter-hegemony?

A final question concerns the role of agency. Gramsci identifies organic intellectuals as political actors with a high degree of agency, understood as the power to consciously reformulate political life. Foucault, however, refuses to discuss agency in much of his work, focusing instead on webs of discursive power. These webs, in his view, are the truly pervasive forms of power in society, "yet, it is often the case that no one is there to have invented them, and few who can be said to have formulated them" (Foucault 1978, 95). How are we to reconcile these two understandings of power in a way that builds, rather than takes away from, the insights of both of these influential thinkers?

These questions cannot be answered at a solely theoretical level. So, I bring them along into the empirical chapters of this book, addressing them when the material offers relevant insights. I conclude Chapter 7 by showing how the case study presented in these pages actually allows us to bridge some of the theoretical differences between Gramsci and Foucault.

2
The Biotech Bloc

The significance of the Cartagena Protocol can be understood only in the context of the larger story of agricultural biotechnology in the 1990s. This story is best described, in Gramscian terms, as the gradual establishment of a new historical bloc in the global agricultural and food system. Although not the product of the interests of only one group in society, the techniques of genetic engineering became the material foundations for a hegemonic project spearheaded by specific actors in the agri-chemical sector in the late 1970s and early 1980s. Together with its state and academic allies, the companies at the heart of the biotech bloc worked over ensuing decades to foster for a new direction in global agriculture.

In this chapter and the next, I document some of the key strategies adopted by these actors to build consent for their project, especially for the commercialization of GEOs in the 1990s. I begin by looking at the realm of material production. This is where the agri-chemical corporations that led the bloc employed genetic engineering to create products that were innovative not only for the traits they embodied but also for the relationships they built with others in the food system. I then examine strategies adopted to build supportive organizations and institutional structures, and how, as popular resistance to GE crops and foods grew, the bloc worked through civil society bodies such as universities and non-profit organizations, as well as regulatory arms of supportive states, to establish consent for agbiotech. Business and state arms of the biotech bloc also operated in a variety of international fora, including the WTO and the OECD, to institutionalize industry-enabling policies during the 1980s and into the 1990s. Strategies adopted by the bloc in the third field of political activity, that of ideas, is my focus in Chapter 3.

This discussion of the biotech bloc is largely focused on the United States and Canada. It was in the United States that the first successful genetic engineering experiments took place, in the early 1970s, and it is where GE crop varieties were first commercialized in the 1990s. What happened in

Canada is of interest given the role Canada would eventually play in the Biosafety Protocol talks. Still, it is important to note that the biotech bloc has been transnational from the outset. Corporate and state participation in the biotech revolution in Argentina and Canada closely mirrored that of the United States. And much of the early research in genetic engineering, including developments in agricultural applications, also took place in Europe. Despite the popular assumption that Europeans are not supportive of biotechnology, governments, research centres, and corporations in France, the United Kingdom, Switzerland, Germany, and Belgium, in particular, had all been involved in genetic engineering since the 1970s. This is an important part of the GE story, shaping the politics of regulation in Europe and eventually setting the context for significant intra-European divergences over genetic engineering in the Biosafety Protocol talks. China also developed its own commercial agricultural biotechnology sector in the late 1980s and early 1990s.

Origins

The biotech bloc emerged in the 1970s at the intersection of three sets of material forces: the biopolitical drive to develop biology into an engineering science; a globalizing food and agricultural system centred on the North American livestock industry; and state-sanctioned commercialism in US academic science. The story begins with the discovery of the techniques of genetic engineering, and these were no accident. GE techniques were the product of years of diligence and hard work on the part of teams of scientists around the world; they were also the product of particular social relations of power. Specifically, the techniques of genetic engineering emerged towards the end of a postwar period of heavy investment in the sciences by governments in the United States and the United Kingdom, the two principle centres of molecular biological research activity. Both of these countries experienced sustained and rapid economic expansion from the end of the Second World War to the late 1960s, and both responded by promoting the advancement of science, especially "big" science such as nuclear weapons, nuclear power, and synthetic chemicals (Wright 1994, 19-20). While both pure and applied sciences were beneficiaries of state largesse, there is no doubt that the underlying motivation for investment was a particular kind of technological progress. In the field of biology, which since its inception a century and a half earlier had focused on describing the underlying mechanics of life, the focus in the second half of the twentieth century was clearly on understanding these mechanics in order to control them. As Kay's work shows, in the United States, the majority of the funding that led to the discovery of the genetic code came from the Department of Defense and the Atomic Energy Commission (Kay 2000, 10-11). It was in this context that the first successful genetic engineering of a strain of bacteria took place, at Stanford University in California in 1973.

The second part of this story concerns the structure of the food and agricultural system in the 1970s and early 1980s. Along with other areas of the global political economy, the transnational historical bloc was growing in its influence on the food system. The result was a global food system increasingly centred on transnational corporations based in the United States. This was in marked contrast to the period from the 1930s to the 1970s, when agriculture in the United States, as in most industrialized countries, was dominated by protectionism and a nationally oriented food system. Friedmann's analysis shows how the shift towards neo-liberal food policies and a global outlook began as the three separate and distinct complexes that made up the US postwar food regime – wheat, durable foods, and livestock – production became increasingly integrated (vertically and horizontally) around the livestock industry (Friedmann 1994). The new system had two specific characteristics that created incentives for participants in the food system to invest in genetic engineering. First, it relied on a very limited number of crops (corn, soybeans, wheat, and rice among them) to produce the bulk of the world's food and fodder. This meant that any new beneficial traits developed in these crops would quickly find a global market, if those traits could be controlled or considered proprietary in some way. Indeed, such intellectual property protection over genetic engineering quickly became the norm. A landmark 1980 decision by the US Supreme Court on the case of a patent claim on a genetically engineered strain of bacteria meant that proprietary rights to GEOs were protected by US courts. Second, this system was highly dependent on inputs such as fertilizer and pesticides. As a result, agri-chemical corporations had become formidable players in the food system, and they wanted to protect and build market share.

The third part of this story, that which led to the meeting of a globalizing food system and genetic engineering, is about the growing commercialism in academic science in the 1970s. Significantly, this commercialism was not simply the result of large corporations moving into the field, as one might expect. Rather, it was many of the molecular biologists themselves, in the mid-1970s, who started looking at what they were doing in terms of potential economic benefits. These scientists began biotech start-up companies to research products and techniques. From 1975 to 1983, more than 250 small biotechnology firms were founded in the United States, bankrolled by venture capital (Shand 2001, 223-24). Because of this commercial drive among some US scientists and the willingness of venture capital in that country to finance them, this number was significantly higher than in any other country or region of the world.

While many molecular biologists began commercial ventures on their own, this entrepreneurial spirit was encouraged by the new economic and political climate of the 1970s. Because of economic strains – and the growing influence of neo-liberalism – states on both sides of the Atlantic started to target their funding of science, and they urged industry and financial capital to increase their investments (Wright 1994, 52). Monsanto, DuPont, Pfizer, and other companies in the agri-chemical and pharmaceutical sectors began to invest in biotech in the late 1970s and early 1980s. These companies also started forming partnerships, or undertaking buyouts, of biotech start-ups. This was the period when genetic engineering became the material basis of a nascent historical bloc. The techniques had been adopted by particular sectoral interests (those of the agri-chemical industry, in the case of food and agriculture), and these interests had set their sights on the development of commercial applications of genetic engineering in farming designed to complement their overall business strategies.

The outcome was a bloc, and a particularly powerful one at that, because the industries that had moved into the field of agricultural applications of genetic engineering (just like the pharmaceutical companies that were interested in medical applications) managed to harness and cultivate the close relationships between the state and academia that had given birth to genetic engineering in the first place. As an example, in a groundbreaking 1975 Monsanto–Harvard Medical School agreement targeted at developing medical applications of genetic engineering, Harvard received $23 million over ten years. In return, Monsanto received first rights to all patentable results of the project. By 1982, the emerging biotechnology industry invested $326 million per year in US universities, and investment rates were rising at a rate of 11 percent annually (Wright 1994, 58).

Rather than execute a full retreat, in the early 1980s, industry branches of governments in the United States and Canada retained their presence in the biotechnology field, on the assumption that this would form part of the foundation of a new knowledge-based economy. The Canadian government, for example, released the 1983 National Biotechnology Strategy. The strategy included direct investment in research to promote genetic engineering. Tax incentives were given for corporate investment in research, and matching-funds models were developed, whereby public funds would be given only to those researchers who could find matching funds from industry (AAFC 1996). By the late 1990s, 10 percent of the Canadian government's research budget was devoted to biotechnology. In the fiscal year 1997-98, this amounted to $314 million (White 2000, 14).

By the time foreign genes were first successfully incorporated into the DNA of plant cells, in the early 1980s, then, the social relations of genetic engineering had changed substantially from ten years earlier. What was new about this constellation, in the context of agricultural research, was the strength of the tripartite relationships among academics, governments, and industry, and the way in which the research projects financed by these forces was so narrowly determined by the interests of industry. For instance, it was not university researchers but scientists working for Monsanto in St. Louis, Missouri – drawing on work from academic scientists that both Monsanto and the US government had sponsored – who were the first to cultivate genetically transformed petunia cells into thriving plants (Charles 2001, 5).

At first, the position of the agri-chemical industry at the heart of the biotech bloc meant that more funding was available to undertake a wide range of basic research. These companies encouraged an exploration of the possibilities of the new engineering science in what Charles terms the "golden age of biotechnology." By the mid-1980s, however, agendas came to be disciplined by the drive for profit. Researchers learned that although they could use these techniques to develop virus-resistance in certain plants, for example, these ideas would not be pursued for commercialization. Instead, the research agenda would be dominated by the search for traits that could bring major economic returns, such as herbicide-resistance (Charles 2001, 31-40).

As the engineering capabilities of molecular biology were pursued, other kinds of research were also left aside. Scientists critical of genetic engineering argue that little funding was made available during this period to those who wanted to undertake research on the ecological or health implications of GEOs (Crouch 1991). Although rarely acknowledged by the governments and companies at the heart of the biotech bloc, this knowledge gap would begin to be filled only when European consumer concerns in the late 1990s spurred on more extensive ecological and health studies.[1]

The Production of GEOs

While there were incentives for agri-chemical companies to move into the field of genetic engineering, a widespread transition to GEOs was no guaranteed success. Testimony can be found in a case documented by Charles, that of the first US company to commercialize a genetically engineered crop: Calgene. During the 1980s, molecular biologists at this small US biotech company (financed through the French chemical giant Rhône-Poulenc) developed the "Flavr Savr" tomato. This tomato contained a new gene that delayed the softening process in the fruit once it had been picked. Despite that this trait appealed to consumers, Calgene and its Flavr Savr were a failure on two fronts. First, at the level of production, the company did not manage to transfer its trait into tomato varieties that would produce high volumes under varied growing conditions; this resulted in poor yields. Second, its own in-house sales distribution system for the ripe tomatoes was poorly managed and expensive compared with conventional distribution channels for unripened fruit (Charles 2001, 126-48). The Calgene story shows how an agricultural revolution based on genetically engineered traits required business models that would work, and alliances with other links in the food chain that would be profitable: it required a material base. How did the agri-chemical and biotech companies in the biotech bloc manage to make genetic engineering profitable for themselves? And how did they manage to build loyalties for their project among other fractions of capital in the food system: the seed companies, farmers, food processors, and distributors?

The most successful strategy adopted by these companies was to introduce traits that would further the sales of their existing product lines of agri-chemicals. This strategy is evident in the most widespread application of genetic engineering in crops – one that continued to account for over 80 percent of all commercial GM seeds planted in 2005 (James 2005) – the introduction of genes that make plants tolerant of broad-spectrum herbicides. Herbicide-tolerant traits were first introduced in cotton and soybean seeds in the United States, then in Argentina and Canada. In the case of Monsanto, the sale of its Roundup Ready seeds enabled the company to extend indefinitely the patent life of its glyphosate-based herbicide, Roundup.

To get seeds bearing these traits to farmers while ensuring maximum return on investment for developing them, Monsanto established a new form of licensing deal with seed companies and farmers. In its agreements with seed companies, Monsanto licensed the right to use the herbicide-resistance genes as long as each bag of seed was printed with the phrase "Roundup Ready." Europe-based Hoechst (later AgrEvo) followed with a similar strategy to promote its Liberty Link herbicide-tolerant soybeans. Despite limited returns for the herbicide-tolerant traits themselves, Monsanto and Hoechst learned that the strategy of developing and marketing the new GE seeds as distinctive was of enormous benefit in extending chemical markets and in building recognition for their agri-chemical products (Charles 2001, 120-21). While farmers had previously identified plant varieties according to a wide variety of traits that they might offer (such as length of growing season and drought tolerance), the new seeds built brand recognition for only one genetic trait and the associated chemical.

The second most important application of genetic engineering – still found in almost 30 percent of globally produced transgenic crops in 2005 – is the generation of corn, potato, and cotton seeds with Bt genes from the bacterium *Bacillus thuringiensis* (Bt) to express proteins that are toxic to a variety of insect pests.[2] Powdered Bt had long been used by organic farmers seeking natural pest control. Now, Bt seeds were designed to replace the use of chemical insecticides used by most conventional farmers. As a business strategy, these seeds allowed the biotech purveyors to take market share from agri-chemical competitors.

In marketing Bt crops, Monsanto and the other agri-chemical firms used this trait as another opportunity to solidify new relationships with seed companies and farmers. Seed companies were granted licences in exchange for tacking on a premium to the price of any seed enhanced with Bt genes. This newly introduced "technology fee" was designed to be just a little less than whatever costs (e.g., of insecticides) the farmer saved by using the new genetic properties of the seed (Charles 2001, 122). This arrangement with the seed companies allowed the biotech firms to keep direct control of the price charged to farmers for *their* genes, as well as the right to enforce their property rights to the seed.

The biotech companies also developed "technology use agreements" (TUAs) that farmers were expected to sign when buying biotech seeds. The TUAs were a contract with the company that owned the gene (as opposed to the company that owned the seed). They stipulated that the farmer would use only that company's herbicides on the crop, even if generic equivalents were available; that the farmer would not save any seed for replanting; and that the biotech company had the right to inspect the farms and grain bins up to three years after the planting (Monsanto 2000).

In practical terms, the TUAs amounted to a new form of contract farming. Since the 1970s, a contract farming model – in which food-processing companies supplied all the necessary inputs and offered a guaranteed price for the output – had been gradually taking over independent farming in various North American agricultural sectors, notably poultry and potatoes (Friedmann 1994). But no inroads had been made into the dominant field crops of soybeans and corn. Along with the hype of genetic engineering, a contract model had now arrived for these field crops, though without even a guaranteed return on the farmer's investment. The only benefit accrued to the farmer was the right to use the gene and associated products.

The introduction of these agreements generated considerable resentment towards the biotech companies from many in the farming community. While producers accepted the need for biotech companies to earn a return on their investment in these new seeds, they also recognized the new contracts as a power grab by the gene purveyors because these reduced, rather than enhanced, the competition needed to eventually lower the price of farm inputs. Still, the contracts did not stop a large number of farmers from buying into GE seeds. In 1996, 1.7 million hectares were planted with commercial biotech crops in the United States, Argentina, and Canada. By 1998, this rose to 27.8 million hectares, mostly planted with herbicide-tolerant soybeans, Bt corn, and Bt cotton (James 2001).

How were the agri-chemical and biotech companies able to generate such strong buy-in among farmers in the United States, Argentina, and Canada, even after adopting strategies that clearly alienated many in the farming community? The answer is complex. To begin with, it is important to note that in the grain sector, farmers in the United States and Canada are direct competitors. So, decisions to adopt a new technology taken by farmers south of the border, especially if it leads to cost savings, carry a great deal of weight in Canada. Argentina was targeted early in the development of GE crops by the biotech industry because that country is where much of the seed used in North America is grown during the northern hemisphere's winter months.

The main reason farmers in these countries chose to adopt herbicide-resistant and pesticidal seeds is because the crops offered immediate advantages. In Argentina, a country known to have major weed problems, herbicide-resistant crops had clear short-term benefits. For farmers in the United States and Canada, many of whom plant or spray at night or on weekends because of their need to hold other, often full-time, jobs, these crops worked and were considered time-savers. GE crops also brought economic benefits to the first wave of producers who adopted them. One US study found that the relative returns for switching from non-Bt to Bt cotton in 1997 were almost as high for the farmers as they were for the biotech company involved. Farmers in regions with lower adoption rates were seen as disadvantaged because they continued to use more expensive technology even as the price for cotton came down (Falck-Zepeda, Traxler, and Nelson 1999).

Furthermore, public resistance to GE crops and foods was limited in these countries in the mid-1990s, especially when compared with what would erupt only a few years later in Europe. When farming, environmental, or consumer groups argued for labelling of genetically engineered foods, for example, it was clear that governments in all three countries stood by the biotech companies, refusing outright to implement anything that might make food containing GE ingredients vulnerable to boycotts. Meanwhile, those farmers who were reluctant to jump aboard yet another technological treadmill found themselves in a difficult position. In the past, they had seen neighbours lose opportunities for bank loans if they did not adopt the latest technologies, so they understood the consequences of being left behind. In the short term, this combination of factors meant strong adoption of genetically engineered crops among farmers in the United States, Argentina, and Canada.

Unfortunately, the history of agriculture suggests that the benefits farmers enjoyed from employing GE crops would not remain the norm over the longer term. Because of the limited number of buyers, the relative glut of food on world markets, and the large number of producers who are willing to work just to break even, most producers have not been in a position to bargain for strong prices since the first half of the twentieth century (except for those working in supply-managed sectors, such as the Canadian dairy sector, where farmers are paid based on a formula that relates to their costs of production; for a history of the supply management system, see Winson 1993, 77-84). In this context, the structural consequence of the introduction of new technologies has typically been shrinkage in the number of farms and farmers, rather than increased net farm incomes. Such consequences can already be seen in Argentina. Since farmers there began growing GE soy in the mid-1990s, commodity prices have declined while reliance on imports of chemicals has increased, overall number of producers has decreased, and average farm size has increased (Lehmann and Pengue 2000). These structural responses to new technologies will persist unless a new political will emerges to ensure stable incomes for small- and medium-size farmers in the global food system.

TUAs, and their enforcement through North American courts, ensured that farmers returned to the seed companies to buy GE seed each year. Recognizing that such legal means of protecting intellectual property may not be as robust in many countries of the global South, one US biotech company, Delta & Pine Land, developed a series of genetic modifications designed to shut off the reproductive ability of a plant, so that the seeds from a transgenic plant cannot reproduce successfully. This was termed the Technology Protection System (TPS), because it "simplifies protection of technology and removes it from the legal arena" (Radin 1999, 24). Significantly, the research on the TPS was funded by the US Department of Agriculture (USDA). The mandate of the USDA had previously been to support research that benefits farmers, but under neo-liberalism, this mandate has been reinterpreted. In the case of TPS, the USDA's stated goal was to "increase the value of proprietary seed owned by US seed companies and to open up new markets in Second and Third World countries" (USDA representative interviewed by the Rural Advancement Foundation International, quoted in Lehmann 1998, 6-7).

Although the US patent for TPS was awarded in 1998, the trait has not yet been introduced in a commercial seed variety, possibly because the proposed trait raised more public backlash than any other. TPS has the potential to have an enormous impact on agriculture. For example, in India, over 80 percent of farmers plant their own farm-saved seeds (Swaminathan 1998). When the patent on the trait was made public in 1998 by the Canadian-based Rural Advancement Foundation International (RAFI; later renamed the ETC Group), environmental, human rights, and development groups around the world criticized the new "terminator technology," holding it up as proof that the emerging biotech industry had set its sights on removing the rights of the world's farmers to save and reseed their crops.

This outcry was so great that late in 1999 the CEO of Monsanto – which earlier that year had made an offer to buy Delta & Pine Land – was forced to state publicly that his company would not commercialize the TPS. This move appears to be an example in which the biotech industry has made concessions because of public pressure. However, in March 2000, insiders in Delta & Pine Land's labs told RAFI that they have never slowed in their development efforts for terminator seeds. RAFI also discovered that more than thirty other patents have been issued in the United States to thirteen institutions describing techniques that control seed germination or the use of inducible promoters to activate traits or performance of GE plants (RAFI 1999). In 2005, the Ban Terminator Campaign, a project spearheaded by RAFI, further revealed that patents on TPS had been granted in Canada and in the EU in October of that year (Ban Terminator Campaign 2005).

Each of the above is an example of how the advent of genetic engineering in seed development and marketing has been used, or will be used, to benefit the agri-chemical companies that took control of this field in the early 1980s. Generally, the first generation of GE crops has been an enormous success for the agri-chemical and biotech industry. The choice of marketed traits played an important role in this. At a time when there was considerable uncertainty about the profitability of genetic engineering in agriculture, herbicide-resistance and Bt genes brought returns by changing the rules of the game in markets that the early biotech purveyors knew well: the herbicide and insecticide markets. These were also traits that would appeal to farmers. Given the controversial strategies developed to retain control of the engineered genes, such as technology use agreements and terminator technology, it was important to be able to get farmers onside. Furthermore, while there was some resistance to the new GE products from environmental organizations in the United States (as described below), the first generation of traits was actually not as dangerous for the environment as it could have been, and this fact played in the biotech bloc's favour. Monsanto's Roundup, for example, has a lower persistence in the soil and is less toxic to wildlife than most other herbicides. If the first GE crops had been made to be resistant to atrazine (a more dangerous herbicide used in corn production), for example, criticism of herbicide-resistant crops would likely have been higher.

The introduction of GE traits was so successful in the Americas that it utterly changed the global seed business. In the early 1990s, agri-chemical companies were licensing the new traits, but by the late 1990s, most had undertaken outright buyouts of seed companies. Monsanto bought out US seed giant Asgrow in 1996, followed by Plant Breeding International, the leading seed company in Great Britain, in 1998. Meanwhile, DuPont bought out the largest seed company in the United States, Pioneer Hi-Bred, in 1999 (Charles 2001, 195, 235). By the fall of that year, five companies (AstraZeneca, DuPont, Monsanto, Novartis, and Aventis) accounted for nearly 100 percent of the global GE seed market, two-thirds of the global pesticide market, and almost one-quarter of the total seed market (Shand 2001, 231).

Despite the success of the GE seeds, and the major consolidations in the global farm input industry that the adoption of these seeds spurred, there are even larger titans in the food system. These are the giant food companies and grain traders. Since the 1970s, the global food-processing, distribution, and retail system has become enormously concentrated through integration (via mergers, acquisitions, and other mechanisms) in the supply chain. For example, in the United States, by 1999, 80 percent of beef cattle and 57 percent of hogs were slaughtered and marketed by the four largest firms operating in the country (Heffernan 1999, 124). The largest food company in the world, Nestlé, had 1997 revenues of US$45.3 billion, surpassing the entire commercial seed industry (US$23 billion) and the entire agri-chemical industry (US$31 billion) (Shand 2001, 229). Even grain-trading giants Archer Daniels Midland and Cargill dwarf agri-chemical companies such as Monsanto. Thus, the extent to which these companies support or do not support the entry of GE crops into global markets figures centrally in the political success of the biotech project.

To date, the reaction from the food giants and grain traders has been mixed. One can document close cooperation between the food processors and traders and the input suppliers in some respects, but in other ways it appears as if the grain traders and processors have maintained a certain distance from the biotech revolution. Shand believes that we are seeing close cooperation between players in the agri-chemical and biotech industry, the pharmaceutical industry, and the food-processing industry. The common interest among all three groups is the potential to develop new food ingredients whose nutritional enhancements would increase shares of the retail food market (Shand 2001, 229). Such products would need their identities preserved from the seed to the dinner plate in order for any genetic "enhancement" to also realize a profit for the companies involved. Along these lines, Heffernan (1999) posits the emergence of three clearly identifiable "food system clusters" in the global food system, with a few large outlying companies that can be expected to form one to three more clusters in the near future. As one example, the Cargill-Monsanto cluster includes the control of genes, seeds, fertilizers, chemicals, and contract farmers by Monsanto; grain collection and processing; the production and processing of beef, pork, turkeys, and broilers by Cargill; and joint ventures with DuPont, Dow, Mitsubishi, and others. These numerous alliances lead to a seamless system, with a fully integrated food system from the gene to the supermarket shelf:

> Within this system, there will be no markets and thus no "price discovery" from the gene to the shelf ... Starting with the intellectual property rights that governments give to the biotechnology firms, the food product always remains the property of a firm or cluster of firms with close working relationships. The farmer becomes a grower, providing labour and often

some of the capital, but never having clear title to the product as it moves through the food system and never making the major decisions. (Heffernan 1999, 131)

While Heffernan (1999) and Shand (2001) provide evidence for cooperation across the whole food system, these clusters have not yet manifested themselves as outright mergers, suggesting an ongoing divergence in material interests between the input suppliers and the food processors. I have already noted how the global food-processing and distribution system is largely centred on the livestock complex. There is little room or need in this complex for identity preservation. Indeed, this system is based on inputs being largely interchangeable. Processing companies thus have a limited stake, at the present time, in ensuring that the genetic engineering revolution is an unqualified success. Most of the large companies in this field could probably survive in markets hostile to GEOs as well as in markets friendly to them, especially since technologies developed in the late 1990s allow segregated, GE-free channels of food processing and distribution. This divergence of interests was made evident when European consumers, followed by some of their North American counterparts, began boycotting GE foods in the late 1990s. At that time, several food-processing companies, including McCain in Canada, and food retailers such as Britain's Marks & Spencer, decided to remove GE ingredients from their products. In the fall of 1996, when Europeans had not yet accepted Roundup Ready soybeans for import, the grain trader Archer Daniels Midland sent out a letter warning grain terminals that they may have to segregate GE soy. A similar scenario played out in early 1999, when Europeans were hesitant about accepting Bt corn imports. In the first case, Europeans did decide to accept the soy, so costly and complicated segregation was not necessary. In the second case, American corn was diverted elsewhere, so segregation was again not needed (Charles 2001, 164, 254).

These examples show that the interests of the food processors and grain traders (each of which appear to be more easily affected by consumer demand) are not necessarily the same as those of the agricultural input industry. To date, the grain companies and food processors have worked within the biotech bloc on some levels – to set minimalist labelling policies, for example – but they have yet to see an immediate economic interest in a successful genetic engineering revolution. The absence of a compelling economic interest on the part of these food titans remains a dangerous scenario for the pushers of GE seeds.

The Organization of Consent
Hegemony is dependent, first and foremost, on the construction of consent in civil society. This is the field in which a hegemonic project must be made

to appear as the new "common sense." What organizational strategies did the biotech bloc employ in its attempt to build consent in civil society? Proponents of genetic engineering had to overcome considerable dissent, including the responses of farmers and others to the TUAs and TPS/terminator technology. Who else was involved in the resistance to GE crops and foods in the heartland of the biotech bloc? How was dissent organized, and with what effect? I explore these questions below. For even more detailed accounts, see Krimsky (1982, 1991), Tokar (2001), and Schurman and Munro (2003).

Resistance to agricultural applications of genetic engineering occurred in three waves in North America. Each wave built on the previous one while drawing in a wider group of civil society actors as they became concerned about GE. The first wave took place from about the mid-1970s to the late 1980s. The critics at the centre of the resistance were a small group of primarily well-educated, middle-class professionals who had come of age in the civil rights, peace, women's, environmental, and sustainable agriculture movements of the 1960s (Schurman and Munro 2003). In the United States, the core group of activists engaged with GE issues was made up of representatives from the Committee for Responsible Genetics (later renamed the Council for Responsible Genetics [CRG]), the Institute for Consumer Policy Research, the National Wildlife Federation (NWF), the Environmental Defense Fund, and the Foundation on Economic Trends, among others (Schurman and Munro 2003). Several Canadians also were involved in early GE activism, including Pat Mooney (the founder of RAFI) and Brewster Kneen, both of whom had been engaged in extended battles over the growing corporate control of seeds and the food system.

Most of these early GE critics could be seen as true organic intellectuals in the Gramscian sense: they had emerged from earlier social movements with the skills, contacts, and confidence to make an impact in GE politics. As Schurman and Munro (2003, 10) write,

> Armed with good educations, literacy in science and law, and professional credentials, such individuals were not afraid to take on the scientific, legal and political establishments; in fact, these were the terrains on which they felt most comfortable, and where their particular skill sets and experience could be used to greatest effect.

These organic intellectuals included Margaret Mellon, a molecular biologist and lawyer who worked with the US National Wildlife Federation (and later with the Union of Concerned Scientists). Mellon would eventually co-author one of the chief scientific critiques of the environmental implications of GMOs in the mid-1990s (Rissler and Mellon 1996). Notably, many in the first generation of biotech critics were not necessarily against the introduction of GMOs to agriculture per se. Mellon's initial work, for instance, was intended to "assure that the technology is developed with the proper environmental controls" (Charles 2001, 95).

Others among these critics took stronger anti-GE positions. Jeremy Rifkin, founder of the Foundation on Economic Trends, questioned the value of genetic engineering altogether, and his critiques were laden with Christian overtones about scientists "playing God" and the end of Creation. Rifkin was a masterful strategist who engaged in a multi-pronged attack on the nascent biotech industry, combining media stunts with litigation to prevent the construction of new biological research facilities (Krimsky 1991, 119).

Despite their differences in ideological positions on whether the science might eventually prove to be of value, the first generation of GE critics had in common a strong belief that technological change needed to be subject to democratic accountability; the key issue uniting them was *how* GEOs were being developed. As Schurman and Munro's interviews with this first generation of activists illustrate, the critics wanted to create space for democratic input into these *life*-altering (in the most literal sense) technological choices that were otherwise being made solely by corporations (Schurman and Munro 2003).

Early GE activists in the United States had various impacts on the development of molecular biology in that country. In 1977, in Cambridge, Massachusetts, scientists and activists (including members of Science for the People, who would later play key roles in the CRG) worked with residents and the city council to create a citizen review commission to oversee gene-splicing experiments at university labs within the city's limits. After the Cambridge law was passed, another dozen US cities followed suit (Krimsky 1991, 101-2; Tokar 2001, 321-22). As another example, in the mid-1980s, the Foundation on Economic Trends won a series of legal challenges designed to shut down genetic engineering experiments that included studies of transgenic animals, field tests for genetically engineered micro-organisms, human genetic engineering experiments, and the US military's biological defence weapons program (Krimsky 1991, 119-24).

Many of the key GE activists first met face to face, along with representatives of European Greens and an Australian organization, GenEthics, in the late 1980s, thanks to funding from a small California foundation, to form the Biotechnology Working Group (BWG) (Schurman and Munro 2003). The emergence of the BWG was the consolidation of the first wave of a North American (and even international) GE movement and the beginning of a second. The BWG would play a pivotal role over the coming years in GE activism by publishing reports on herbicide-tolerant crops, bringing new actors into the fold to work on regulatory issues, and even presenting the first international workshop on GE issues at the Rio Earth Summit in 1992 (Schurman and Munro 2003; interview #2).

While Canadians were not directly involved in the BWG, a second wave of GE activism also emerged in Canada in the late 1980s, around the development of regulatory standards for field trials, and eventually commercial release, of GEOs. This second wave of Canadian activists also consisted of professionals with scientific and legal savvy. Among the actors to get involved at this time were the Canadian Environmental Law Association and the Canadian Institute for Environmental Law and Policy.

By the 1990s, as the commercialization of GE products loomed closer, a much wider coalition of groups came to be active in a third wave of anti-genetic engineering activism. The specific issues that brought these forces together were Calgene's efforts to have the Flavr Savr tomato approved for consumption, and Monsanto's application to have an engineered bovine growth hormone (rBGH, known within the industry as recombinant bovine somatotropin, or rbST) approved for use in the US and Canadian dairy industries. This hormone was designed to increase milk production by as much as 20 percent, but it was also associated with increased incidence of mastitis in cows.

The forces in opposition to the hormone, and to the first generation of commercially produced genetically engineered crops, were many and varied. And in contrast to the first two waves of professional activists, more CSOs with broad public support became involved. In Canada, they included the National Farmers Union (a farm organization dedicated to supporting family farming) and the Council of Canadians (a group with a long history of engagement in environmental, social, and trade issues), as well as environmental groups with large memberships such as Greenpeace and the Sierra Club of Canada. Each of these CSOs opposed GE products for various reasons – corporate concentration in the food system, a desire to see family farms protected, or environmental and health concerns about genetic engineering in general. However, for reasons I explore in Chapter 3, the main focus of their work was on the possible health risks of the GE products, along with calls for labelling foods derived from them. In the United States, for example, Rifkin set up an organization called the Pure Food Campaign to declare war on the Flavr Savr and then rBGH (Krimsky and Wrubel 1996, 234).

Behind their concerns of possible health impacts and labelling were larger concerns about the lack of comprehensive regulatory structures in North America. Canadian GE critics of the early 1990s argued that their regulatory system lacked transparency and privileged a limited range of scientific questions over others. The first questions to ask about a new technology, a 1995 Canadian Institute for Environmental Law and Policy report maintained, are: "What is its purpose? Does it address a legitimate need? If so, does it address the cause of the problem or just the symptoms?" (Mausberg and Press-Merkur 1995, 20). These questions were raised because it appeared as if most of the GEOs that were about to be commercialized targeted issues that could also be solved through less invasive approaches, such as pest or weed problems. In other cases, products such as rBGH were being proposed when there was no identified problem (e.g., a milk shortage) to be solved. Canadian and US regulatory systems appeared to be based on the assumption that GE technologies would be beneficial, or at the very least neutral, in their social and economic impacts.

During this third wave of consumer activism, which continued to grow throughout the 1990s, protests were carried out in front of grocery stores, letters were written to political officials, parliamentary committees were struck, and public debates between proponents and critics of biotech foods were held in town halls across the United States and Canada (see Sharratt 2001; Tokar 2001).

How did the biotech bloc respond to these three waves of activism against genetic engineering in North America? I have already noted that the bloc, since its inception, involved industry arms of governments as well as academic research establishments. Also supportive were the courts and patent offices that accepted and enforced intellectual property rights to transgenic organisms and genetic engineering processes. Each of these actors had a role to play in R&D stages of the biotech revolution. The next stage of the revolution, the commercialization of biotech products and foods in the 1990s, required the participation of a new set of state and civil society actors. These actors consisted of supportive "independent" CSOs, and of supportive government regulators. Together, they found a common purpose in the normalization of an ideology that saw genetic engineering as an important step along a continuum of improvements in plant and animal breeding. They worked equally diligently to try to disassociate genetic engineering from other narratives, such as the story of technological hubris articulated by Mary Shelley that surfaces in the critics' favoured term "Frankenfoods."

At the heart of this new set of actors were the biotech bloc's organic intellectuals. These were the scientists who took it upon themselves to build support for the biotech vision across society. They were organic to the bloc in the sense that they had been cultivated within the bloc's own joint industry-state-academic relationships of the 1970s and 1980s. While these relationships formally constituted a research apparatus, they also served as the bloc's ideological apparatus, and many scientists had built professional lives within this tripartite nexus. Graduate research may have been paid for by joint industry-state funding, and the scientists may have worked as consultants for governments in helping develop regulatory frameworks. Their authority in the public sphere came from the fact that among the hats they wore was the mortarboard and tassel, and this hat carried with it a perception of formal independence and objectivity.

These organic intellectuals included the likes of Donald Duvick, for a time the head of research at Pioneer Hi-Bred. Duvick wrote articles justifying the biotech revolution in academic journals using his Iowa State University address (e.g., Duvick 1995). Roger Beachy was another example. This Washington University professor was key to many of Monsanto's early discoveries in genetic engineering and a major champion of the technology's ability to serve the world's needy (Beachy 1991). Beachy was one of the biotech pioneers who saw his ideals compromised by the emerging realities of a market-driven industry. It was his research on virus resistance that Monsanto funded but then dropped because it would be of little commercial value (Charles 2001, 37). Still, while some became discouraged by the direction in which corporate priorities were taking the science, many in key positions remained defenders of the biotech revolution. This support can be seen most clearly in the defence of biotech by the deans of agricultural colleges in the United States and Canada. When the public debate about agbiotech was at its height in 1999, twenty-six deans of the major US and Canadian agricultural colleges signed the National Agricultural Biotechnology Council's "Vision for Agricultural Research and Development in the 21st Century." This document supported every aspect of the biotech industry's agenda, including, among other developments, the contentious TPS, or terminator technology (NABC 2000, 189).

One role of the academic biotech establishment was to quell dissent within academic ranks. This action was felt most keenly in the agricultural colleges that had become dependent on biotechnology industry funding, and it exacted a clear disciplinary price from the tradition of academic freedom. Following a review of the limited scientific research carried out that casts doubts on the claims made for GEOs, a professor from the University of Guelph wrote:

> Most of the above researchers have met with varying degrees of professional ostracism or sanction, ranging from public ridicule, to withdrawal of laboratory facilities, teaching privileges, or research funding, to vitriolic condemnation by peers and even outright firing. Many of the questions [raised by the research findings] remain unanswered, due to the absence, scarcity, or actual withdrawal of funding – or even job loss. Is this kind of reaction consistent with reasoned, scientific discourse among peers, institutes, journals and funding agencies? Or is something more than academic rivalry at stake? (Clark 2000)

Besides speaking through academics to trumpet the biotech revolution, the biotech bloc undertook a second key organizational strategy to influence public perception: the creation of "independent," non-profit organizations with a mandate to "educate" the public on this issue. Foremost among such organizations in Canada were the Food Biotechnology Communications Network (FBCN), based in Guelph, Ontario, and the National Institute of Nutrition (NIN), based in Montreal. During public debates, the FBCN actively enlisted and paid independent experts to present "balanced, science-based facts about food biotechnology" at local town-hall debates on genetic engineering across Canada. However, these "experts" were far from independent: they included nutritionists trained by Monsanto through a Dietitians Network that the company organized in 1995 (L. Stewart 2002a). This lack of independence should be no surprise, though, given that the vast majority of the FBCN's funding came from only a handful of its members: the major agri-chemical and agbiotech corporations. Significantly, the FBCN also received $750,000 in federal government funding between 1997 and 2001 (L. Stewart 2002b).

For its part, the NIN sponsored research on biotech-related issues in Canada. Once again, however, this research was not really independent. The board of the NIN was made up of executives from Monsanto and Nestlé, and administrators from Canada's agricultural colleges (NIN 1998). The NIN was also sponsored by the Canadian government. Both the NIN and the FBCN are examples of what Stauber and Rampton (1995) term "astroturf" organizations: they position themselves as representing the public interest but have no real base of popular support. These kinds of front groups, and the expertise they were able to marshal, played an important role in the North American biotechnology debates in the 1990s. But a final group of actors figured even more centrally in the struggles to gain consumer acceptance of GMOs: government regulators.

It is often assumed that the regulatory system brought into place for GE crops and foods in the United States in the late 1980s and early 1990s was the result of a government initiative to respond to public concern. However, as Eichenwald and his colleagues (2001) show, the impetus for the regulatory system actually came from industry. Specifically, it was representatives from Monsanto and Calgene who first sought government regulation in order to assuage public fears about genetic engineering. And because of the fiercely anti-regulatory climate in the Reagan administration, these companies faced an uphill battle. In response to their requests for regulatory oversight, key administrators in the US Food and Drug Administration argued that the larger biotech companies wanted regulation only in order to keep smaller competitors, who could not afford the expense of sending a product through regulatory hurdles, out of the market (Miller 1997).

The ideal regulatory system, from the perspective of the biotech companies, would have to meet three conditions. First, for the purposes of building public confidence, it would have to give a government stamp of approval for their products, certifying that they are safe to introduce both into the environment and into people's diets. The industry had done enough market research to know that the general public trusted claims made by governments about a product's safety more than if those same claims came from a private company, especially if the government claims could be corroborated by university scientists. The second condition sought by the industry was that any new regulatory system should not cause significant delays to the commercialization of biotech products, nor require onerous new research. The third condition was that the products of biotechnology would not be singled out in any way. The companies wanted to see GEOs treated the same as other types of new food products, and governed by the same legislation. They also wanted to avoid any sort of GEO labelling regime, as called for by some of the critics of genetic engineering. Many in industry recognized that labelling could end up stigmatizing, rather than enhancing, their products.

When the case for regulation, as pressed for by the biotech industry, was finally heard by the Reagan administration, the resultant framework did meet all three conditions set out by the industry. The United States set in place a system of voluntary notifications rather than mandatory authorizations, rooted in the idea that regulations should address only risks to public health and the environment that are "real and significant, rather than hypothetical and remote" (OSTP 1986; 1992, 6761-62; US FDA 1992). Since there was little evidence of significant harm caused by GEOs to date, and given that most GEOs could be considered essentially "familiar" and "substantially equivalent" to related non-GEOs that were already "generally recognized as safe," the new regulatory system called for little new research into the health or environmental impacts of GEOs before they would be given approval. (I revisit these regulatory concepts and their implications in Chapter 3.) Furthermore, because of the substantial equivalence of engineered and non-engineered plants and foods, this regulatory system would not require the labelling of biotech foods unless they were expected to pose a health risk.

In Canada, the federal government took a slightly more hands-on approach to the regulatory system developed in the early 1990s for GE seeds and foods, institutionalizing a mandatory regulatory approvals process, rather than a voluntary one. Still, this approach was closely related to that of the United States, in that it assumed the essential manageability of any risks posed by genetically engineered plants and foods.

In the United States, Canada, and Argentina during the 1990s, the government stamp of approval on biotech products before their entry into the marketplace, for the purpose of building public consent, became the central argument in the public relations campaigns undertaken by industry and academic molecular biologists. The approval of genetic engineering within these arms of the state also firmly established government regulators as active allies of industry in ensuing struggles to ensure that genetic engineering came to be accepted as the new norm in global agriculture. These alliances would have profound implications, especially in the context of intergovernmental negotiations about genetic engineering safety, such as those undertaken for a biosafety protocol, because the state was now fully committed to the biotech revolution in all three countries.

The response of the Canadian government to anti-biotech activism in 1999 reveals just how tight the relationships between the regulatory arms of the state and industry had become in this country. Documents made available through an access to information request reveal that in April of that year, the Canadian minister of agriculture, Lyle Vanclief, convened an extraordinary roundtable meeting. This meeting included representatives from the Prime Minister's Office; the Canadian Food Inspection Agency (CFIA), the main government regulators of GE crops; CEOs from major biotech companies such as Novartis; and the head of the Food Biotechnology Communications Network (FBCN). Also invited was one journalist, Anna Hobbs, the associate editor of the magazine *Canadian Living*. The purpose of this meeting was to "highlight the need for immediate co-ordinated action to deal with this crisis at hand." Outcomes of the roundtable included more federal money for the FBCN and a plan to distribute, through *Canadian Living*, an insert produced by the CFIA that would promote the safety of food biotechnology at a cost to taxpayers of $300,000 (L. Stewart 2002a).

Another series of events in Canada during this same period, and involving many of the same actors, also demonstrates close industry and government coordination: the effort to prevent mandatory labelling of GE foods. In the late 1990s and into the early 2000s, polls consistently showed that over 90 percent of Canadians wanted the government to mandate GE labels (e.g., Council of Canadians 2000). However, a growing public backlash against labelled GEOs in Europe suggested that labels could become targets of anti-GE campaigners. The joint industry-government response in Canada, made public in September 1999, was the formation of a Canadian General Standards Board (CGSB) committee to develop a standard for food labelling. The CGSB initiative, established at arm's-length from government and dominated by industry players, was designed to divert attention away from the GE regulators on the labelling issue. Once the CGSB committee was established, all references made to labelling by the Canadian government referred to the work of this committee. In keeping with industry interests, the CGSB committee's mandate focused on voluntary labelling, instead of the mandatory labelling that critics had been calling for. And while the committee was charged to consider the labelling of "foods derived from biotechnology" – a mandate even industry groups initially accepted (FCPMC 1999) – its first decision was to adapt the scope of its work to deal with the issue of labelling of "foods obtained or not obtained through genetic modification" (M. Humphreys and N. Nishikawa 1999, 1). This move allowed the committee to focus on creating onerous standards for any company that would want to label its product as "non-GEO" instead of placing new expectations (and costs) on food manufacturers using GE ingredients.

Close coordination among industry and regulatory arms of the state in the biotech bloc, and a strong shared vision, meant that the bloc was in a good position to undertake the process Gramsci calls "transformism": the piecemeal co-optation of opposition movements (Forgacs, 2000, 430). In the case of labelling debates, for example, one of the best allies of the biotech industry became the Consumers' Association of Canada (CAC). Despite that a vast majority of Canadian consumers supported the call for mandatory labelling, and that provincial chapters of the CAC also took this stance, the anti-labelling position of the CAC's national office was frequently invoked by Minister of Agriculture Vanclief to justify the government stance on the issue. Lee Ann Murphy, spokesperson for the CAC, argued that the CAC "[did] not feel that the label is a place for public debate" (L. Stewart 2002b, 9). Not surprisingly, the CAC received $1.3 million in government funding, through Industry Canada, after it adopted this stance in 1998. Several trade associations also supported the CAC financially. In 2001, Monsanto Canada hired Murphy as a director of public affairs (L. Stewart 2002b, 9).

The Canadian examples presented here demonstrate the close institutional relationships between arms of the state and civil society that exist in a fully hegemonic formation. This is not to say, however, that the biotech bloc managed to achieve all of its aims. It did have to make concessions to critics to gain this position. An instructive example is the fate of rBGH in Canada.

In the United States, citizen activism delayed governmental approval of rBGH for several years. Still, it was eventually approved in 1993. In Canada, by contrast, the groups that mobilized around this issue were ultimately victorious. In early 1999, Health Canada rejected Monsanto's application to have rBGH approved, on the grounds that the health of cows could be placed in jeopardy by this veterinary drug (Health Canada 1999). While animal health was the official reason for the rejection of Monsanto's application, it was clear to observers that this regulatory outcome was the combination of several factors, including that Canada's supply-managed dairy industry saw no major advantages to the increased yields that rBGH promised, while public concerns about the growth hormone represented the real possibility of reduced milk sales (see Sharratt 2001 for more detail on this case).

By rejecting Monsanto's application, the actions of the Canadian government demonstrate that the dynamics within a hegemonic formation cannot be taken for granted. The agri-chemical and biotech industry needed strong alliances with the regulatory arms of nation-states to achieve the gains it made in the 1990s, but these relationships offered no guarantees, and concessions towards critics and allies alike were sometimes required that would have material implications for the class fraction at the heart of the historical bloc: the agri-chemical and biotech industry.

Despite these concessions, the agri-chemical industry and its allies did manage to establish a climate favourable to the majority of the GE applications that were being developed in the United States, Argentina, and Canada. While rBGH was stopped at the Canadian border, it did come to be employed by farmers in the much larger market of the United States. And the first genetically engineered varieties of cotton, squash, and potato were also approved for commercial production in the United States in 1994. A year later, GE canola was approved in Canada, and a steady trickle of genetically engineered crops and foods followed. By 2002, fifty-one varieties of "novel foods" and "plants with novel traits" – about three-quarters of which were genetically engineered – had been approved for consumption in Canada. And, in all three countries, any demand to label these foods has been successfully resisted by the relevant government agencies and departments to date. Because of these material gains, and the organizational and ideational relations of force they relied on, it would be appropriate to say that the hegemony of the biotechnology industry's project was successfully consolidated in these countries during the 1990s.

Consent and Coercion on the Global Stage

Agricultural applications of genetic engineering, such as rBGH and genetically engineered crops, were first introduced into farming in three countries of the Americas. Despite these localized introductions, this revolution was intended to be global from its inception, in part because the food distribution system was already globalized in the 1990s. GE crops raised in one country would almost inevitably find their way around the world within weeks of harvest, and the global grain-trading system relied on the bulk movement of foodstuffs that were differentiated by grade, and not necessarily by variety or even origin. These realities meant that the agri-chemical and biotech corporations' hegemony in only a few (albeit critical) states did not provide stability in any meaningful sense. What the bloc needed was a truly global regime of consent for genetic engineering. Such a regime would require the active participation of state authorities, and it would need to be accepted by civil society. It would require shared norms and common understandings of the relevant issues that would be internalized in the practices of the biotech industry and its critics.

The biotech bloc's efforts to establish a global regime supportive of the GE project got underway in the late 1980s and early 1990s, and were concentrated in two areas. First, actors in the bloc worked to develop supportive international policies for intellectual property rights (IPRs) over genetically engineered organisms. Second, they tried to internationalize minimalist regulatory requirements for environmental and food safety.

With regard to intellectual property rights, the biotech bloc adopted a combination of coercion and consent to build international support for American-style patent protection. As was the case in the Americas, proponents of patenting began by using the hype associated with genetic engineering to try to convince farmers and lawmakers around the world of the common sense of removing the rights of farmers to save transgenic seeds. This stance was then backed by coercion. As Purdue (1995) demonstrates, as early as 1988 the US government was using the threat of trade sanctions as a device for ensuring that countries sign, and then enforce, bilateral agreements protecting IPRs of all types. In 1994, the precedence of these coercive actions enabled trade negotiators from the United States to get international consent for the GATT (General Agreement on Tariffs and Trade) agreement on Trade-Related Aspects of Intellectual Property Rights (TRIPS) (Subraminian 1990). As one of the central coercive projects of neo-liberalism, the TRIPS represents an international effort to institutionalize Lockean-style property rights through the GATT/WTO. The most recent TRIPS agreement requires among its provisions that countries create systems for protecting IPRs over new micro-organisms and plants developed through genetic engineering through a domestic patenting system, or an effective *sui generis* system. Most developing countries were to have implemented such systems by 1 January 2000, with the least developed countries following suit by 2006.

The TRIPS demand for patents clearly serves the interests of the transnational agri-chemical corporations at the heart of the biotech bloc. In fact, Shiva quotes executives from these corporations admitting that they actually drafted many parts of the agreement themselves in the early 1990s (Shiva 1998). It is notable, however, that in order to have the TRIPS accepted, the proponents of global IPR rules had to make concessions in the final text. The mention of *sui generis,* meaning "of its own kind of class," was included to appease the governments of some developing countries that were not comfortable with the idea of patenting living organisms. What this means in practice remains at the heart of the ongoing international debates over IPRs in genetic engineering.

In the late 1980s and early 1990s, the biotech bloc's organizational efforts to build an international food safety and environmental regulatory regime supportive of agbiotech were primarily focused in Europe. Europe was a crucial target for three reasons. First, some European governments and companies had been active participants in the biotech bloc from the outset, and the biotech industry wished to establish the same kind of supportive relationships that were emerging in the United States within the European context. Second, a trans-Atlantic consensus on the technical aspects of regulation, if achievable, could usually be expected to be accepted by the rest of the world. Third, for reasons I explore in Chapter 3, European publics and their governments appeared to be more skeptical towards genetic engineering than publics in the Americas, so gaining support in Europe was seen by the biotech bloc as crucial for building global consent.

The main strategy the biotech bloc adopted in Europe was to try to generate support from above, through official regulatory channels. This strategy involved three main mechanisms: it was undertaken by North American regulators – now firmly embedded in the biotec bloc – who promoted the benefits of the regulatory models in place in their countries to European regulators; it was undertaken by North American and European industry arms of the biotech bloc, which lobbied the EU for regulations that would lead to the commercial approval of genetically engineered crops and foods; and it involved coercive activities on the part of the US government against European countries.

The promotion of genetic engineering by North American regulators took place through international regulatory "harmonization" processes. Even though consensus was the official objective, it would be more accurate to say that these processes were directed at universalizing the permissive regulatory models developed in the North American heartland of agbiotech, because they took place at a time when the US regulatory system was already established while the European one was not. The initial avenue for such talks was the Organisation for Economic Co-operation and Development (OECD), which has a history of supporting trade liberalization by promoting regulatory harmonization among developed countries. Beginning in the mid-1980s, the OECD brought together GE experts and government representatives to try to develop a common perspective on how to assess the risks of GEOs (see OECD 1986, 1993). In the early 1990s, similar discussions on the regulation of GE foods also took place under the auspices of the Food and Agriculture Organization (FAO) of the United Nations and the World Health Organization (WHO) (FAO/WHO 1991, 1996). A consensus was always achieved in these talks, but the results were actually highly ambiguous regulatory definitions for terms such as "substantial equivalence," which could then be interpreted very differently in the various regulatory cultures. Still, through the 1980s and into the early 1990s, regulators in the biotech bloc were able to use the OECD and FAO/WHO processes to entrench, at the international level, concepts and terms originating in the permissive US regulatory system.

The biotech bloc also worked through its transnational industry networks to build a European regulatory environment hospitable to the introduction of genetic engineering. These activities included the formation of a working group of the TransAtlantic Business Dialogue, an organization of European and North American businesses that informally develops trade policy proposals for streamlining regulatory hurdles in the United States and Europe. By the late 1990s, this working group put forward a proposal to set up a

European counterpart to the US Food and Drug Administration to speed up and harmonize the approval of new products (Levy and Newell 2000, 18).

As Gramsci would expect, the biotech bloc's actions to encourage a favourable regulatory environment through lobbying and consensus building were supplemented by coercive pressure on European countries to import GEOs. In an example described by Charles (2001, 186), when US farmers started growing Roundup Ready soybeans in the spring of 1996, and these were not yet approved for import into Europe, the United States sent Special Trade Representative Mickey Kantor (later a Monsanto board member) to argue that if Europe denied imports of soy, the United States would consider an appeal under the WTO. Even though Greenpeace and other CSOs tried to block shipments of GE soy in the fall of 1996, the EU gave in to these US pressures and allowed the imports.

Initially, the combination of international regulatory harmonization activities, business lobbying, and coercion was successful in Europe. Most important, from the perspective of the bloc, it helped spur on the development, adoption, and then favourable application of regulations that allowed for the introduction of GE crops and foods to the European food system. Under these regulations, eight different genetically engineered crops were approved for commercial use in Europe between 1996 and 1998. These included three varieties of canola, four varieties of corn, and one of soybean (Birchard 2000). Despite this apparent victory, there is much more to this story. There were some important differences between the European regulatory model and the regulations developed in Canada and the United States that suggest that European regulators were taking a more cautious approach to the evaluation of the possible risks of genetic engineering than their North American counterparts. Furthermore, the European regulatory structure was set up in a way that would allow for greater political intervention in the approvals of genetically engineered crops and foods than was possible in North America. While initially insignificant, these differences would become very important once there was a large-scale public backlash against genetically engineered foods, in 1998 and 1999.

Each of the examples presented here shows how proponents of genetic engineering worked through the regulatory arms of states and through international institutions to try to move the biotechnology revolution forward at a global level in the early 1990s. At the same time, international fora were also being used by critics of biotechnology to try to slow this revolution. A number of CSOs, for example, including Genetic Resources Action International (GRAIN, based in Spain) and RAFI, made the call for "no patents on life" the primary focus of their activities in the 1990s. Working with state partners, primarily in the global South, these groups also used multilateral fora to entrench their positions on intellectual property rights. As a result, one outcome of the 2001 International Treaty on Plant Genetic Resources for Food and Agriculture, passed under the auspices of the FAO, ensures that certain traditional crop varieties will be excluded from patenting (FAO 2001). This move may place the new FAO agreement in conflict with the TRIPS agreement.

Conclusion

The biotechnology revolution that took place in agriculture during the 1990s had strong material foundations; it was about the application of the new genetic engineering techniques in the food system. However, a close look at this revolution confirms that technological developments do not in and of themselves produce social and political change. These technologies were harnessed in the context of particular interests: scientific, governmental, and industrial.

The central organizational structure of the biotech bloc was established early. It incorporated a strong state and university research infrastructure, primarily (but not exclusively) based in North America, and was spearheaded by agri-chemical corporations that sought to consolidate a stronger position in the global food system. These corporations were able to decide on the direction of genetic engineering, in part because of the neo-liberal shift towards market-driven research in the public sector. The novelty of this agricultural research constellation meant that the majority of applications of genetic engineering in this field would be designed, first and foremost, to further the interests of agri-chemical companies.

The corporations at the head of the biotech bloc brought farmers onside in the United States, Argentina, and Canada by developing crops with traits that would appeal to them, such as herbicide resistance and pesticidal properties. The cost to these farmers was their indenture in a new contract farming model that saw them lose their rights to save and plant seeds in exchange for the privilege of using genetically engineered varieties. Should it be commercialized, the TPS/terminator technology will have the same effect as a strong patenting system reinforced by technology use agreements and the courts, even though its disciplines are biological rather than legal. While these contracts and technologies have had limited impact in the three countries at the heart of the biotech bloc (where seed saving is practised only in a limited way), they will have profound effects when adopted in areas of Southeast Asia and Africa, for example, where seed saving remains an integral part of contemporary farming.

In addition to such social implications for farming, the marriage of the genetic engineering revolution to the agri-chemical industry can also be expected to have adverse environmental implications. Already in Canada there are cases of wild mustard weeds growing in Prairie towns that are resistant to the three most common herbicides used in farming. These are the product of multiple crosses with herbicide-tolerant canola varieties. While the herbicides to which they are resistant, such as Monsanto's Roundup, are considered an environmental improvement over earlier products, farmers must now resort to more toxic chemicals (such as the herbicide 2,4-D) to deal with this new "superweed" problem.

Generally, the biotech bloc has been successful in consolidating a hegemonic formation in the United States, Canada, and Argentina by building solid links through civil society organizations and the state, especially universities and regulatory authorities. It has been the authority of these "independent" bodies that has assuaged the fears and concerns of the public. An examination of these links has shown that the premise of independence really is a charade in the case of a hegemonic formation such as the biotech bloc. The material interests of the class fraction at the head of the bloc cannot be untangled from those organs of the state and civil society that have genuinely come to adopt these private interests as the "public" or "national" interest.

The biotech bloc has made enormous gains, but a hegemony rooted in these three countries is far from stable, despite concessions towards some critics. This instability is because of the global nature of the food system. In this system, food processors and grain traders are the major players. By the late 1990s, the agri-chemical companies did manage to have their presence felt in the global farm input industries, and the result was a consolidation of seed and chemicals into a handful of companies. Were the biotech revolution to provide plant traits with immediate appeal to consumers, the input companies might even merge with food processors such as Nestlé, leading to even greater vertical integration in the food system and a concomitant reduction in control and choice over food by consumers and farmers. This increased integration has yet to occur, however. There are identifiable food system clusters that appear to work cooperatively from the level of the seed to the grocery store, yet no outright buyouts of input companies by the giant corporations that are closer to the consumer in the food chain have taken place. The biotech bloc's revolution is not yet global, and its fortunes could still turn quickly should a major food scare involving genetically engineered products cause food processors or grain traders to see GE foods as a liability.

To globalize its revolution, and to give it more stability, the biotech bloc worked to develop an international regime supportive of genetic engineering. This regime included the GATT/WTO and its TRIPS agreement. It also involved regulatory harmonization efforts under the auspices of the OECD and the FAO/WHO and efforts to construct an acceptable regulatory framework in Europe. Still, a successful regime necessarily involves more than institutional dimensions; it has cognitive and normative dimensions that entail that the relevant actors have a shared understanding of a given problem and how it should best be solved. Such cognitive and normative congruence did not yet exist in the biotechnology regime in the mid-1990s, despite the best efforts of the biotech bloc to create it. Instead, divergent framings of genetic engineering, the problems it poses, and how they should be solved existed in this regulatory field, and these only grew as the decade progressed.

These divergences point to the fact that the emerging genetic engineering regime was not solely influenced by actors in the biotech bloc. Forces critical of genetic engineering were also working to shape the governance regime with the intention of placing controls on the biotech revolution. The FAO Treaty on Plant Genetic Resources was one such institutional manifestation. Aspects of the Convention on Biological Diversity (CBD), the parent convention to the Cartagena Protocol on Biosafety, also appear to have been shaped by forces critical of genetic engineering in an effort to try to reign in the biotech bloc.

Before I turn to the CBD and the Biosafety Protocol, there is a crucial piece of the story of the politics of genetic engineering that remains to be explored. The material and organizational aspects of the biotech bloc presented in this chapter have shed light on the politics of biotechnology, but there is much that is not explained here about the nature of the consent and dissent at the heart of biotechnology politics: Why was the wide range of concerns that critics raised about genetic engineering reduced to a regulatory debate about environmental and health risks? How were biotechnology companies able to secure patents to genetically engineered organisms, when plant breeders were never able to exercise this degree of legal control over their work? How could the regulations established in North America be touted to the public as offering a satisfactory safety assessment of GEOs when there were no long-term data upon which to make such judgments? What was different about the European response to genetic engineering from that in the three countries at the heart of the bloc, and what were the implications of this difference? These questions all deal with the realm of ideas, the third arena of political activity in my theoretical framework. Ideas are shaped in the context of material, organizational, and institutional factors, but they are also forces in their own right. In the next chapter, I examine the central ideational currents in the politics of biotechnology up to the mid-1990s in order to bring to light their nature, their sources, and their effects in the global politics of genetic engineering.

3
The Ideational Politics of Genetic Engineering

What is genetic engineering? Is it simply another tool, on a par with plant and animal breeding, or is it a risky game of dice being played with the building blocks of life? The public acceptability of biotechnology depends on which of these ideas is accepted as the norm. On the one hand, this means that those groups that succeed in establishing the definition of genetic engineering have a great deal of power to determine the course of political events. On the other hand, it means that an ostensibly technical definition may be politicized in both overt and subtle ways.

Four sets of ideas have been particularly influential in shaping the politics of genetic engineering: that of the gene, the environment, liberalism, and risk. Together, these concepts have provided the basic ideational framework within which the major struggles over GEOs have taken place, including debates over the regulation of GEOs. Where did these concepts come from, and what are their effects as political forces in their own right?

The Gene

Foucault describes life's entry into history as a distinct object of expert inquiry and government preoccupation in the nineteenth century. In the 1950s, the molecular gene made its own entry into history. For nearly a century, biologists had recognized that many phenotypic traits were related to functional units of heredity. In Mendelian genetics, these were termed genes, but little was understood about what these units were composed of or how they worked. With the discovery, in the 1950s, of deoxyribonucleic acid (DNA), which is found in the nucleus of each cell, genes came to be understood as specific segments of DNA that carry the information needed by an organism to construct the amino acids and proteins that perform biological functions. Although there were still many questions about how the Mendelian genes (linked to phenotype) related to the molecular segments of DNA, experiments suggested that direct correlations between both types of genes could be found in many instances. The assumption that this would eventually be true in all cases set the context for much of the work that followed. This presupposition is summed up in Watson and Crick's "central dogma" of molecular biology: an organism's genome, its full complement of DNA, should fully account for its characteristic assemblage of traits (Crick 1958).

The notion that genes are bits of information led to a host of projects aiming to isolate and identify genes. Eventually, by the 1970s, these projects provided the tools for transferring segments of DNA from one organism to another, giving birth to the new applied science of "gene splicing" or "genetic engineering." This latter term, which was already in use among molecular biologists in 1970, was defined as the deliberate and controlled modification of the genetic makeup of living things (Sinsheimer 1970).

The idea of transferable molecular genes that took hold both in science and in the popular imagination is an example of the way scientific theories are also discursive phenomena. More than simply a description, the gene is a social construction that emerged in the context of biopower: the particular set of discourses and practices that aims to bring human beings, through investment in their bodies, under rational control and management. In this case, molecular genes were constructed as "information," as part of a genetic "code" or "blueprint" that could be translated with the right tools. Once this translation was complete, authorities would have the tools needed to achieve corporate and public policy objectives through "genetic enhancement." Such objectives could be envisioned at both poles of the axis of biopower, from the treatment of the "genetic illnesses" of individuals to new approaches to population control.

Keller (2000, 143) refers to this linear, reductionist discourse of genetics as "gene talk," a term I adopt here. This discourse can be traced to particular historical narratives and social relations of power. The metaphor of the "genetic code," for example, combined centuries-old theistic discourse of the "book of life" – which many scientists saw themselves as engaged in deciphering – and the more immediate historical context, in the 1950s, of the nexus connecting genetics research, military funding, and the emerging computer age. This combination fostered the command-and-control view of genetics that would take hold in popular consciousness (Kay 2000, 11).

Haraway (2000, 92) refers to the fixation on the transferable gene as "genetic fetishism," with fetishism understood in at least two distinct ways. Following Marx, Haraway contends that the gene is a fetish that is defined by its value as a tradable commodity (that can be moved from one organism to another), rather than by values that recognizes the social (and natural) processes through which it was produced. Haraway also refers to the gene as a "cognitive" fetish. Drawing on the ideas of Alfred North Whitehead, Haraway sees cognitive fetishism as a further productive mislocation: a mislocation of the abstract in the concrete. In the case of gene discourses, the gene is tied to notions of blueprints, programs, and codes, each of which is itself an abstraction. In popular discourse, however, "the layers of abstraction and processing that have gone into producing notions of code and program are then simplified and mistaken for the real" (93).

The fallacy of cognitive fetishism was revealed early in the development of molecular genetics. For example, Kay (2000, 14) shows that efforts by scientists using computer analysis, information theory, cryptanalysis, and linguistics to "break" the genetic code in the 1950s yielded no results. In hindsight, it is clear that there were no results because these scientists were not dealing with a code, in the conventional sense of a linear script, at all. This "genetic code" is little more than a powerful metaphor for the correlation between nucleic and amino acids. Despite these pitfalls, informational and scriptural representations of genetics still took root and proliferated. Why? Gene talk took hold because of its cultural resonance and because of its role in validating the academic project of genetic engineering. Eventually, by the early 1980s, the biotech bloc had solidified around specific commercial goals, and it had an active interest in adopting and cementing gene talk, as a justificatory ideology, into popular consciousness.

The idea that molecular biologists were on the verge of discovering and solving a host of genetic "defects" in humans and other species drove research in biology, pharmacy, agriculture, and related fields throughout the final two decades of the twentieth century. The ultimate expression of this drive was the Human Genome Project (HGP). This US$3 billion project, undertaken between 1990 and 2001, set out to read the entire human genome of approximately 3 billion nucleotides. The HGP was expected to yield molecular translations of approximately 100,000 genes. Once the molecular structures of these genes were available, the work could begin to match these genes to specific desirable or undesirable traits so that these could be altered, transferred, or switched off through genetic manipulation.

In the context of agriculture, gene talk was employed in two main ways to further the interests of the biotech bloc. It was used to demonstrate first, the precision, and thus safety, of genetic engineering, and second, the unique importance of these techniques vis-à-vis traditional breeding techniques. As an example of how gene talk helps demonstrate the safety of genetic engineering, consider the definition of the "new developments in biotechnology" (the term "genetic engineering" is avoided completely) included in a 2000 industry brochure. These techniques, the brochure states, simply involve the "precise" identification and transferring of "the specific gene that creates a desired trait in a plant" to another plant (CBI 2000, 3). This portrayal is designed to maximize public trust by minimizing any uncertainties or complexities involved in the science of moving genes from one organism to another. It extrapolates from gene talk the notion that GEOs are no different from non-engineered organisms because they are simply the sum of well-known parts.

I call this particular use of gene talk the idea of GEO "equivalency," as it is related to the regulatory term "substantial equivalence." The designation of substantial equivalence is the standard that most GE foods must reach in order to be considered safe enough for human consumption. In theory, substantial equivalence between GE and non-GE products can be demonstrated by comparison of DNA structure, gene expression, proteomic analysis, and secondary metabolite profiling (see S. Barrett et al. 2001, 187-89). In practice, however, this technical threshold, like the wider idea of equivalency that reinforces it, is often taken for granted (i.e., accepted as fact without evidence of a detailed technical comparison). The use, and implications, of substantial equivalence as a regulatory concept are discussed in further detail later in this chapter.

An article in *The Economist,* commenting on consumer concerns about the health implications of GEOs, provides a second example of the idea of equivalency in a popular text:

> After all, the process works by transferring single genes, designed to do particular jobs, from one species to another. If products of these genes are harmless ... it is difficult to conceive of any way in which human health might be damaged. (Economist 1999, 19)

These examples from the industry brochure and *The Economist* actually show how the idea of equivalency functions in two different ways to help normalize genetic engineering. The first example is clearly a case of the active deployment of this idea, as an ideology, to further industry interests. The news article excerpt, however, shows how a journalist's analysis can unwittingly reproduce the idea of equivalency as an assumed truth (yet one that is far from value-neutral). This is equivalency functioning as a *discourse,* defining what makes sense, and what does not, in the public debate over GE safety.

While reinforcing assumptions about the equivalence of GEOs to non-GEOs, gene talk has also enabled an emphasis on the distinctiveness of the new techniques of genetic engineering by framing genes as transferable commodities. From this perspective, plants and other organisms essentially became gene reservoirs while the techniques of genetic engineering allowed the full value of these genes to be realized where they were needed. The fetish of the transferable, molecular gene (as a piece of information), along with the novelty of the techniques that brought it into being, came to be used by the proponents of genetic engineering to argue that genetic engineering was a form of "invention" that deserved to be accorded patents, something hitherto not possible with traditionally bred plants and animals.

Underpinning both equivalency of GEOs and the uniqueness of the techniques of GE, gene talk served as an ideological resource for the biotech bloc. At the same time, however, gene talk functioned discursively as a site of resistance. This happened in two ways. First, the actual development of molecular biology eventually provided data that called into question the assumptions of the reductionist model of genetics, thereby creating the footing for a reverse discourse focused on genetic "complexity." Second, the biotech bloc's use of gene talk to emphasize the uniqueness of genetic engineering, in the context of patent claims, also came to be juxtaposed against the bloc's insistence that the safety considerations associated with GEOs were essentially equivalent to those of non-GEOs.

The notion of genetic complexity, or what we might call "complexity talk," acknowledges the importance of genes. However, it holds that gene-organism-environment relationships are far more complicated than the linear model of gene talk allows. Proponents of complexity talk, including Commoner (2002), point to the experimental evidence that shows that a single gene can affect more than one trait – a phenomenon known as pleiotropy. Others, such as Holdredge (1996) and Rissler and Mellon (1996, 38), argue that the position of a gene on a chromosome can also affect phenotypic outcome in the positional effect, and that genes have been shown to silence the expression of otherwise "un-related" genes (a phenomenon known as "epistasis"). Advocates of complexity talk also focus on the role of non-genetic factors, such as the organism's environment, and the embryonic environment inherited from the mother in sexual reproduction, as elements shaping the DNA-organism relationship (e.g., Keller 2000, 146). Together, all these factors help explain why, in many experiments, the "same" gene functions very differently in different organisms.

As Commoner points out, the results of the Human Genome Project in 2001 dramatically reveal the need for models that can account for complexity. Although intent on finding 100,000 genes, the HGP yielded only about 30,000 identifiable genes (Commoner 2002). In seeking to explain how so few genes could be responsible for the complex array of traits exhibited by human beings, one major theory holds that as many as four thousand human genes may be "alternatively spliced." This proposal is significant because alternative splicing disrupts gene talk. It suggests that a gene's original nucleotide sequence could be divided into fragments that are then recombined (by a spliceosome) in order to encode a multiplicity of proteins, each different in its amino acid sequence from the others and from the sequence that the original gene, if left intact, would encode (Collins and Gutherie 1999 in Commoner 2002, 6).

As with gene talk, complexity talk gains its power from its status as a truth claim, combined with its having been shaped and reinforced by a community that extends well beyond the field of molecular biology to include environmental activists and others. These political actors share a distrust of genetic engineering, and complexity talk offers a scientific tool for challenging the presumed safety of GEOs. For reasons I explore below, the debate over whether or not it is acceptable to introduce genetically engineered organisms into the environment and into people's diets has come to hang, to a large extent, on the degree to which the effects of the genetic engineering process can be predicted. The positional effect, alternative splicing, pleiotropy, and epistasis all suggest that unanticipated effects of genetic engineering are not only possible but likely. The proponents of GEOs were particularly vulnerable on this subject because molecular biologists did not know (in most cases) exactly how many copies of a new genetic construct were inserted into a host genome, nor where on the chromosomal structure those constructs took hold. Sometimes they were not even sure what it was they introduced. For example, in 2000, Monsanto informed Health Canada that a variety of Roundup Ready soybeans had "extra" pieces of DNA that were not reported, and thus not specifically examined for their health impacts, in its initial submission for safety approvals in 1994 (OFB 2000).

It is important to point out, however, that no discourse is inherently resistant. Under current circumstances, complexity talk appears to undermine some of the gene-talk–related premises of genetic engineering safety, but in Chapter 7, I discuss ways in which proponents of GE have also begun to adopt complexity talk in order to counter their critics.

The second way that gene talk provided a site of resistance is revealed in the apparent contradiction between the biotech bloc's use of gene talk in property rights discourse and its use of it in regulatory debates. While a gene fetish helped shape an understanding of genes as a distinct form of intellectual property, the assumption that underpins this view – that genetic engineering is indeed distinctive – also took hold in the popular imagination. This time, however, genetic engineering was interpreted as a distinct *threat* to food and environmental safety. As Charles (2001, 30) writes,

> Scientists and biotech entrepreneurs sometimes cursed ... terms ... that created mental images of forms of life that were different in some fundamental way ... and blamed Rifkin and his allies for promulgating them. In fact, the enthusiasts of biotechnology had mostly themselves to blame. It was the genetic engineers, after all, who called the splicing of one or more genes into a cell "genetic transformation." This choice of words revealed their pride of authorship and implied that they had indeed "transformed" a plant,

creating something new, different, and (with a wink toward Wall Street) uniquely valuable. Inevitably, those who were wary of biotechnology would turn this notion on its head, sensing a threat where others perceived only promise.

This desire of the proponents of agbiotech to frame GEOs as both unique and equivalent to non-GEOs remained a thorn in the foot of the biotech bloc in ensuing decades.

Environmental Governance

Foucault identifies biopower as the constellation of discourses and practices focused on the management and control of human life at the level of individuals as well as populations. In recent years, social theorists have noted how the wave of concern for the environment since the 1960s, coupled with associated environmental governance strategies, have extended biopower to the living world as a whole. Luke (1995) refers to this inclusion of the environment as a field of biopower as "geo-power," whereas Rutherford (1999) calls it "ecopolitics." I have chosen to refer to this set of discourses and practices as ecopower.

Ecopower recasts nature in biological, chemical, and physical terms that are assumed to be fully amenable to rational management, intervention, and increases in production. Like biopower, ecopower is driven by the joint interests of capital (to colonize and commodify) and the state (to manage for the sake of the perceived public good). Ecopower has two poles. At one end, discourses and practices of the environment are directed towards large and complex bodies as objects of management: lakes, forests, ecosystems, cities, the biosphere, and the atmosphere. At the other end, considerable attention is given to the human individual as a subject, presented either as a destructive organism that needs to be tamed through scientific management, or as a consumer who must be taught to make more environmentally responsible consumption decisions. Within ecopower, best management practices are imposed from the outside by the state, and they are internalized through voluntary initiatives taken by industry or by individuals following their environmental consciences.

Environmental discourses have not always embedded assumptions about the total management of human-nature relationships. In fact, before 1965, the term "environment" was itself rare in policy discourse (Luke 1995). In the early 1960s, advocates for nature, including Rachel Carson, drew the world's attention to the adverse effect that insecticides used to kill specific agricultural pests, among other industrial products, were having on other species in complex food chains (Carson 1962). Discursively, Carson's arguments were framed within a purity model of nature. This notion of a pristine natural world (or what Carson called the "web of life") that is put into jeopardy by human actions would become a central concern of the nascent environmental movement of the late 1960s. Also clear in Carson's work is the framing of nature's contamination in catastrophic language. Pesticides such as DDT would not just harm some birds that preyed on agricultural pests, they would result in the total silence of a spring devoid of all birds and their calls.

Such descriptions of nature despoiled by pollution, coupled with a revival of Malthusian framings of population–resource use dynamics, had, by the early 1970s, led many in the environmental movement to embrace a zero-growth philosophy. This movement argued that resource consumption and other human incursions into nature had to be drastically reduced and brought to a steady state for the sake of the long-term health of the planet and its human population. That the Club of Rome's 1972 report, entitled *Limits to Growth,* could sell more than 2 million copies illustrates the appetite at the time for such critiques (Meadows et al. 1972). In discursive terms, the zero-growth framing embraced a pristine view of nature and a portrayal of human actions as catastrophic if they continued along their current path. To these elements were added assumptions about the possibility of global management of the human population and/or consumption patterns in order to curb an impending environmental apocalypse. Significantly, however, this management was framed as a biopolitical project. It was directed towards human beings rather than towards a presumed management and control of the planet as a whole.

The contemporary configuration of ecopower was fully consolidated only in the mid-1980s, when rallying cries of zero growth and nature protection were subsumed and reoriented in the governmental discourses of sustainable development and ecological modernization. For its part, sustainable development, as framed by the United Nations World Commission on Environment and Development (WCED), purports to reconcile environmental goals with the economic needs of the world's poor (WCED 1987). As understood by the WCED, the environment is no longer a separate entity (as "nature"), and ecological catastrophe will be avoided only by more, not less, economic growth. Such growth can be made sustainable, it is assumed, by relying on ecological expertise. Luke (1999, 138) shows how the WCED's framing of sustainable development expresses the central logic of contemporary ecopower, which is that

everything [the WCED] stipulates can be known – how to define aspirations of a better life, what constitute basic needs, when to manage economic growth, why to improve technology, where to organize environmental resources, who is to judge the ability of the biosphere to absorb human pressures – is known, or is, at least, knowable. And since these eco-knowledges exist, all that existing state regimes need to do ... is operationalize this knowledge.

Ecological modernization embeds similar environmental values, though it places less importance on the goal of distributive justice. Originally developed by a group of Dutch and German social scientists in the 1980s and into the 1990s, ecological modernization emphasizes the need for continuous improvement in environmental practices in order to avoid conflicts between environmental growth and environmental protection (e.g., Janicke 1985; Hajer 1995). Discursively, ecological modernization does not hold assumptions of a pristine nature, seeing the natural world instead as primarily an object of expert knowledge. While this perspective reflects many elements of the *Limits to Growth* arguments of the 1970s, it has higher expectations of the potential for industrial ecologists, ecological designers, and other experts to refashion the technological and natural artifacts of contemporary existence so that they might be compatible with scientific notions of carrying capacity (Cohen 2001, 7).

Even as the idea of a pristine nature continued to motivate many environmental activists, the official strategies for ecological governance that emerged in the late 1980s and early 1990s, as found in governmental programs such as the Dutch National Environmental Policy Plan of 1988 or Canada's Green Plan of 1992, were strongly influenced by the discourses of sustainable development and ecological modernization. I have identified some of the elements of these discourses, but I have not yet explained why they became hegemonic. Such a discussion requires an examination of the social relations of environmental regulation at that historical juncture, including, in particular, the rise of the transnational historical bloc and its ideological emphasis on liberalism.

Liberalism

In Chapter 1, I noted that liberalism has been an important force in the global political order since at least the early nineteenth century. This is true for liberalism both as an ideology and as a discourse. As an ideology, liberalism holds forth a set of assumptions about social relations, combined with a prescription of how these relations should be organized to maximize the freedom of individuals, thereby achieving the best for society as a whole. Like their predecessors, today's acolytes of liberalism, including von Hayek, argue for a "freeing" of "market forces" as the way to maximize the social good. They suggest that the primary role of government is the establishment and enforcement of the rule of law, the purpose of which is to assign and protect the rights of individuals to property and "economic freedom" (von Hayek 1976, 101-4). The wave of liberal prescriptions for the global economic and political order evident since the early 1970s, often termed "neo-liberalism," is rooted in a particular historical configuration of social relations: the rise of the transnational historical bloc (Gill 1991). Planning groups such as the World Economic Forum gather the organic intellectuals of this historical bloc and actively work to universalize neo-liberal norms across all sectors of society. When these norms have not been fully internalized on a voluntary basis by nation-states, corporations, and other actors, they are implemented through the coercive arms of the transnational historical bloc. The World Trade Organization, World Bank, and International Monetary Fund all have means at their disposal for forcing states to lock-in neo-liberal reforms.

A central feature of the contemporary discourse of neo-liberalism is its emphasis on jurisdictional competitiveness among nation-states. This competition is an object of truth – neo-liberal political economists present it to us as a simple fact – that functions to discipline the regulatory activities of nation-states. The pervasiveness of this discourse has led many states to internalize the goal of providing what Gill (1998) terms the three Cs: *credibility* and *consistency* in public policy designed to inspire the *confidence* of investors.

It was during the rise of the transnational historical bloc, with the growing influence of neo-liberalism on government policy, that academic and public environmental concern reached its peak in industrial countries, in the late 1980s. This confluence of forces helps explain the emergence and implementation of the concepts of sustainable development and ecological modernization. Ideologically, these concepts can be seen as attempts, by the transnational historical bloc, to accommodate environmental movements (located both inside and outside this state and civil society complex) while maintaining the bloc's hegemonic position in the global political economy. That transnational business interests consented to the adoption of these environmental discourses was clear in the run-up to the 1992 United Nations Conference on Environment and Development in Rio de Janeiro. At that time, a particularly strong voice for the concept of sustainable development was the World Business Council on Sustainable Development (WBCSD) (see Schmidheiny 1992). The WBCSD included some of the world's largest environmental offenders, Mitsubishi, DuPont, Shell, and Monsanto among them (Chatterjee and Finger 1994, 121-36). These corporations were prepared to adopt the ideology of sustainable development – rather than zero growth, for example – because its acceptance of ongoing development allowed for a shift from questions about excessive consumption, global inequities in wealth distribution, and the threat of tough environmental regulations to the more manageable goals of increased efficiencies, industry self-regulation, and the purveyance of "green" product lines. For these corporations, the implications of sustainable development and ecological modernization were essentially the same: they were vehicles for channelling environmental critique in a way that would benefit, or at least have minimal impact, on their bottom lines.

While these reasons may explain why the transnational historical bloc adopted sustainable development as an *ideology,* sustainable development also had effects as a *discourse* of contemporary ecopower, some of which appear to have been at odds with the bloc's goals. As with governmentality in general, which was originally the product of both liberalism and *raison d'état,* as a set of discourses and practices, neo-liberal ecopower embraces a mix of freedom, ongoing state intervention, and expert knowledges and modes of administration outside the direct control of the state. On the one hand, in the field of international environmental politics, the rise of the ideology of neo-liberalism has been accompanied by the entrenchment of economic expansion, free trade, market forces, and incentives as a means to address global environmental issues (Bernstein and Cashore 2002, 213). On the other hand, neo-liberal ecopower invokes a very intrusive role for state and non-state experts alike. Consider the norms of liberalized trade instituted through the World Trade Organization.

The 1994 WTO Agreement on the Application of Sanitary and Phytosanitary Measures (SPS Agreement) allows states to assess the risks of products such as new pesticides for their impact on the environment and on human and animal health before their approval for use in those countries (WTO 1994a). And according to outcomes of SPS Agreement disputes, states are permitted to set the regulatory bar (the level of harm deemed unacceptable) as high as they like (WTO 1998a). These import-restrictive measures are a far cry from free trade, yet they are promoted as acceptable practices by the proponents of neo-liberal globalization. Such practices are accepted as legitimate forms of green protectionism if – and these are the disciplinary boundaries of neo-liberal governmentality – they can be justified by appeals to scientific assessments of the risks, be non-discriminatory, and be the least trade-restrictive means to achieve the specific environmental or public health goal.

The ideas of sustainable development and ecological modernization, then, emphasize a reduced role for the state in some respects, but remain highly interventionist in others. Significantly, the locus of this intervention has shifted from the state, narrowly defined, to wider state and civil society networks and their expert views on how to manage the planet and its inhabitants. This reality sets the context for many of the key struggles in contemporary environmental politics, which are centred on expert knowledge claims and, more often than not, on competing interpretations of environmental risks and how they should be managed. I now turn to considering these political dynamics.

The Politics of Risk

What are the productive effects of contemporary ecopower, as embodied in the discourses and practices of sustainable development and ecological modernization? First, the adoption of these discourses has led to an internalization of environmental norms across many industries and governments. This is evident in government and corporate green plans, for example. But such processes are far from uniform, as they occur in different ways in different corporate and political cultures. This lack of uniformity leads to the potential for conflict within sectors and among governments, as later chapters demonstrate in the case of biosafety. Second, these discourses have clearly empowered those with technical resources in this area of truth, according governments, corporations, intergovernmental bodies, and universities with the financial resources and skills needed to undertake scientific research, an important role in determining the general thrust of governmentality in this field. It is their interpretations of environmental risks, and how these risks should be managed, that are the first to be embedded in regulatory practices. Third, this configuration of ecopower has resulted in new sites of conflict and new modes of resistance.

The case of sustainable development provides a useful example of a new site of resistance. While sustainable development, as an *ideology*, was deployed by the transnational historical bloc to further its interests, it did originally emerge out of a process of accommodation to social movements critical of the status quo, so the *discourse* of sustainable development remains contested. Since its arrival in official policy parlance in the mid-1980s, environmental and social activists have continually tried to (re)define sustainable development in order to have their values embraced in the environmental strategies adopted by industries and governments. Critics of the World Business Council on Sustainable Development's vision of sustainable development, for instance, have argued that serious attempts at sustainability must address intra-generational (and not just inter-generational) social, political, and economic justice concerns to be viable.

In the example of sustainable development, the term and its definition provide a site of struggle among alternate normative framings. More often than not, however, debates within the context of ecopower revolve around whether or not certain practices *are* really sustainable. In other words, normative positions are couched in scientific truths. This dynamic is precisely what Foucault and his interlocutors observe in other biopolitical struggles: questions of value become debated as questions of fact.

When it comes to contemporary environmental issues, Beck has developed a useful sociological interpretation of these political dynamics in modern industrialized countries. In his works on the risk society, Beck (1992, 1995) argues that the central current of politics is no longer concerned with the distribution of the benefits of industrialization but with the distribution of its social and environmental costs. According to Beck, the struggle for one's daily bread has lost its urgency for many in these countries, while the potential costs associated with modern industrial processes (such as nuclear power, the use of industrial chemicals, and genetic engineering) are increasingly global in scale, complex, and potentially catastrophic.

In the risk society, political struggles often take the form of controversies over the nature and extent of risks. These controversies are seen in cases when official development decisions, such as a municipality's decision to give planning approval for an intensive hog operation, are challenged on the basis that the true costs of those decisions, in terms of possible impacts on the environment and society, have not been adequately considered in the decision-making process. The individuals and groups that bring such challenges forward may have various reasons for their positions (e.g., concerns about animal rights or property values), but the question of risk becomes the centre of their argument. For both sides, the apparent objectivity of risks allows normative assumptions to remain hidden, while the authority of truth sides with them. As a result, governments devise accounts of risks as manageable. Relying on risk analysis techniques, which define a risk as the product of the magnitude of a possible hazard multiplied by the likelihood that it will occur, estimates are produced and management strategies are devised to demonstrate that there is no real cause for concern. On the other side, critics adopt the language and tools of science to show the limitations of official risk estimates, uncertainties in the data, and flaws in interpretation, all of which clearly demonstrate the need for greater caution or the pursuit of alternative courses of action.

This discussion shows how the idea of risk analysis is much more than simply an ideological tool for state or industry interests, despite frequently being wielded this way. As a discourse, it also enables forms of resistance that are powerful in their own right. Experts affiliated with the state and corporate interests are empowered in this field, but so are resistance movements when working with those whom Irwin (1995) calls "counter-experts." Given that these resistance movements often lack economic clout, the power they can exercise through expertise represents an important dimension of contemporary environmental politics.

Beck (2000, 42) speaks to these dynamics of power and resistance in the risk society in terms of "involuntary politicization":

> Perceived dangers appear to prise open firmly bolted mechanisms of social decision making. Things which used to be negotiated with no attempt at justification must now suddenly have their consequences justified in the

biting wind of public debate. Whereas the execution of particular legislation once seemed to take place automatically, those responsible now appear
in public, and, when the pressure is on, may even admit to mistakes or
mention the alternatives that were once rejected. In sum, the risk technoc-
racy unintentionally produces a political antidote as a result of, and in opposition to, its own way of handling things. Dangers become publicly known,
even though the relevant authorities claim to have everything under control, creating new leeway for political action.

Beck sees this involuntary politicization as a step towards greater democratic accountability in post-industrial societies. While this analysis is tinged
with a Habermasian reading of communicative rationality that Foucault
would resist, the latter would certainly agree that these dynamics show that
ecopower does not extinguish the possibilities of resistance; rather, it enables particular forms of resistance. A Foucauldian analysis would also add
this warning: the ability of some activists to function effectively in the field
of ecopower often comes with a disciplinary price. The terms of the debate
they engage in tend to assume that the management of all environmental
issues is ultimately both desirable and possible. This can be frustrating for
environmentalists who use the language of risk to criticize but whose concerns are still not fully addressed by the possibility of simply more "scientific" management of nature.

The Biopolitics of Genetic Engineering

Over the past three decades, the genetically engineered organism has become a new realm of biopower. This field is composed of a constellation of
material and discursive practices that have recast biology in the terms of
engineering, risk assessment, and property rights. At one end of this new
biopower axis we find the production of proprietary traits in individual
organisms (and their progeny) by splicing foreign genes into their genetic
blueprints. At the other end we see the emergence of new approaches to the
assessment and management of the risks of transgenic organisms to consumers, agriculture, and the environment by state regulatory agencies and
their civil society partners. Whereas biopower has sex at its axis, sex has
now been supplemented with the laboratory-based recombinant DNA (rDNA)
techniques. Foucault saw the goal of producing docile, useful bodies through
a technique of overlapping objectification and subjectification as critical to
making the human sciences, as central components of both governmentality
and biopower, historically possible (Dreyfus and Rabinow 1983, 160). Today, the goals of creating a docile, productive nature, and of shaping consumers into condoning subjects of genetically modified food, have given
rise to new realms of government and biopower that reveal the importance,

safety, or potential risks of transgenic organisms. However, as we see in other fields of biopower, the realm of the GEO is rife with strongly contested constructs. Critics are continually challenging and reinterpreting the ideas that become hegemonic in this field, and outcomes remain far from certain.

The remainder of this chapter presents a detailed examination of the way struggles in the field of biotechnology have taken place at intersections among the discourses of the gene, the environment, liberalism, and risk. I focus on two particular sites of struggle: the ownership and control of genes and transgenic organisms, and the discursive politics of regulating agricultural biotechnology, with a view to protecting public health and the environment from the risks they may pose.

Genes as Property

The emergence of the biotech bloc in the early 1980s was predicated on the ability of the agri-chemical industry to foresee returns on their investments in the genetic engineering of living organisms. These returns would not have been possible without an important shift in North America that allowed genes to be recognized as private property. This shift, first formally realized in US Supreme Court and US Patent and Trademark Office rulings, was enabled by the growing influence of neo-liberalism coupled with an acceptance that genes really are independent, transferable commodities, as propagated by gene talk.

Patents are normally awarded only when three conditions are met. The product or process must be an innovation (involving a novel step or producing a hitherto unknown product), it must be non-obvious, and it must have industrial applications. By awarding patents, states ensure that innovative methods and products become public knowledge and that they are applied within the country. In exchange, the state guarantees the inventor the exclusive right to manufacture the product, to use the process, or to license others to carry out these activities, for a period of fifteen to twenty years (depending on the country and type of product). In the case of genetic engineering, patents (or similar forms of intellectual property rights) have been awarded for GE processes and for whole lines of transgenic organisms. Patents have also been awarded for specific genes, since the act of isolating and replicating DNA outside an organism's own body has been accepted as an innovative step under many patent laws.

While the patenting of genetically engineered micro-organisms and plants has become the accepted norm in most industrialized countries today, before the 1980s, companies that developed new seed varieties had only limited rights to these organisms, because plant breeding was recognized as different from invention. The germplasm of most domesticated plant species had been cultivated and selected by farmers and breeders for millennia, and this germplasm was seen as common, rather than private, property. As a result, even when the 1970 and 1978 US plant variety protection (PVP) acts gave seed producers the exclusive rights to market seeds of a plant line they had developed, this right was balanced with the "farmer's privilege" to replant seed, as well as the rights of other breeders to develop varieties from the protected line (US Supreme Court 1980; Ghijsen 1998). That farmers could buy seeds and then reproduce their own seed for ensuing years also meant that the PVP acts provided little commercial incentive for companies to invest heavily in developing seeds of non-hybrid crops, such as soybeans or wheat. (Hybrid crops, such as corn, were an exception. Hybrids are developed by crossing two distant lines of a crop. The first generation of a hybrid crop tends to demonstrate the strongest qualities of its parent lines, but ensuing generations do not. Because the parental lines are held as trade secrets by the companies that sell hybrid seeds, farmers must buy hybrid seeds annually.)

The shift that would open the way to the patenting of seeds came in 1980 when the US Supreme Court ruled that a transgenic micro-organism could be assigned a patent in the five to four *Diamond v. Chakrabarty* decision (US Supreme Court 1980). This ruling was followed by a 1983 US Patent and Trademark Office judgment that allowed transgenic plants to be patented. In 1987, this same office ruled that a transgenic mouse designed to develop cancer at a predictable rate, Harvard University's OncoMouse, could be awarded a patent. What was it that made the patenting of living organisms the new norm in the United States?

In his majority decision in the *Chakrabarty* case, Chief Justice Burger wrote (US Supreme Court 1980, 3, emphasis mine) that any matter that has been subject to useful improvement should fall under the Patent Act, as originally authored by Thomas Jefferson in 1793, since that act is based on the idea that "ingenuity should receive a *liberal* encouragement." However, as King and Stabinsky (1998/99) point out, Jefferson himself was a plant collector and breeder, yet plants and animals were not included in his original patent law. Mr. Justice Brennan, who wrote the dissenting opinion in the case, took note of this detail, and of the fact that over time the United States had developed PVP acts that gave breeders much more limited protection over biological innovations (US Supreme Court 1980). The closeness of the 1980 vote in the US Supreme Court suggests that this ruling was about more than applying simple logic to the question of property rights to transgenic organisms, as the majority decision suggests. To understand this shift, we need to examine the social relations and ideational moves that enabled it.

In terms of social relations, the late 1970s and early 1980s saw the nexus of the biotech bloc forming in the United States. Industry was beginning to invest heavily in biotechnology, and the United States government was increasingly active in this field. Thanks to the growing influence of the ideology of neo-liberalism, it was increasingly accepted that states were in competition for investment in new technologies. Consequently, the Carter and then the Reagan administrations actively supported mechanisms, such as taxation strategies, that would spur innovation and promote the transfer of knowledge from universities to private industry (Wright 1994, 57). While these administrations did not directly control the Supreme Court, its decision must be viewed in light of the growing emphasis on competition and legal structures that would support innovation. That US biotech start-up stocks soared in the fall of 1980 is evidence of the immediate economic importance of the *Chakrabarty* decision (Charles 2001, 11).

The growing strength of gene talk was another ideational factor that made these rulings possible. Seed companies had tried for years to extinguish the farmer's privilege to replant seeds, but had been unable to do so. Gene talk, with its construction of organisms as gene reservoirs waiting for the high-precision science of molecular biology to "invent" new genetic combinations, fundamentally shifted earlier views of plant research. GEOs would be the first beneficiaries of patents, in the early 1980s, but this move was eventually followed, in 1991, by new PVP laws that would strengthen plant breeder rights to all types of new seeds, whether GE or not.

The five-four decision in the *Chakrabarty* case shows that while there were forces pushing the idea that individuals and companies should be able to gain exclusive patent rights over specific genes, this was not yet widely accepted as common sense in the early 1980s. (In fact, IPR over genes remains a central issue of contention in the global politics of agricultural biotechnology today, as noted in Chapter 2.) The impact of the recognition of genes and transgenic organisms as intellectual property in the United States was felt around the world. Since the early 1980s, variations on these US decisions also have been made in Europe, Canada, Japan, and many other industrialized countries. These countries then cooperated in the development of the 1994 WTO TRIPS agreement, which would become the primary disciplinary vehicle for globalizing US-style patenting. Yet, despite the globalization of patenting norms for living organisms, there is considerable variation among countries in how far one can go in patenting life forms. Furthermore, discourses of resistance that focus on contradictions between IPRs and sustainable development, and on the rights of communities to the traditional seeds and germplasm developed by them, have also gained footholds in the debate.

As an example of the variation in patenting norms, although Harvard's OncoMouse received patent protection in Australia, Japan, and several European countries, in Canada the OncoMouse patent claim was rejected. In December 2002, the Supreme Court of Canada denied the OncoMouse patent on the basis that only the Canadian Parliament, and not the Patent Office, should undertake a step as radical as permitting the patenting of animals and other "higher life forms" (Supreme Court of Canada 2002). Significantly, the Canadian Parliament has never debated the question of intellectual property rights over living organisms, and the Supreme Court decision has not yet resulted in efforts to bring forth legislation on this issue.

Despite this outcome in the OncoMouse case, Canada has not rejected patents on life. As if to reinforce this point, a second Supreme Court of Canada decision released in 2004, in the case of *Monsanto Canada Inc. v. Schmeiser,* reaffirmed the rights of companies with patents on genes found in transgenic plant varieties (as opposed to the plants in their entirety) to control how those genes are used.

Schmeiser, a Saskatchewan farmer, had been sued by Monsanto for growing patented Roundup Ready canola without paying royalties to the company. Schmeiser admitted that there were Roundup Ready genes in the canola on his farm but argued that these genes were unwanted contamination caused by seeds falling from passing trucks or from pollen drifting in from neighbouring fields. He argued that he had never signed a contract with Monsanto and had never applied Roundup to his crop. His lawyer further argued that because Schmeiser did not actively use the proprietary trait, Monsanto's right to enforce its patent protection should not extend to this particular farmer (Anonymous 2004). Despite these arguments, Schmeiser lost his first case in a Saskatchewan Federal Court in 2001. This decision sent shockwaves around the world as farmers started to realize that contamination by proprietary GE genes could cost them the right to save their own seeds for replanting. Schmeiser appealed the lower court decision and the case was heard by the Supreme Court of Canada, receiving prominent attention as a David versus Goliath story until its conclusion. The Supreme Court of Canada decision of 2004 upheld the Monsanto patent on the Roundup Ready canola genes, but did not award Monsanto the royalty of $15 per acre because Schmeiser "earned no profit" from having the genes in his field (Supreme Court of Canada 2004).

Together, the OncoMouse and Schmeiser cases, and the apparent contradictions between the two, have led Canadian CSOs to argue that a full public debate on the issue of intellectual property over living organisms and their parts is long overdue (Swenarchuk 2003). These cases brought the issue of intellectual property rights over living organisms to the public's attention and raised new concerns among farmers and environmentalists in particular. One of the immediate concerns for environmentalists is that, in the Schmeiser case, Monsanto appears to be rewarded for genetic pollution caused by its Roundup Ready genes, rather than punished. Interestingly, concerns over IPR issues among mainstream environmental activists represent a significant shift from the early days of GE patents, a shift that relates to the way that the discursive framing of so-called environmental issues has changed over time.

The *Chakrabarty* case galvanized the small group of activists and critics that made up the first wave of the anti-genetic engineering movement (Schurman and Munro 2003), but its overall impact on building resistance to GEOs was limited. Many of the early activists in this movement were concerned that the patenting of a micro-organism was the first step towards the patenting of human beings. In response to the *Chakrabarty* decision, for example, the Council for Responsible Genetics released the first "No Patents on Life" declaration, in the early 1980s (interview #2). However, larger North American environmental organizations did not mount campaigns on the issue of patents. Mooney (1996) believes that this lack of interest was because Chakrabarty had developed a strain of bacteria engineered to help clean up oil spills. Because the mainstream environmental movement saw Chakrabarty's work as an environmentally friendly technological development, it did not jump aboard the anti-patenting campaign. Like the case of Roundup resistance versus atrazine resistance, this situation again demonstrates how the end uses of a particular genetic transformation can help mitigate public dissent for the biotech project as a whole.

By the mid-1990s, however, a much wider array of activists, including the larger environmental groups, did get involved with the patenting issue as part of the third wave of GE resistance. As genetically engineered organisms moved towards commercialization, and as the implications of proprietary rights over whole lines of GE seeds became clearer, a global movement emerged around the call for No Patents on Life, spearheaded by the CRG, Canada's RAFI (led by Pat Mooney), and India's Research Foundation for Science, Technology and Ecology (directed by the public intellectual and activist Vandana Shiva). Significantly, the activists now challenged IPRs over living organisms within new discourses that had gained currency in global environmental politics.

Most important, in terms of the discursive currents being discussed in this chapter, activists used three major lines of argument to demonstrate the contradiction between the patenting of GEOs and the (by then widely accepted) goal of sustainable development. First, they drew attention to the fact that in countries such as India, over 80 percent of farmers plant their own farm-saved seeds. These resource-poor farmers were not in a position to buy new seed each year, so the provision of patented seeds would effectively lock them out of agricultural development. Second, they argued that the arrival of patented seeds could wipe out traditional landraces in many countries. Landraces are locally adapted varieties bred by farmers, and are key components of the world's agricultural biodiversity. The pressure on farmers to adopt genetically engineered seeds – from which companies could earn a profit each year – would be intense, leaving little incentive for the preservation of local seed varieties, even though they might be better suited for dealing with climate variations or water scarcity, for example. As a result, genetic engineering would lead to yet more monocultures, leaving farmers, regions, and countries vulnerable. Third, these activists pointed out that the acceptance of patents on seed meant that companies with access to genomics (the techniques of sequencing and interpreting the genomes of living organisms) are in a position to patent material that has actually been used by farmers and indigenous communities for generations. Where the biotech industry saw the gathering of exotic germplasm as "bioprospecting," these activists saw it as "biopiracy." As an example, US patent laws allowed US-based W.R. Grace & Co. to procure a patent on the well-known Indian medicinal plant neem in 1994 (*Financial Express* 2005). Similarly, Texas-based RiceTec, Inc. registered US patents on varieties of basmati rice (a type of rice historically associated with Pakistan and India) in 1997 (Shiva 1998). These concerns about biopiracy, along with critiques of the TPS/terminator technology (when this was uncovered by RAFI in 1998 [RAFI 1998]), tied anti-patenting campaigns to larger struggles against globalization and corporate control in the 1990s, effectively bringing young environmentalists, labour unions, and others to the anti–genetic engineering camp.

Discursively, these arguments are interesting because they draw on the idea of sustainable development, a discourse initially supported by the biotech purveyors such as Monsanto, to criticize the impact of IPRs in developing countries. In another clever strategic move, the CSOs mobilized the discourse of rights over genetic material against the biotech purveyors by promoting the idea of "farmers' rights" to traditional germplasm. This concept built on the notion of the farmer's privilege to plant farm-saved seeds established in early PVP laws. In the context of the TRIPS, these groups noted that this international agreement allows for *sui generis* systems other than patenting. Such systems, they argued, could be designed to protect the rights of farmers to save and use seed varieties, and the rights of indigenous communities to compensation for stewardship of plant varieties and their genetic material, as a complement to the protection of genetic engineers' claims on invented traits that the agreement was originally designed for (Louwaars 1998, 13).

Clearly a discourse of resistance rooted in farmers' rights is enabled in, but subverts, the discourse of proprietary rights to genetic material. It has had an enormous impact in the global politics of agbiotech. Through the gradual acceptance of this discourse, we are seeing *sui generis* laws enacted in some countries of the global South that recognize the rights of traditional plant breeders and farmers to register plant varieties used and developed by them or their communities as part of IPR legislation. One example is India's

Protection of Plant Varieties *and Farmers' Rights* Bill of 2001 (CSE 2001, emphasis added).

The notion that indigenous people and farmers should retain rights to the uses of genetic material developed in their communities has even had an impact in the United States and the EU. In the United States, criticism of RiceTec's patent application for basmati rice, from both the Indian government and Vandana Shiva's Foundation for Science, Technology and Ecology, led to a re-examination of the patent claim by the US Patent Office, in 2000, and eventually a request to RiceTec to withdraw seventeen of its twenty patent claims (Clement 2004). In the end, the US Patent Office upheld only three of RiceTec's patents for varieties the company had genetically engineered, and the patents granted on these varieties referred to them by number rather than the name "basmati." This move was hailed as a victory by Shiva's organization, because it "prevents the potential use of the basmati patent against farmers growing traditional 'Basmati' rice and ensures the economic status quo of traditional basmati rice farming" (quoted in Clement 2004, 16). In the EU, similar concerns led to the European Patent Office's revocation of a joint United States Department of Agriculture and W.R. Grace & Co. patent for a fungicidal product developed from seeds of the neem tree, originally granted in 1994 (*Financial Express* 2005). Following up on complaints lodged by Shiva's foundation along with two other environmental organizations concerned about IPRs, the European Patent Office revoked the patent in 2000. It further upheld this decision against a challenge by the patent applicants in 2005. The grounds for revoking the patent application, were that there was no novelty and inventiveness in the neem product, since farmers in India had already long recognized, and utilized, the fungicidal properties of neem (RFSTC 2005).

As these examples reveal, the debate about property rights to living organisms has evolved in recent years, as ideas about IPRs, sustainable development, and farmers' rights become better defined. However, it is important to recognize that, as a discursive phenomenon, this debate has resulted in unanticipated productive effects for forces on both sides of the struggle. For activists against patents on life, the adoption of the discourse of farmers' rights has given them a foothold in debates about IPRs, but only at a cost. The notion of farmers' rights, like that of patents, is ultimately a practice of ecopower that accepts plant lines and their genes as commodities to be owned, bought, and sold. It also does not inherently negate the possibility that companies may claim property rights to "invented" varieties, even as communities secure their claims over traditional ones. By adopting a discourse that accepts the notion of organisms as property, then, groups advocating for farmers' rights may have implicitly accepted a framing of the human-nature relationship that does not entirely fit with their wider social and political agendas. Meanwhile, for proponents of genetic engineering, the emphasis they placed on the distinctiveness of the techniques of genetic engineering, while important in the context of gaining IPR protection, became fodder for those who argued for tough assessments in the face of these radically new technologies. I now turn to examining the political dynamics of these debates about the regulation of GEOs for their potential environmental and health impacts.

Defining the Regulatory Problem: Containing Laboratory Hazards

A discourse focused on the regulation of the effects of genetically engineered organisms first emerged in the United States in the mid-1970s. At that time, regulations aimed at containing the potential hazards of genetic engineering took shape at the interface between discourses of the gene, the environment, and liberal norms of self-regulation.

In the late 1960s and into the 1970s, the mixing of genes from more than one species of micro-organism through cell fusion – the technique that was the immediate predecessor to genetic engineering – was widely heralded as an advance that had enormous promise for medicine and agriculture. However, this period also saw the rise of environmentalism as a broad-based social movement with its focus on nature protection and on the reduction and elimination of pollution. As the science of molecular biology developed, and genetic engineering came within reach, concerns were voiced about organisms that included such unnatural complements of genetic material. In an example that demonstrates the way apocalyptic themes enter into the genetic engineering debate, Rifkin (1983, 74) argued that the price of genetic engineering would be the "human soul," because creation would no longer be sacred. Once we undertake genetic engineering of other life forms, he wrote, "we are responsible for nothing outside ourselves, for we are the kingdom, the power and the glory, for ever and ever." Critics also voiced concerns about a host of potential economic, social, public health, security, laboratory safety, and environmental threats that could be posed by micro-organisms engineered (whether intentionally or not) to be pathogenic. Genetically engineered micro-organisms (GEMs, the first generation of GEOs) were framed as potential biological pollutants with the awesome ability to reproduce and multiply. Drawing on the catastrophe discourses that had become commonplace in environmental parlance, Rifkin (1983, 80) also argued that "genetic engineering could very well pose as serious a threat to the existence of life on the planet as the bomb itself."

In July 1974, the molecular biology research community responded to the controversy raised by genetic engineering. A National Academy of Sciences committee headed by Paul Berg, then a leading genetic engineering researcher, published a letter to molecular biologists calling for a moratorium on further experiments until the hazards of GEMs could be properly assessed (Wright 1994, 136-40; Krimsky 1982). The result was a voluntary moratorium that continued until 1975, after the second of two conferences held in Asilomar, California, at which these issues were debated by the scientists involved. Following the Asilomar conferences, a voluntary regulatory approach was institutionalized that advocated laboratory practices designed to contain genetic engineering experiments so that GEMs would not escape into the environment. This approach was formally constituted in laboratory safety guidelines by the National Institutes of Health (NIH) in the United States in 1976. The Canadian manifestation can be found in the *Guidelines for the Handling of Recombinant Molecules and Animal Viruses and Cells* brought in by the Medical Research Council of Canada in 1977 (PHAC 1996; an updated version, Health Canada's *Laboratory Biosafety Guidelines,* continues to govern genetic engineering laboratory and greenhouse-based research in Canada).

The NIH guidelines were premised on the idea that genetic engineering experiments should be small in scale and undertaken under controlled and contained conditions. In principle, no genetically engineered organisms should be released to the environment, and if one did get out, it should not survive. Furthermore, the more dangerous an experiment was perceived to be, the more cautious should be the conditions under which it would be carried out. Containment would be achieved by the use of physical and biological barriers. Depending on the nature of the experiment, physical barriers could range from the standard techniques used in American microbiological laboratories to those used in facilities designed to handle the most lethal of pathogens, such as anthrax and the smallpox virus. Biological containment depended on the use of experimental organisms that were unlikely to survive outside the laboratory environment (Krimsky 1991, 100-1; Wright 1994, 168-69). For the purposes of implementation, the NIH guidelines called for the establishment of institutional biohazard (later biosafety) committees to ensure that laboratory facilities were properly equipped and that procedures were being followed. Nationally, committees known as study sections reviewed research proposals and decided on the safety procedures that should be adopted for specific experiments (Wright 1994, 168). While adopting the committees' recommendations was theoretically voluntary, a lack of compliance on the part of researchers could see them lose all government funding – a heavy sanction for university-based scientists at the time.

The establishment of a regulatory system narrowly designed to address only the safety concerns associated with genetic engineering – rather than the much wider set of social and ethical issues being raised by critics – represents a very particular response to the genetic engineering problem. What factors enabled this discourse to become the norm, and what were their implications? To answer these questions, we need to revisit the social relations of genetic engineering of the time. In the mid-1970s, corporate interest in genetic engineering was still limited, and there were no GEOs on their way to commercialization. Instead, these techniques were in the hands of the researchers, whose primary interest was in seeing research continue and ensuring that any regulation of their work remained in the hands of the scientific community (Wright 1994, 140). As Wright demonstrates, it was these scientists, and the Berg committee in particular, that established this narrow framing of GEO risks in their original letter of 1974. And the voluntary moratorium proposed by these scientists was a public relations coup. It demonstrated that the molecular biology community exhibited a healthy degree of caution in exercising its new-found powers over life, as critics and a cautious public were demanding. This apparently responsible attitude, combined with the way that liberal norms of governance deferred to expertise, ensured that industry self-regulation became the accepted norm until the early 1980s.

As part of a regulatory discourse, the framing of genetic engineering in terms of hazards needing containment was a significant departure from past experience in at least one respect. Serious attention to the regulation of pesticides, for example, was given only long after they were in widespread use and found to be harmful. In the case of genetic engineering, regulatory efforts were being made *before* the introduction of *potentially* hazardous organisms into the environment. That scientific researchers were prepared to police themselves to this degree reveals how distinctive the debate over GEOs was, even in its early days. This emphasis on the prevention of harm within ecological discourse foreshadows the development of precautionary discourse in the 1990s. The whole rationale for the containment approach was, according to the Berg committee, because "there are few available experimental data on the hazards of ... DNA molecules" (Berg et al. 1974, quoted in Wright 1994, 179). The recognition of this kind of scientific ignorance would eventually resurface in arguments for institutionalizing a precautionary response to commercial releases of GEOs into the environment.

Despite the novelty of containment before evidence of harm as an approach to regulation, this case still had much in common with previous regulatory efforts. In fact, it reproduced already existing norms of ecopower in at least three ways. The discourse of hazards effectively separated technical issues from social issues related to genetic engineering. It placed the genetic engineering research community of "independent" experts, rather than the state, in the driver's seat when it came to defining and finding solutions to the genetic engineering problem. And, this regulatory discourse functioned as a tool for normalizing genetic engineering. Building acceptance for genetic engineering by suggesting that its potential hazards were fully managed would become a recurring theme in ensuing decades.

Beyond reproducing norms of ecopower in this new field, the focus on genetic engineering hazards had further productive effects on the politics of agbiotech. Specifically, the acceptance of this discourse meant that public skepticism was not yet quelled when GEMs were brought out of the laboratory for field tests and commercialization. In other words, the regulatory emphasis on GEMs as potential hazards that worked in the biotech research community's favour in the early 1970s may have helped lay the foundation for some of the resistance to the biotech industry in the 1980s. For example, public skepticism was intensified by news footage showing field tests of a GEM designed to inhibit frost formation (an organism called ice minus) on California strawberry fields in the mid-1980s. This footage showed applicators wearing "moon suits" to avoid contamination (Krimsky 1991, 132). These suits may have been appropriate given accepted procedures for dealing with potentially hazardous micro-organisms, but they were exactly the wrong image for a nascent commercial biotech industry. The eventual development of commercial GEOs to be used outside of the lab would require a very different set of regulatory discourses and practices.

The regulatory focus on hazards channelled the genetic engineering issue into a narrow technical debate, as Beck and Foucault might have predicted. Nonetheless, a wider set of issues continued to motivate activism and public concern. In fact, the same kinds of catastrophe arguments voiced by Rifkin in the 1970s were revisited in subsequent years. Consider, for example, the apocalyptic title of Canadian food activist Brewster Kneen's 1999 book, *Farmageddon*. Concerns about patenting and corporate concentration also resurfaced through the 1980s and 1990s, even though these were formally excluded from "science-based" regulatory decision-making processes. These issues continued to motivate critique, but the focus of that critique, in the regulatory debate, remained on the potential hazards of the new technologies, whether to human health or the environment.

Redefining the Problem: Managing the Risks of Biotechnology

The second major discursive moment in the politics of biotechnology regulation came in the early 1980s in the United States. At that time, several factors came together to displace the discourse of GEOs as inherent hazards with a discourse focused on the management of the risks of biotechnology. Organizing genetic engineering in terms of risk was an important shift in the governance of agbiotech and reflected the growing emphasis on risk management within ecopower. What difference did it make? First, the implicit cost-benefit calculation had changed. It was now accepted that GEOs had to be proven dangerous, rather than this being taken as a given. Second, the risk framing suggests that the potential harms of GEOs could be easily characterized and calculated. This move assumes sufficient baseline data and scientific understanding of how GEOs will act in the environment. Third, GEO risks would now be framed in terms of their management, rather than their avoidance. Finally, the risk framing implied that costs and benefits of GMOs could and would be fairly traded and that no distinction needed to be made between benefits and costs to the producer of the technology versus those that might befall the consumer.

What enabled this shift in the way the dangers of GEOs were framed? The official narrative within regulatory circles states that a decade of experience with GEMs had led molecular biologists to believe that the hazards of genetic engineering were not as significant as had first been imagined (e.g., NRC 2000). This narrative is attractive for its simplicity, but it is problematic because it assumes that the new discourse was based on facts alone. Wright (1994, 337-82) points out that while there had indeed been no major environmental catastrophes associated with the release of GEMs at that point, there were also few facts about the hazards posed by GEMs in the environment and the extent to which these could be managed. One cannot assume that the scientific picture of the hazards of genetically engineered organisms was finally complete in the early 1980s. Evidence does show, however, that the GEO hazard picture was being looked at from a new perspective by the main policy actors involved: the molecular biological community. By the late 1970s, many of the researchers undertaking genetic engineering had become employed by the fledgling American biotech industry. This shift in position meant that their sights were now set on field tests and on the commercialization of GEOs, rather than simply on laboratory research. For this group, the cost-benefit calculation regarding the value of in situ experiments had changed, and these scientists were still the most influential actors determining regulations in the biotech sphere.

Wright (1994) provides a compelling argument for the idea that the move to frame GEOs in terms of risks rather than hazards in the United States in the 1980s (followed by the adoption of a similar regulatory discourse in Canada in the late 1980s and early 1990s) involved a combination of scientific evidence and shifting interests on the part of the genetic engineering community. I believe that it also depended on developments in the four major ideational currents: the gene, the environment, liberalism, and risk.

First, starting in the late 1970s, actors in the biotech bloc made a deliberate effort to avoid singling out genetic engineering as being different from traditional plant and animal breeding practices. The primary mechanism employed in this effort was the inclusion of the new rDNA techniques in a broader narrative of biotechnology. For example, one industry-sponsored booklet defines food biotechnology as "diverse activities[,] from the use of yeast in brewing or bread-making to advanced plant-breeding techniques," with a history that dates back to 1800 BCE (CBI 2000, 3). This notion of a biotechnology continuum is an example of the ideology of equivalency in action. It would have a significant impact on official regulatory policy governing genetic engineering in the United States and Canada in the 1980s. In Canada, biotechnology came to be defined as the "applied use of living organisms, or their parts, to produce new products" – a definition that includes, but does not mention by name, GEOs (Agriculture Canada 1993, 1). The adoption of this definition was an important victory for proponents of genetic engineering, because it pointed towards a regulatory system that would treat the products of genetic engineering in the same way that the products of traditional plant breeding had been treated: with fairly minimal oversight.

This minimalist position was reinforced by the growing strength of neo-liberalism during the 1980s. In American regulatory debates, for example, evidence provided by Wright (1994, 451) shows that the genetic engineering problem was framed in terms of a "race" between North American corporations and their European competitors, in which the controls on GEO hazards represented a major "handicap." Arguments that "overly cautious" regulation could result in "delays in achieving benefits" were taken very seriously by the Reagan administration in particular (Wright 1994, 275). A similar set of arguments was made in Canada in the late 1980s and early 1990s as Canada created its regulatory system for field-testing biotech plants. In 1991, Mulroney's Conservative government established a Regulatory Affairs Directorate under the Treasury Board, with a mandate to ensure that regulation does not impede innovation (Stanbury 1992, 58). According to guidelines published by this directorate, departments had to demonstrate that for existing or proposed regulations,

> a problem or risk exists, government intervention is justified, and regulation is the best alternative; the benefits of the regulatory activity outweigh the costs; and steps have been taken to ensure that the regulatory activity impedes as little as possible Canada's competitiveness. (Stanbury 1992, 2)

The notion of a biotech continuum, together with the growing strength of neo-liberalism, led some in the North American regulatory community to argue against any regulatory control that "unfairly" targeted genetically engineered organisms (Miller 1997). This position faced opposition on two fronts. On the one side, the nascent commercial biotech industry argued that a regulatory system for GEOs was actually necessary to win public approval of its products. On the other side, academic ecologists, who also carried expert authority in the field of environmental risks, started raising concerns about the potential impacts of GEOs.

Krimsky (1991, 133-51) documents an important "paradigm conflict" between ecologists and molecular biologists in North America throughout the 1980s and into the 1990s on the subject of the risks of genetic engineering. These ecologists were part of this second wave of genetic engineering critics. Rather than adopt the apocalyptic and anti-corporate rhetoric of Rifkin and Mooney, they shaped their arguments in the context of the discussion about hazards, and the management of risk, that had taken hold at the centre of this field of governmentality. Their positions took two main forms. First, they argued that tangible hazards would result directly from the genetic transformations taking place. With modified crop plants, these included the possibility of increased weediness of transgenic crops resistant to herbicides, gene flow to wild relatives of modified crops, impacts on non-target organisms from crops with pesticidal properties, and impacts to overall crop-plant genetic diversity (Rissler and Mellon 1996). Second, they argued – from a position that can be seen to be located in complexity talk – that there was ultimately no simple framework for predicting the effects of genetically engineered organisms on the environment. Exemplifying this kind of argument was an influential 1989 report from the Ecological Society of America that cited cases of secondary phenotypic effects resulting from a single genetic alteration, with some effects expressed only in certain environments (Tiedje et al. 1989).

Together, the two sets of arguments the ecologists raised were part of the initial shaping of a reverse discourse, within the context of risk, that would later come to revolve around the notion of precaution. Like the hazard discourse espoused by the genetic engineers in the 1970s, this perspective was based on a combination of scientific knowledge, norms, and interests. There was no strong evidence of major environmental harms caused by GEOs, but the past experience of ecologists with other environmental pollutants such as DDT, combined with a proclivity towards seeing nature in complex, rather than simple, terms, led them to be extremely cautious about the presumed benefits of GEOs.

Given their ability to operate within the discourse of hazards and risks, these critics had some impact on the governmental outcomes of the regulatory debates they engaged in. Consider the case of Bt refuges. When first proposed, Bt seeds raised important concerns. Specifically, bacterial forms of Bt had historically been used as natural pesticides, and ecologists and some agricultural scientists worried that widespread production of Bt toxins by plants would result in the rapid development of insect resistance to this valuable pesticide, which they believed was a public good that should not be squandered. The biotech companies argued against this perspective but were unable to counter the argument that Bt would lose its effectiveness if not used in moderation. Eventually, the critics won, and the industry was forced, first by the US Department of Agriculture and then by similar government regulatory actions in Canada and elsewhere, to accept a refuge strategy that would see farmers plant up to 20 percent of their fields with non-Bt plants of the same species to slow the development of insect-resistance to the Bt toxin (NRC 2000). This case is an interesting example of how the agrichemical and biotech industry was forced to make real economic concessions towards its critics in order to win both state and civil society consent.

Aside from their influence through such formal government channels, the ongoing scrutiny of environmentalists also led to the normalization, in the biotech bloc itself, of the need to consider and address concerns voiced by the general public. This normalization resulted in the smooth public relations strategies of the commercialization period. It also resulted in the internalization of certain environmental norms in the self-governance practices of the genetic engineering community. That Ciba-Geigy never developed atrazine-tolerant corn, although it had isolated the appropriate gene and considered the endeavour, as noted by Charles (2001, 156), is an example of self-governance in the industry in the context of intense CSO and media scrutiny.

These cases illustrate how the more tangible concerns raised by ecologists had an impact on the governmentality of GEOs in North America. But what became of complexity talk and its narrative of uncertain risks? These concepts were part of the debates over GE plant regulation, but to what extent did this position influence the formal regulatory structures that were emerging for genetically engineered organisms? To address these questions, a more fine-grained analysis of the GEO regulatory system is necessary.

Normalizing the Commercial Release of Plants with Novel Traits

Undertaking risk analysis normally depends on historical data of hazards and their likelihood of occurring. At the very least, it requires a sense of what those hazards could be. In the case of genetically engineered organisms, there was still a great deal of debate in the 1980s about what the real hazards of GEOs were, and there were few experimental data upon which to base likelihoods of occurrence. Given this, how could GEOs be assessed for their risks in the United States and Canada, and which risks were considered?

In the United States, the Environmental Protection Agency, the Department of Agriculture, and the Food and Drug Administration are the primary regulators of genetically engineered plants and foods. In Canada, this role is mainly undertaken by the Canadian Food Inspection Agency and Health Canada. There are some important differences in procedures for evaluating GE plants and foods in these two countries. In the United States, the regulatory process for GE foods is a "consultation" process in which the developer needs to show the Food and Drug Administration that it is confident there are no unmanageable risks associated with the new GE food (US FDA 1995). In Canada, by contrast, the process is more formal, with a new food needing formal approval from regulators before coming to market (Health Canada 2004). Nonetheless, the overall approaches to GEO safety assessment in the United States and Canada are similar, so I examine these systems together. In particular, in both countries there is a reliance on three specific regulatory concepts to facilitate the conclusion that GEOs are safe for the environment and human consumption. These are the concepts of novelty, familiarity, and substantial equivalence.

With definitions of biotechnology rooted in the notion of a biotechnology continuum, Canadian and American governments do not single out GEOs as a distinct category of plants and foods. Instead, they are considered part of a larger category of "plants with novel traits" (PNTs, in Canada) and "novel plants" (in the United States). In Canada, PNTs are defined as

> plant varieties/genotypes possessing characteristics that demonstrate neither familiarity nor substantial equivalence to those present in a distinct, stable population of a cultivated species of seed in Canada and that have been intentionally selected, cultivated or introduced into a population of that species through a specific genetic change. PNTs include those derived from both recombinant DNA technology and plants derived through traditional breeding. (PBO 2000, 1)

This definition of PNTs clearly depends on the definitions of familiarity and substantial equivalence. These are as follows:

> Familiarity is defined as the knowledge of the characteristic of a plant species and experience with the use of that plant species in Canada. Substantial equivalence is defined as the equivalence of a novel trait within a particular species, in terms of its specific use and safety to the environment and human health, to those in that same species, that are in use and generally considered safe in Canada, based on valid scientific rationale. (PBO 2000, 2)

Familiarity and substantial equivalence are more than simply aids for defining PNTs. These concepts, when elaborated with reference to specific variables that can be empirically observed and tested, are both the triggers and the endpoints of the risk analysis process. The concepts are triggers in that a lack of knowledge about a plant's familiarity or substantial equivalence is what sets a data review in motion. The concepts are endpoints in the sense that once a PNT has been determined to be familiar and substantially equivalent to a similar crop already in the food system (a "comparator"), with any new traits falling within acceptable boundaries or not posing unmanageable risks, the PNT is considered safe and needs no further assessment (PBO 1995).

Tests to determine familiarity and substantial equivalence consider specific sets of characteristics of the plant or food in question. For environmental assessments, regulators look at previous experience with the plant species, the new trait, the trait introduction method, and the cultivation practices that will be used. The assessors then evaluate the PNT against its counterpart for altered weediness potential, gene flow to related species, altered plant pest potential, potential impact on non-target organisms, and potential impact on biodiversity (PBO 2000, Appendix 3). And as noted earlier in this chapter, to assess the substantial equivalence of a novel food to a conventional counterpart, a set of molecular, compositional, and nutritional data is evaluated for the modified organism and its comparator.

These practices suggest that genetically engineered plants and foods (along with other novel plants and foods) undergo a thorough scientific assessment of their ecological and food safety risks, as the American and Canadian governments contend. Unfortunately, it would be more accurate to say that the concepts of novelty, familiarity, and substantial equivalence actually allow risk assessors to bypass a comprehensive assessment of the full range of possible health and environmental implications of GEOs.

Consider the concept of novelty. First and foremost, the term "novelty" shifts attention away from genetically engineered organisms as a potential source of unique risks. As S. Barrett and his colleagues (2001, 177-80) point out, the idea of novelty comes from the traditional breeding process, where years of crossing and selecting plants with specific characteristics allow traditional breeders to develop lines that are distinguished by novel traits such as increased disease resistance or altered oil profiles. Within these programs, the relative genetic uniformity of the material used means that interactions of the breeding-derived trait with other parts of the genome are assumed to be of no functional significance. Should a negative impact be created, it is expected to be discovered during the breeding process, and these plants are simply weeded out of the new line.

In traditional plant breeding, then, the assumption is that the novelty is located entirely in the new trait, and that the rest of the plant can be thought of as generally the same as others of the species. When novelty is deployed in relation to the GEO, it brings with it the assumption that the only significant difference in the GEO is the engineered trait. This assumption allows risk assessors to sidestep the possibility that there may be unanticipated effects of the genetic engineering process that could also harbour new hazards. In other words, novelty reproduces the linear and reductionist reading of genetics found in gene talk.

When it comes to assessments of the impacts of GEOs, familiarity and substantial equivalence reinforce the normative biases embedded in novelty. For example, even though the environmental risks being examined are some of the risks that ecologists articulated in the 1980s and early 1990s, these reviews do not embody the ecologists' emphasis on complexity and caution. Instead, reviews for familiarity and substantial equivalence assume that knowledge of the parent species, the introduced trait, and the vector used, are sufficient to allow predictions about the implications of the GEO. This approach would be valid if the new organism is nothing more than the sum of its parts, as the discourse of equivalency suggests, but this assumption does not hold in all cases of genetic engineering, as advocates of complexity emphasize.

In those cases when new studies are actually undertaken to determine possible toxicological impacts on non-target organisms or possible allergic reactions by consumers, these are often lab-based (rather than in situ) and depend on bridging studies (tests of samples of the new proteins, such as the Bt protein introduced in Bt corn, produced in bacterial cultures). Again, these studies are suitable if one assumes that the transgenic plant and the transgenic bacteria produce exactly the same protein, and that no other toxic or allergenic protein or secondary metabolite is inadvertently produced by the transgenic organism. But these are reductionist assumptions that can be verified only through experimentation, and such experiments are not routinely undertaken in Canada or the United States.

This general overview of the Canadian and US regulatory systems shows that while attention was paid to the concerns of ecologists and other critics of GEOs in the organization of these systems, the concepts of novelty, substantial equivalence, and familiarity all reproduce the assumption that there is sufficient knowledge to authorize the release of GEOs into the environment and the food system because these organisms are unlikely to be more than the sum of their parts. Under the guise of scientific objectivity, these systems appear to ignore the possibility that gene-organism-environment relationships could be more complex than gene talk assumes. A lack of any formal peer review mechanisms also means that these limitations are nowhere second-guessed in the formal regulatory process.

The stamp of approval that followed such regulatory reviews became a key tool for the biotech bloc in normalizing genetic engineering in the United States, Canada, and Argentina, despite its shortcomings. Government assurances that the "new foods developed through biotechnology [undergo] thorough safety assessments ... before they can be sold in grocery stores or in the marketplace" was a key argument of the biotech bloc in the 1990s (CFIA 2000).

At the same time, the move to commercialization was still accompanied by a spike in public interest and skepticism, more so than at any point in the 1970s and 1980s. Why had the genetic engineering issue finally hit a public nerve? There are two answers to this question. First, the activist groups had clearly worked hard to get the word out. Rifkin's Pure Food campaign, for example, had actively built skepticism towards genetically engineered foods among the general public. He did not argue that the Flavr Savr tomato and other products of genetic engineering were necessarily dangerous. Instead, Rifkin argued that they had not been proven safe (Charles 2001, 134). Given the workings of the regulatory system outlined above, this argument was clearly true. The idea that there could always be other, as

yet undiscovered, risks is the Achilles heel of any proponent of a new technology in the risk society. Still, the work of Rifkin and his fellow campaigners does not fully explain the spike in public skepticism in the 1990s. For a
more complete picture, we need to think carefully about what was happening in this emergent field of ecopower.

Struggles within the relations of biopower, Foucault (1983, 211) suggests,
are often "immediate struggles" in which "people criticize instances of power
which are closest to them, those which exercise their action on individuals." At one level, GE crops are targeted towards the management of nature.
Through their food, however, people were also becoming visceral subjects
of this new domain of ecopower, and the immediacy of this fact was the site
upon which the backlash to biotech was built. Thus, in the United States
and Canada, followed by other countries, it was during the debates over
rBGH, and then over the commercialization of genetically engineered crops
such as canola, potatoes, soybeans, and corn, that active resistance came to
be centred on the interests and rights of individuals as consumers of food.

A public opinion poll published in 1994 shows that Canadians were "very
concerned about the safety of milk from rBGH-treated cows and 96 percent
of those surveyed stated that they wanted this milk labelled" (Mausberg
and Press-Merkur 1995, 26). The way the poll results are framed in this
quote from an environmental CSD shows that two discourses became central when GE food was at issue. The first was the discourse of food safety;
the second was the discourse of labelling and consumer choice. What is
interesting about this second discourse is that it is rooted in ideas close to
the heart of the biotech bloc. The argument for labelling brings together
the liberal tenet of freedom of choice with the notion that GEOs are indeed
unique (although uniquely risky, in this case). This argument, rooted in the
liberal tenet of consumer rights, is strong. In the United States, Canada, and
Argentina, however, labelling of genetically engineered foods has yet to be
mandated by central governments. Why? The lack of labelling is a good
example of the discursive power of equivalency. These governments do not
directly challenge the notion of choice, which they recognize would be a
tough argument to win. Instead, they state that there is no scientifically
justifiable reason for singling out GEOs from their non-engineered counterparts, because they are "substantially equivalent" to one another (Health
Canada 2002). This debate exemplifies the way that an issue that many
perceive to be political is ostensibly decided on the basis of a "scientific"
judgment in the context of the risk society and ecopower. In this case, that
scientific judgment is clearly far from neutral.

The Advent of the Genetically Modified Organism
In the United States and Canada, regulatory practices that relied on the concepts of novelty, familiarity, and substantial equivalence became critical

tools for normalizing genetic engineering in the 1990s. These assessment practices were the product of material, organizational, and ideational politics in these countries, where the biotech bloc had deep roots in academia, industry, and the state, and they reflected these dynamics. My analysis suggests that risk assessment for GEOs might have taken a different form were other priorities to shape this field. In fact, the evidence shows that alternative models of risk assessment for GEOs were emerging in other countries in the early 1990s. In the EU, in particular, there appears to have been a more cautious adoption of the risk discourse employed in North America. Norway and India also provide interesting examples of alternatives to the North American regulatory model.

In the EU, 1990 Council Directive 90/220/EEC was the first major piece of pan-European legislation governing the deliberate release of genetically engineered organisms. Like its North American counterparts, 90/220 was premised on the notion that regulators should assess the risks of transgenic plants before their release into the environment. This reveals the pervasiveness of the risk framing, by the late 1980s, in the molecular biological community on both sides of the Atlantic. However, 90/220 differs from the North American regulatory model in several important ways that reveal a more cautious reading of the possible risks of genetic engineering. For example, 90/220 explicitly names "genetically modified organisms" (GMOs) – a term rarely used in North America at the time – as the central object of concern (EEC 1990). The GMO is defined as "an organism in which the genetic material has been altered in a way that does not occur naturally by mating and/or natural recombination" (EEC 1990 Article 2, para. 2). This definition lines up with the view, voiced by many ecologists, that transgenic organisms may represent unique hazards to the environment and public health when compared with traditionally bred plants and animals. The directive also notes that GMOs "may reproduce in the environment and cross national frontiers thereby affecting other Member States [and] the effects of such releases on the environment may be irreversible" (EEC 1990 Preamble, para. 6). This notion of irreversible harm, along with the need for "preventative action" and the right of states to "avoid" (rather than manage) adverse effects, suggests a more cautious stance than we see in North American legislation. Another important aspect of 90/220, in terms of events that would transpire years later, is Article 16, also known as the safeguard clause. This article states:

> Where a Member State has justifiable reasons to consider that a product which has been properly notified and has received written consent under this Directive constitutes a risk to human health or the environment, it may provisionally restrict or prohibit the use and/or sale of that product on its territory. (EEC 1990, Article 16, para. 1)

Perhaps most significant is not what is in directive 90/220 but what is absent. Whereas American risk assessors were already employing the concepts of familiarity and substantial equivalence to evaluate novel plants, this directive gives no specific parameters for what would constitute "evidence for safety" or "environmental harm" (von Shomberg 1998). Rather, 90/220 establishes flexible procedures for defining these terms in practice. This move should be read, in part, as a rejection of the North American model, with which European regulators were well acquainted.

Does the European Council's regulatory directive 90/220 represent a significant departure from the North American perspective, one rooted in very different value and policy judgments about genetic engineering? Schweiger (2001, 365-70) suggests that the negative European response to GEOs in the late 1990s was indeed the result of deeply rooted cultural values, values that include (1) an aversion to technological "tampering" with food ingredients, which for many Europeans is bound up in traditional food cultures and national cuisines; (2) an affinity towards agriculture that is in harmony with the rural environment; (3) a preference for organic food; (4) anti-American sentiments; and (5) a view that "it is just wrong." I accept that each of these values may have played a role in the resistance to GEOs in individual European countries, but it is important not to overemphasize these factors as being intrinsic to only Europe, nor to generalize them across all European cultures. Austria, for example, is actively committed to organic agriculture, but elsewhere in Europe organic farming has also been characterized as an unnecessary, even backwards, step in relation to the larger project of ecological modernization. Similarly, while anti-American sentiment and a penchant to celebrate local cuisine was mobilized against companies such as Monsanto in France in 1998 and 1999, this reaction did not dominate the debate earlier in the decade.

Consider the United Kingdom, where the British company Zeneca used Calgene's Flavr Savr tomato (from California) in its tomato paste in 1995. This was the first GE food to hit the grocery store shelves in Europe, and Zeneca actually promoted the tomato paste as "genetically altered." Far from generating a public backlash, Charles (2001, 168) notes, Zeneca's tomato paste outsold its competitors in test markets. But within three years, by early 1998, some British supermarkets were announcing that they would keep "Frankenfoods" off their shelves. Had the British found their deeply rooted cultural values in the intervening years? The story is more complex than this, involving a confluence in the late 1990s of several factors within all three sets of relations of force: material, ideational, and institutional (as I examine in later chapters). It would be fair to say that there was a skepticism towards genetic engineering in many European cultures during the early 1990s, and that this skepticism was likely influenced by the factors suggested by Schweiger (2001) as well as by the characteristics of the risk society identified by Beck (1992), but these factors did not (yet) result in a wholesale rejection of genetically engineered foods.

Through the 1980s and into the 1990s, European countries were in the process of developing an alternative policy response to risk controversies: the precautionary principle. As mentioned in the Introduction, in Europe, this principle had its immediate origins in the German concept of *Vorsorgeprinzip,* which states that society should seek to avoid environmental problems by careful forward-looking planning that blocks the flow of potentially harmful activities (Jordan and O'Riordan 1999). The *Vorsorgeprinzip* was first employed in international policies governing North Sea pollution in the late 1980s, though it can be seen to have influenced European policy leanings as early as 1981, when Europe banned US and Canadian beef that had been fattened with the use of growth hormones. This was a precautionary move because there was no solid evidence at the time that the beef represented a health risk.

Despite the history of the precautionary principle in Europe, it would be a mistake to think of this approach as distinctly European. US biotechnology activists point out that a precautionary response lies at the heart of numerous American regulatory practices, including ones instituted only recently. The United States Department of Agriculture's prohibition on the use of rendered protein from ruminants in feed for other ruminants (based on the possibility that such uses could transmit Bovine Spongiform Encephalopathy, a disease that was not yet identified in the United States at the time the prohibition was enacted) is one example. Another is the US Food Quality Protection Act of 1996, which covers the way that the Environmental Protection Agency (EPA) sets pesticide residue limits on food. This law requires the EPA to be "reasonably certain of no harm" to infants. In the absence of data that provide "reasonable certainty" that residues are safe, the EPA requires an additional tenfold safety factor in setting safe exposure limits (Consumers International 2002). These examples demonstrate a willingness to take stronger measures to avoid potential risks than might be called for based on the best available science. Such US applications of precaution have led Wiener and Rogers (2002) to conclude that the EU and the United States have both been proponents of precaution at different points in history and in different policy issues, depending on, among other things, the specific political and economic contexts of the issue in question.

Internationally, the idea of precautionary action was mobilized by the US EPA in the debate about protecting the ozone layer in the mid-1980s (before the discovery of the Antarctic ozone "hole") (Litfin 1995). A precautionary perspective also had a major impact on the climate change debates of the late 1980s and into the 1990s (before the Intergovernmental Panel on Climate Change developed a consensus on human-induced climate change). In 1992, a version of the precautionary principle was formally iterated in Principle 15 of the Rio Declaration. It states:

> Where there are threats of serious or irreversible damage, lack of full scientific certainty shall not be used as a reason for postponing cost-effective measures to prevent environmental degradation. (UNCED 1992b)

Significantly, the notion of precaution is rooted in the discourse of risk. A scientific assessment of risk, or of potential (even only hypothetical) risk, and many of the values underpinning this assessment, still figure centrally in any precautionary response. However, one of the key features that distinguishes the precautionary response in the cases of hormones in beef, the ozone layer, and food safety is that this framing of risk assumes the need to err on the side of protecting the public when a strong scientific account is lacking, even if such action means missed opportunities for producers. This approach has much in common with the hazard discourse of the 1970s, which also emphasized the inherent riskiness of some laboratory practices, and thus the need to err on the side of caution through risk avoidance or containment. Because of the embeddedness of the precautionary framing and of the more permissive manageable risk framing (as found in North American GEO regulatory policies) in the shared discourse of risk, these two perspectives hold certain common assumptions. For one, they both accept the centrality of scientific assessments of potential health and environmental impacts of GEOs in defining the GEO "problem." At the same time, these two positions point to two very different solutions to that problem.

In keeping with the two ways of conceptualizing the power of ideas adopted in this study, these framings can be thought of as two contrasting ideological positions, or currents, each with its own arguments and adherents, within the shared discourse of risk. These positions are ideological in the sense that they are each grounded in a set of arguments for their defence as the appropriate, or common sense, response given the overarching (i.e., already widely accepted) discursive framework of risk. At the same time, in terms of their social function, each of these positions works ideologically as the glue that holds together various political actors who come to share and accept those arguments – even if only because the arguments fit with the worldviews of those actors as informed by countless other interests and concerns – as their own.

It is important to recognize that it was not at all clear in the early 1990s how exactly precaution, as a policy principle, should relate to risk analysis methodologies. Some scientists accepted precaution as a general call for proactive environmental policy but did not believe there was a direct connection to the practices they used to obtain scientific evidence for risk management and mitigation (Gray 1990; Jordan and O'Riordan 1999). Meanwhile, sociologists of science who advocated the adoption of precaution argued that this principle would require fundamental changes to the way risks of all types (environmental and socio-economic, for example) would be assessed, as well as a radical rethinking of the nature and implications of scientific uncertainty in risk decision making (e.g., Wynne 1994). For these advocates, precaution was expected to colour all aspects of the risk assessment process, though what resultant methodologies would look like was still the subject of vigorous academic discussion.

Growing acceptance of broad norms of governance centred on the idea of precaution was influential in the GEO debate in Europe through the 1980s and into the early 1990s. From the perspective of the agri-chemical industry, even just the possibility of more burdensome regulatory structures rooted in precaution led some European companies, such as Germany's Hoechst, to set up on American shores and launch their first biotech applications there in what they regarded as a biotech-friendly regulatory environment. Once formal regulatory structures were established, precautionary norms were evident in directive 90/220's mention of the need to "avoid irreversible damages," as well in its lack of any mention of the North American concepts of substantial equivalence and familiarity. The term "precaution," however, is not found in the 1990 directive at all. This demonstrates that precaution had *not* coalesced as a distinct policy framing of the GEO risk issue with direct implications for regulatory practice by the early 1990s, at least not in pan-European governance of GEOs. As a result, while the European regulatory framework did suggest that attention be paid to the uncertainties inherent in genetic engineering processes, in practice, risk assessors working under this framework still came to assume, like their North American counterparts, that risks could be identified and managed based on limited field trial and nutritional data provided by the biotech companies that compared GEOs to non-GEOs. That EU risk assessors made these same assumptions was clear when, by 1996, European scientists looking at the same data sets as the North Americans arrived at the same conclusions that would permit a number of GE crops to be planted in Europe.

Rather than present a clear alternative framing of genetic engineering, then, 90/220 initially functioned, like the North American regulatory frameworks, as a tool for normalizing genetic engineering in Europe. It did contain elements that were the product of a greater skepticism towards GEOs, and these elements left open the possibility of more stringent risk assessment practices for GEOs in particular (rather than a wider class of "novel" organisms). However, this higher level of regulatory scrutiny is not how 90/220 was translated into practice until after 1998.

It is instructive to consider two other examples of national regulatory frameworks that put forward still other readings of the genetic engineering issue. In Norway, the regulation of genetically engineered organisms was strongly influenced by the discourse of sustainable development. Whereas in North America and Europe the assessment of GEOs was seen as a purely technical exercise, in Norway, non-scientific criteria were included as part of the risk assessment process. Specifically, Norway's Gene Technology Act of 1993 requires that "the production and use of genetically modified organisms take place in an ethically and socially justifiable way, in accordance with the principle of sustainable development and without detrimental effects on health and environment" (Norway 1993, chap. 1-1). This move appears to place an onus on the developer of a GEO to show how it will benefit sustainable development, rather than simply assume that it represents a manageable risk. In practice, Norway's gene act has been stringently applied. In June 1997, the Norwegian government announced, in a distinctly precautionary move, that it would prohibit the production, import, and marketing of GEOs containing antibiotic-resistance marker genes because of their potential health impacts. Such antibiotic-resistance marker genes are commonly introduced into GEOs so that the transformed cells can be separated from the non-transformed cells. Whether or not they represent a health risk remains hotly debated. By prohibiting their use, Norway effectively excluded almost all GEOs on the global market to this day. As a result, Norway has approved only one application of genetic engineering, a crop plant not altered with the use of antibiotic markers (Ivars 2002, 194).

Readings of the GEO regulatory problem that attempted to incorporate the social dimensions of the biotech issue also emerged in the global South. In India, an interesting example discussed by Gupta (2000b, 27), regulators argued that assessments of biotech crops "must of necessity include not just ecological and human health considerations, but also responsiveness to the needs, constraints and priorities of Indian agriculture, since these cannot be divorced from biosafety evaluations." As a result, regulations require that GEOs be both "environmentally safe and economically viable" (DBT 1998, 6). These expectations led the Indian government to announce that it would not allow terminator seeds to be used in India in 1999 (Gupta 2000b).

The assessments Norway and India developed for GEOs began to reflect the type of comprehensive assessments of genetically engineered organisms called for by many of the activist CSOs involved in GE issues in the early 1990s. These organizations saw the risks associated with GEOs as only one dimension of the genetic engineering problem. The intellectual property right issue, the move to a contract farming model, and increased vertical integration with the food system were all reasons to be concerned about genetic engineering. While the approaches of India and Norway can be seen to have reflected this broader vision, these were isolated exceptions in the emerging global regime for the governance of genetic engineering based on a narrow definition of the risks of GEOs and how these are to be managed.

Conclusion

By the time states and other actors began negotiating a biosafety protocol, in 1996, the general contours of the GEO debate were well established. Gene talk was the strongest ideational current in the political field of agricultural biotechnology, and this largely favoured the interests of the biotech bloc. The fetish of the gene, conceived as an independent, transferable unit, enabled the bloc to win the right to patent genes and certain genetically engineered organisms, first in the United States and then globally through the WTO TRIPS agreement. At the same time, the gene, seen as a unit of information that functions in an equivalent manner regardless of the organism in which it is found, allowed the bloc to frame genetic engineering as a precise and thus safe exercise. This understanding was embedded in the regulatory systems that emerged for biotech crops in the United States, Canada, and Argentina in the late 1980s and early 1990s, and the approval of GEOs by these regulatory systems was critical to winning public consent for agbiotech in those countries.

In hindsight, the discourse of equivalence, which emerged from gene talk, may have been particularly important in the context of the labelling issue. A 1997 European Commission regulation that mandated labels for GE foods and ingredients appears to have been a pivotal moment in the movement to resist GE foods on that continent (European Parliament 1997a; this regulation was amended in European Parliament 1997b and 1998). Through this decision, GE foods were made visible to consumers and could become the targets of boycotts. In the United States and Canada, however, the call for labels, which was widely supported by the public, thanks to its fit within a broader discourse of consumer choice, was successfully resisted by the biotech bloc on the basis of substantial equivalence. It is difficult to say exactly how important an absence of labels was to the biotech bloc's hegemony in North America. However, that almost half of Americans surveyed in a 2000 study thought their groceries were "free of biotechnology" suggests that it likely played a role in establishing consumer consent (IFIC 2000).

While a source of power for the biotech bloc in the early 1990s, gene talk also created a site of resistance. Critics emphasizing the complexity of gene-organism-environment relationships were able to portray genetic engineering as an inherently risky endeavour. Activists had no concrete examples of catastrophic results from genetic engineering, but the idea that these practices compromised nature's own boundaries, with unpredictable results, served as an ideological mooring for a wide array of actors who saw the application of genetic engineering in agriculture as undesirable. Mobilizing the biotech bloc's own arguments that GE was distinctly different and important, these critics shaped a discourse of "genetically modified organisms" that placed these organisms in a category of their own in terms of the risks they might harbour. While resisted in North America, this language was normalized in regulatory policy through the European Council's Directive 90/220/EEC, where it was influenced by a growing emphasis on precaution in environmental risk controversies. In other jurisdictions, including Norway and India, the debate about the regulation of GMOs was framed by the discourse of sustainable development.

Initially a product of accommodation between the transnational historical bloc and its environmental critics, conversations about agbiotech's contributions to sustainable development brought other sets of issues to the fore. In this debate, leaders of the biotech bloc, such as Monsanto's CEO Robert Shapiro, argued that genetic engineering would "save our planet" through the development of transgenic plants that required less use of dangerous herbicides and pesticides while providing greater yields (Shapiro 1999, 28). At the same time, the discursive context of sustainable development enabled critics to bring a wide range of social and economic concerns about the potential implications of genetic engineering on the structure of agriculture, and on the conservation of genetic resources and traditional landraces, to the regulatory debate in these countries.

Still, by the early 1990s, these types of concerns had yet to have an impact in the United States, Canada, and even the EU, each of which defined regulation in narrower, science-based terms. That the EU was adopting North American regulatory ideas is clear: the very same 1997 regulation on "novel foods" that required mandatory labelling of GEOs in Europe also contained an article that allowed for food "derived from, but no longer containing, genetically modified organisms" to pass through a "simplified notification procedure" (European Parliament 1997a, Article 5). This notification procedure meant that GE foods (such as processed canola oil) could avoid more detailed risk assessments (and labelling) if the notifier could provide a scientific justification that the product was "substantially equivalent" to existing foods (in terms of composition, nutritional value, metabolism, intended use, and level of undesirable substances).

How did these discursive dynamics affect the biotech bloc's desire for a supportive international governmental regime by the mid-1990s? Generally, from the perspective of genetic engineering proponents, there was progress, but it was slow. GEOs were making their way through the European regulatory system, and the first GE crops would be approved for commercial planting by 1996, despite public protest. At the level of international harmonization efforts, such as those being carried out under the auspices of the Organisation for Economic Co-operation and Development (OECD) and Food and Agriculture Organization and World Health Organization Joint Consultation, differences among domestic framings of the genetic engineering issue created challenges for the creation of internationally acceptable norms for assessing transgenic organisms, but North American regulatory framings appeared to be making inroads. In the OECD talks, for example, North American regulators tried to have the concepts of novelty, substantial equivalence, and familiarity recognized internationally, but the European delegates were not comfortable with the use of these concepts. Regarding the concept of familiarity, one European Community regulator quoted by Levidow and his colleagues argued that "familiarity breeds contempt" (Levidow et al. 1996, 140). In the end, familiarity was adopted in a bid to avoid an impasse, but only after it was given a definition designed to distance the word from its American usage. According to the OECD compromise, familiarity "is not synonymous with safety." It means "having enough information to be able to judge the safety of the introduction" (OECD 1993, 7, 28).

This definition of familiarity meant that all sides could go home believing that their regulatory approaches were reflected in the international compromise. Becoming familiar with a GEO could be interpreted as the demand for a significant knowledge base for the complex array of possible implications of this new organism before it is released into the environment, as those cautious of GEOs sought. But the term could also be understood in the way it was used in North America – to suggest that familiarity with the species and the new trait (before their recombination) meant that the new organism can be assumed to be safe. Through such processes, then, North American regulatory concepts were slowly being internationalized, but the emerging international regime for biotech regulation did not yet exhibit a shared understanding of the problem being dealt with.

How did the discursive dynamics described in this chapter affect the critics of genetic engineering? Basing their arguments in the possible risks of GEOs to health and the environment, in the liberal discourses of consumer choice, and the discourse of sustainable development, critics were able to slow the biotech revolution through the 1980s and 1990s, and to exact accommodations from the biotech bloc. The challenge for the resistance movement, however, was the disconnect between the discourses they had adopted with success, and, for many, their overarching belief that genetic engineering was undesirable, unnecessary, and perhaps even just wrong. Specifically, the discourses that critics were able to mobilize within most effectively were those that centred on possible health risks and the demand for labelling. Even if they were fully successful with these arguments, however, more stringent assessments of GEOs would do little to deal with the problems of vertical integration in the food system, or the possible introduction of the terminator technology, for example. Labelling, for its part, would not prevent the genetic pollution that many organic farmers were concerned about; it would only allow knowledgeable consumers to avoid GEOs. Each of these discourses, then, was based on assumptions that did not necessarily speak to the full set of concerns motivating the activists. By adopting – and having some success with – arguments based on possible health concerns and consumer choice issues, were the activists simply making strategic choices, or were they redefining their own positions in a way that would, ultimately, see them become consensual participants in the GE revolution? It is important to keep this question in mind as we examine the efforts of critics in proposing, and then giving shape to, an international institution designed to place controls on the genetic engineering revolution: the Cartagena Protocol on Biosafety to the Convention on Biological Diversity.

4
Biosafety as a Field of International Politics

The biotech bloc was becoming a formidable force in the global food and agricultural system in the early 1990s. This was the result of years of hard work directed at building acceptance for a very particular genetic engineering agenda. Given this progress, the next part of this story represents a major turnaround. Rather than diminish in strength, during the 1990s, the challenges to GEOs grew internationally, eventually coalescing around the call for a precautionary regulatory framing of agbiotech. Although not the only locus for the elaboration of a precautionary approach (the other main one taking place in the EU), the Cartagena Protocol on Biosafety to the Convention on Biological Diversity provided the key international institutional context for the elaboration of this new discourse of GEO governance.

What does a precautionary framing of the GEO issue, as established in the Cartagena Protocol, set out? Under the protocol, precaution has come to mean that even while there has been no clear evidence of harm to the environment or public health caused by "living modified organisms" (LMOs; defined narrowly as a subset of GEOs), nation-states are expected to give explicit consent, in advance, for the initial import of LMOs intended for introduction into the environment. Such decisions should be made on the basis of scientific assessment of the risks posed by the LMOs. In cases where there is insufficient scientific evidence to make a determination about the risks posed, countries may bar imports on a precautionary basis. Such decisions should also conform to the basic rules of international trade by, for example, being taken on a case-by-case basis and being non-discriminatory (i.e., applying equally to domestic goods and imports).

This view of what precaution vis-à-vis GEOs means, as set out in the clauses of the Cartagena Protocol, was accepted by more than 130 countries and a host of industry and environmental lobby groups when the negotiations on the protocol were concluded in January 2000. Given its widespread acceptance in that forum, and its growing international influence since 2000 (as documented in Chapter 7), the "Cartagena" discourse of precaution seems common sense to most people when told that this is how the world's countries treat GEOs today. However, the sensibleness of this view was not at all self-evident in the early 1990s, at the outset of the negotiations on a protocol. States closely allied with the biotec bloc thought that a set of guidelines instructing officials on how to undertake risk assessments for the products of biotechnology (defined broadly to include the yeasts used in bread-making) was more than adequate, especially given the lack of demonstrated harm attributable to these GEOs. Meanwhile, representatives of industry argued (and continue to argue) that

> all of the critiques and complaints ... get the facts exactly wrong. The concrete experience we have on the ground with crops improved through biotechnology and the foods derived from them are that these crops are more efficiently produced, they consume fewer inputs, they are more time-friendly in the agronomic practices that they require. They lead to higher quality products produced more economically at cheaper prices ... and with much smaller environmental footprints. (Interview #3)

While either of these alternative positions, as just two examples of a wide divergence of views, could have become the understanding of GEO risks found in the protocol, they are not institutionalized in this treaty. Instead, we have the version of precautionary governance outlined above in the field of biosafety.

How did this precautionary framing of the GEO issue emerge through the Cartagena Protocol negotiations, and what are its implications? This question is the focus of the remaining four chapters of this book. In these chapters, I argue that the institutionalization of this framing in the protocol did not occur at once but was made possible through four critical discursive moments, three of which took place before the protocol was even negotiated. I define a discursive moment as a fundamental, even if only subtle, shift in the widely accepted view of the issue in question. These four discursive moments gradually undermined the biotech bloc's minimalist interpretation of the risks of GEOs, and each was enabled by the overall relations of force at particular points in time.

The first moment took place from the late 1980s through the early 1990s, before, during, and after the negotiation of the 1992 Convention on Biological Diversity (CBD). During this period, the various actors active in the field of GE, drawing on discourses of the gene, risk, biodiversity, and sustainable development – each of which was the product of past struggles, as well as the site of ongoing debate – attempted to shape an emergent regulatory field of biosafety in ways that would further their vision of GE. The establishment of this field enabled further negotiation on an international instrument of biosafety. It also set in place the key boundaries of the debate.

Still, within these boundaries there were widely divergent currents, especially about the legal status of a new biosafety instrument (e.g., voluntary or
legally binding), the kind of risks it should address (e.g., only immediate
biological hazards or also socio-economic risks), and even the class of organisms it should encompass (e.g., GEOs or a wider class of novel organisms).

The second and third discursive moments took place in the process of
defining the terms of reference of a biosafety protocol in 1994 and 1995.
During this period, proponents of the view that genetic engineering was
simply part of a biotechnology continuum and thus not deserving of specific regulatory attention compromised on terms of reference that would
focus narrowly on GEOs. Shortly thereafter, countries closely allied with the
bloc came to accept the call for a legally binding international regulatory
framework, rather than simply a voluntary instrument.

The fourth discursive moment took place in 1999 and early 2000, during
the final stages of the negotiation of the Cartagena Protocol. It was during
this period that the operational language of the protocol came to embed
the particular precautionary approach to the risks of GEOs described above.
Between 1996 and 1999, the biotech bloc worked through industry and
allied states to have a minimalist framing of the risks of biotechnology (as
embedded in US and Canadian regulatory practices) accepted by other nations of the world. In the end, however, the balance of material, ideational,
and organizational relations of force in 1999 and early 2000 led them to
make significant concessions in key areas. While the final text of the
Cartagena Protocol includes room for interpretation according to different
interests, there are clear advances for GE critics.

How were each of these discursive moments realized? In this chapter, I
look at the first three discursive moments, while the final moment is the
subject of Chapters 5 and 6. I am not suggesting that each of these discursive moments inevitably led to the next, but rather that each of these moments created the space, both in terms of ideas and institutions, for ensuing
shifts that culminated in the precautionary protocol of January 2000.

An International Instrument for Biosafety

The first moment in the elaboration of the discourse of precaution in the
area of biosafety took place from 1989 to late 1994, before, during, and
following the Rio summit on the environment and sustainable development in 1992. During this period, the idea of biosafety came to be organized through the discourses of the gene, risk, sustainable development, and
biodiversity. The outcome was international acceptance of the need for a
new biosafety instrument under the CBD.

Before Rio, biotechnology was discussed in two separate fora: the Intergovernmental Negotiating Committee for a CBD (INCCBD), and the Pre-
paratory Committee (PrepCom) meetings for the non-binding sustainable

development action agenda, Agenda 21. In each case, the subject had been first raised by states from the global South in regards to technology transfer. In the INCCBD talks, a Sub-Working Group on Biotechnology was formed in late 1989 to "prepare terms of reference on biotechnology transfer" (Bernstein et al. 1994, 1). As Munson (1993, 501) points out, it was at these meetings, as well as at subsequent meetings of the PrepCom for Agenda 21, that Malaysia, supported by the Philippines, proposed adding considera-tion of "the safety of release or experimentation on genetically modified organisms." Framing their position within the discourse of risk, and with specific attention to GMOs as the object of concern, these delegations ar-gued that while genetic engineering could result in benefits when trans-ferred to the South, this was also a potentially hazardous activity. The area of concern the Malaysians raised would soon come to be known as "biosafety," though this term itself is not defined in any official CBD or PrepCom document. It appears as if biosafety was chosen as an abbreviation for biodiversity safety (biodiversity being the object of concern of the CBD) or biotechnology safety, with biotechnology, in the reading of the Malaysians, meaning GEOs in particular. Before its use in this context, the term "bio-safety" referred to laboratory safety practices employed for managing the risks posed by potentially pathogenic micro-organisms. Significantly, this historical use does not single out GE organisms from other kinds of patho-genic micro-organisms.

The Malaysians' framing of the genetic engineering issue in terms of pos-sible risks, although not new in the field of GE regulation, was fairly new to the field of international development. At the time, the literature on this subject emphasized, almost exclusively, the expected benefits of genetic engineering for the future of global agriculture and medicine. One US del-egate of the early 1990s believes that the Malaysians raised the GE safety issue

> purely as a political ploy ... so as to prevent the Convention from evolving in a direction that might actually compel nations who are threatening biodiversity through the degradation of habitat ... to change their behav-iour. (Interview #3)

While it is certainly possible that the issue of biosafety was raised as part of such a political strategy, it would have gone no farther if the argument had no discursive footing. As it was, through the INCCBD and PrepCom processes, countries from the South and their CSO allies, including the US-based Council for Responsible Genetics, were able to put forward a credible case for some sort of regulatory instrument as part of a biodiversity package.

What was the basis of their case? In the early 1990s, ecologists in North America and Europe were becoming increasingly involved in the debate over the possible risks of GEOs to the environment, and they had been able to establish a strong argument for, at a minimum, a regulatory system that would assess new GEOs for their potential risks to the environment and human health. Given that such regulatory systems were being established in the North, it was argued that countries of the South could become the testing ground for technologies not yet approved in the North. One widely cited example took place in Argentina in 1986. At that time, a genetically altered rabies vaccine was tested by a US research organization without the knowledge or consent of the government (Rajan 1997, 179). Put in the context of the pesticide DDT, the trade in hazardous waste, and other issues of environmental risks that were disproportionately impacting on the developing world – all of which were under discussion in the lead-up to Rio – these groups had a powerful argument.

The strongest voices on the other side of the table came from the US delegation, which rejected any call for a new international instrument. This position was rooted in the idea of equivalency that had taken hold in regulatory circles in that country, as is evident in the statement of a US delegate who argued that "whether an organism has been manipulated or not does not bear on risk to the environment" (Munson 1993, 499). Although the United States was itself moving towards a risk-based regulatory system for novel organisms, its version of risk analysis for biotech products had fairly minimal demands, since it assumed the substantial equivalence of GEOs with non-GEOs. As there was no guarantee that a similar understanding would emerge from international negotiations under the auspices of the CBD, the US position, in concert with that of the agri-chemical and biotech industry, was to argue that such an instrument was entirely unnecessary. Citing the possibility that GE crops could lead to biological solutions to agricultural problems, the US delegates argued that biotechnology should be seen as an inherently beneficial tool for sustainable development. As a demonstration of their unbridled faith in biotechnology, they insisted that every reference on safety be removed from working documents. At PrepCom 3, for example, they tried to have the phrase "monitoring and evaluating the effectiveness of the safety" of biotechnology replaced with "evaluating the success" thereof (Munson 1993, 499).

Given the history of the biotech bloc, the unified position of the United States and the biotech industry in the INCCBD and PrepCom talks of the early 1990s should come as no surprise. In fact, US delegations included several key organic intellectuals of the biotech bloc, among them Dr. L. Val Giddings, a scientist who had worked with the US Department of Agriculture and its Animal and Plant Health Inspection Service (APHIS). In his post-doctoral research, he had developed and applied molecular biological techniques to the study of natural populations. Holding the interpretation of sustainable development widely espoused by individuals and corporations in the transnational historical bloc, Giddings saw himself as a passionate environmentalist. He believed that genetically engineered crops would lead to increased yields in agriculture and provide alternatives to the spraying of yet more chemicals to protect crops. His own research led him to believe that the techniques of GE could be used in understanding, and thereby conserving, genetic diversity (BIO 2002). In subsequent years, Giddings would become a prominent figure in the Biosafety Protocol negotiations, first as a US negotiator, then, starting at the second meeting of the Biosafety Working Group (BSWG-2), as the vice-president for food and agriculture of the US-based Biotechnology Industry Organization (BIO), and eventually as the spokesperson for the Global Industry Coalition (GIC).

Another important US delegate in the early 1990s was Terry Medley, also of APHIS. Medley was one of the chief architects of the American regulatory system. By the late 1990s, Medley would be head of Biotechnology and Regulatory External Affairs for DuPont (CAST 1999). That US representatives such as Giddings and Medley had one foot in industry and the other in the regulatory system (a relationship some activists have referred to as the "revolving door") shows the unity of purpose that existed in the American heartland of the biotech bloc in the 1990s.

The centrality of industry interests in the position of the US government was most evident in its reaction to the final text of the CBD, which contained provisions that access to and transfer of biotechnology was to be "provided and/or facilitated under fair and most favourable terms, including on concessional and preferential terms" (UNEP 1992, Article 16, para. 2). In response to this, US biotech interests (organized through BIO) were able to convince then-president George H.W. Bush not to sign the CBD at Rio because of fears that this and other clauses would affect the status of American intellectual property rights claims (Lane and Schweiger 1995). These forces were equally powerful in the Clinton administration. Clinton signed the CBD in 1993 only after presenting a set of "interpretive statements" stating that the United States "does not feel compelled to apply the provisions of the Convention to any private entity under US jurisdiction" (Lane and Schweiger 1995).

Despite the arguments put forward by the US delegation and its allies (including Japan), in the end, both the CBD and Agenda 21 did make reference to an international biosafety instrument. Article 16 of Agenda 21 calls for international technical guidelines, while Article 19.3 of the CBD states:

> The Parties shall consider the need for and modalities of a protocol setting out appropriate procedures, including, in particular, advance informed agreement, in the field of the safe handling and use of any living modified organism resulting from biotechnology that may have an adverse effect on the conservation and sustainable use of biological diversity. (UNEP 1992)

How did the countries from the global South that wanted a new biosafety instrument manage to win consensus from the rest of the world on this issue? To answer this question, we need to look at what was happening within each of the three sets of relations of force.

In terms of material forces, it is important to recognize that Northern countries allied with the biotech bloc did not have the upper hand when it came to having their will imposed in this area of international law. Instead, it was countries from the South that possessed the biggest bargaining chip. They were the holders of the raw resource – it was well understood that the bulk of the world's biodiversity is situated in hot climes – and the principle of national sovereignty over biodiversity was becoming accepted in international law. The North's desire for continued access to these resources, which it had hoped to entrench through the CBD, meant that the South had an unusual degree of clout in its efforts to use this same agreement to impose checks on the actions of Northern biotech developers.

In terms of organization and ideas, one needs to pay careful attention to the framework of the biodiversity talks leading up to the 1992 Rio conference. The framework was significant in at least three ways: it shaped a reading of biodiversity in terms of a set of broad social and technical goals brought together under the umbrella of "sustainable development"; it privileged a scientific reading of biosafety issues in relation to the protection of centres of diversity (those places in the world that contain the highest concentrations of biological diversity); and it allowed biosafety to be linked to other environmental concerns being dealt with at the level of trade, such as pesticides and other hazardous chemicals.

The Rio conference was solidly embedded in the discourse of sustainable development, even as competing understandings of what this term meant held sway in the various official and unofficial fora that were part of it. The discourse of sustainable development, with its attempt to address both environment and development priorities, shaped the way issues of nature (conceptualized as "biodiversity") conservation came to be understood in policy and scientific terms. An important document in this process was the World Resources Institute's (WRI) 1992 *Global Biodiversity Strategy: Guidelines for Action to Save, Study and Use Earth's Biotic Wealth Sustainably and Equitably*, which publicized a detailed strategy for global biodiversity conservation, just four months prior to the Rio Summit. A clear example of expert-driven policy in contemporary relations of ecopower, this document was the result of extensive research and consultation among scientists and policy makers undertaken in collaboration with the International Union for the Conservation of Nature and Natural Resources (IUCN) and the United Nations Environment Programme (UNEP). Both of these organizations also played central roles in the creation of the CBD (Bernstein, Chasek, and Goree 1993).

As ecopolitical programs, the CBD and the WRI Strategy are somewhat unusual because they not only highlight the need for expert environmental management strategies for protecting biological diversity but also call for distributive justice. This last is evident in the three sustainable development policy principles enshrined in the objective of the CBD (and which also figured centrally in the WRI Strategy). These policy principles are conservation, sustainable use, and the equitable sharing of benefits.

The framing of global biodiversity policy in terms of use, conservation, and equity provided the footing for Southern countries to first raise the subject of genetic engineering at the CBD meetings. Initially, when biotechnology was introduced in terms of technology transfer and its usefulness in conservation of biodiversity (through gene mapping, etc.), these moves were welcomed by the proponents of genetic engineering. However, the three principles of the CBD meant that the possible negative impacts of genetic engineering could also be brought to the table. These issues included the equity dimensions of genetic engineering (as raised in discussions about gene patenting), the potentially negative socio-economic implications of adopting GEOs in the global South, and possible risks of GEOs to the environment and human health.

Although both social and scientific issues were raised as possible reasons for being concerned about biotech in the context of the CBD, it was the scientific framing of biodiversity, as an objective truth, that enabled calls for biosafety regulation to maintain their foothold in these talks. The scientific framing of biodiversity, as defined in the CBD (and before it the WRI), includes the totality of genetic, species, and ecosystem diversity (WRI 1992). In the field of biology, the concept of biodiversity is closely tied to the notion that there are a number of centres of diversity around the world. For most major agricultural crops, these centres are in the global South. This meant, it was argued, that countries in the South, and their genetic heritage, were particularly vulnerable to the potential negative impacts of GEOs through, for example, "increased uniformity of plant and animal varieties" (WRI 1992, 47). When it came to possible ecological risks, most scientists from the North – all but the most ideological defenders of GEOs – could at least understand the logic of the argument that the South could be at risk, especially if those countries lacked any capacity to assess and manage the risks of GEOs. It was through this argument, then, that many technical experts from around the world came to see the need for further consideration of a biosafety instrument.

The final way that we see the impact of the context of the Rio Summit is in the reference to advance informed agreement (AIA) in Article 19.3 of the CBD. AIA means that informed consent is given prior to the import of certain products. This idea was brought to the biosafety discussions from concurrent negotiations on the Convention of the Prior Informed Consent

(PIC) Procedure for Certain Hazardous Chemicals and Pesticides in International Trade. The PIC treaty had been negotiated in 1989 as a voluntary procedure to help states make informed decisions on the import of chemicals that have been banned or severely restricted in other countries. In the lead-up to Rio, through the PrepCom for Agenda 21, negotiators were considering the possibility of calling for mandatory participation by all trading states in the PIC process. Many of these same negotiators were working on text dealing with biotechnology, and they drew the parallels between dangerous chemicals and the possible hazards of GEOs. However, the United States, representing agbiotech interests, rejected these comparisons and, while it could not have the reference removed entirely (it was isolated on this issue), finally insisted that prior informed consent language proposed by developing countries be altered to reflect a more benign AIA (Gupta 1999, 6). This combination of relations of force – material, organizational, and ideational – led to the inclusion of Article 19.3 in the CBD.

Most analysts of the biosafety negotiations believe that this article represented the critical juncture, before the actual negotiation of the protocol, for the international regulation of GEOs under the CBD. For example, Gupta (1999, 5) states that it "signalled an entrenchment, in an international legal document, that the potential risks posed by use of genetic engineering merited separate regulatory attention." This article was certainly an important step in the direction indicated by Gupta, but closer attention to the language of Article 19.3 shows that the CBD was not quite the watershed she describes, for two reasons: The article called only for "further consideration of the need for a protocol"; and it does not actually single out genetic engineering for separate regulatory attention.

Article 19.3 refers to "living modified organisms" (LMOs). Like AIA, this term was introduced by the United States in order to shift attention from "genetically modified organisms," the term that was gaining currency in Europe (Gupta 2000a, 208). "Living modified organisms" was a phrase designed to be consistent with the North American regulatory consensus that GEOs should not be treated as a distinct set of organisms with unique risks. The term "living" was introduced to acknowledge that many of the concerns about GEOs, such as risks of invasiveness, might equally apply to organisms developed through traditional breeding (Glowka et al. 1994, 45). Significantly, this understanding of LMO would be fully consistent with the convention's negotiated definition of biotechnology, which includes "any technological application that uses biological systems, living organisms, or derivatives thereof, to make or modify products or processes for a specific use" (UNEP 1992, Article 2). According to the CBD, LMOs include GEOs, but they could also include organisms produced by traditional breeding techniques. As a sign of further debate to come, this possibility led Malaysia, which understood LMO to mean "*living* genetically modified organisms" – with the "living" simply a further acknowledgment of the biological nature of the convention's subject matter – to make its understanding that LMOs were "genetically modified organisms" explicit in a statement when it signed the CBD (Gupta 1999, 5).

Article 16 of Agenda 21, which also deals with biotechnology, does not specify GEOs either. It calls for "further development of internationally agreed principles on risk assessment and management of all aspects of biotechnology," but also uses a broad definition of biotechnology as "the integration of the new techniques emerging from modern biotechnology with the well-established approaches of traditional biotechnology" (UNCED 1992a Article 16.29).

Gupta (1999) sees the term "living modified organism" as an example of a "boundary-ordering device": a term that allows temporary compromise by being open to more than one interpretation (Shakley and Wynne 1996). As a discourse, we might say that the LMO was an ongoing site of power and resistance; it was a site of struggle. A discursive interpretation would also draw our attention to the way that the adoption of a term such as "living modified organism" could enable particular interpretations to appear as logical later in the Biosafety Protocol discussions. Such a move did indeed happen when the concept of an LMO eventually came to be used as a tool for preventing non-living products of GEOs from being included under the protocol.

In the end, nothing in either the CBD or Agenda 21 singled out GEOs as worthy of a higher degree of assessment than traditionally bred plants or animals, and neither text definitively called for a new instrument. As with the OECD compromises (which were being negotiated around the same time), this meant that there was space within the discourse of biosafety for the equivalency position held by the biotech bloc and for the more critical readings of GE. The lack of consensus on these issues came to the fore in the events that immediately followed the signing of the CBD.

In late May 1992, the director of UNEP called for an expert panel to consider the issue of biosafety based on Article 19.3 of the CBD. As this was one of four panels, it came to be known as Expert Panel IV. Its thirty-three members met three times in 1992 and 1993, and its two chairpersons would become key figures in all ensuing discussions. These were Veit Koester of Denmark, who would eventually become the chair of the Biosafety Protocol negotiations, and Tewolde Egziabher, a scientist and director of the National Herbarium of Ethiopia. Egziabher would become the chief spokesperson for African countries (and later for most countries of the South) during the Biosafety Protocol negotiations. He was, without a doubt, the strongest and most outspoken critic of GE throughout the process.

Reflecting the tensions in Article 19.3 itself, Expert Panel IV was not able to arrive at a consensus. As a result, the report it published in April 1993 included a majority and a minority opinion (UNEP 1993). The majority opinion supported the idea of a legally binding biosafety instrument under the framework of the CBD. In the first formal reference to "precaution" in the biosafety debate, the majority report argues that any future cooperation should be based on the precautionary approach articulated in Principle 15 of the Rio Declaration on Environment and Development (para. 32). The report also argues that a protocol should be specifically targeted at "genetically modified organisms," rather than at "biotechnology" defined more broadly. Drawing on the language of the European regulations of 1990, the panel defined GMOs as "organisms in which the genetic material has been altered in a way that does not occur naturally by mating and/or natural recombination" (para. 57). The majority opinion goes on to make a case for a stringent protocol, one that would assess GMOs for environmental, human health, and socio-economic impacts. Framing the issue of socio-economic impacts in terms of "risks to biodiversity," the discursive context of the CBD, the report states that

> the conservation and sustainable use of biological diversity, especially in the case of domesticated plants and animals, is dependent on socio-economic conditions of the people who have been maintaining it. It is, therefore, essential that the socio-economic risks posed by the use of GMOs be evaluated and that any probably adverse effects be mitigated. (para. 85)

Such socio-economic risks include the possible impacts on traditional agricultural exporters were biotechnology to be used for import substitution in the North (i.e., potential genetically engineered substitutes to traditional commodities such as coffee and sugar). The report argues that import substitution could result in the discontinuation of agricultural systems and thus genetic erosion. It concludes that "when the use of a GMO is not clearly seen to offer an advantage, it would make sense that the traditional technologies and systems continue" (para. 86).

The Expert Panel IV report's minority opinion was firmly grounded in the equivalency framing. According to this perspective,

> biotechnology is a set of methodologies or techniques that can be applied to a wide range of organisms to generate a variety of products with many uses. In its broadest sense, biotechnology includes all organisms, and parts of organisms, that humans use. (UNEP 1993, para. 123)

From this point of view, it made perfect sense that "no regulation can easily address the trading patterns of all the potential product types" (UNEP

1993, para. 124). Furthermore, a protocol with a narrower focus on GMOs, as the majority was advocating, was completely misdirected:

> There is no scientific evidence linking the specific techniques of modern technology in biotechnology to adverse impact on biological diversity, still less to quantifying such speculative risks; a protocol would, for no clear purpose, divert scientific and administrative resources from higher priority needs and delay the development of techniques beneficial to biological diversity, and essential to the progress of human health and sustainable agriculture. (UNEP 1993, para. 42)

An international protocol focused on GMOs "stigmatises the techniques and increases public concerns, hence diverting resources, political attention, and delaying innovative and beneficial developments" (para. 42).

The minority opinion articulated in Expert Panel IV's report was clearly that of the biotech bloc. According to inside sources, it was held by only two panel members, the US member and the representative from the OECD secretariat (Khor 1993). The participation of the OECD secretariat in putting forward this position must be seen in the light of that institution's ongoing efforts in the early 1990s to universalize the minimalist American regulatory framework for the products of biotechnology among its member countries.

Even though this minority position was voiced by only two of thirty-three members on the panel, the fact that the panel did not achieve consensus was used by the United States as reason for refusing to allow passage of a motion tabling the report as an official background document for the first Conference of the Parties (COP-1) to the CBD (Lane and Schweiger 1995). Thus, while the majority opinion in Expert Panel IV's report would remain important as an ideological resource for GE critics – laying out for the first time in an internationally negotiated text the basic elements of a legally binding protocol rooted in precaution – it officially disappeared from the UNEP and CBD discussions, and no follow-up was undertaken in UNEP. The power of the US action in blocking this document was evident in the 2002 reflections of the then director of biodiversity programs for UNEP, Hamdallah Zedan. Zedan (2002, 30) saw the lack of follow-up on the panel's report as an indication that "the international community might not have been ready to embark upon a further round of treaty negotiations so soon after the conclusion of the Convention itself." That some key participants took away this impression was clearly the intent of the United States, the OECD secretariat, and their industry allies. It certainly did not reflect the view of many others at the time, including a large majority of the panel itself.

When the Expert Panel IV report did not become a background document to COP-1 of the CBD, another background paper was written by staff of the interim secretariat of the CBD. The introduction of this document reveals the institutional power that lies in the hands of UN secretariats to shape discussion. This overview of the biotechnology issue is remarkable for how differently it was framed from the Expert Panel IV's majority opinion. It reiterates the definition of biotechnology found in the convention, stating that "biotechnology is ... a continuum of technologies, ranging from long established and widely used technologies essentially based on the use of micro-organisms, through to recombinant deoxyribonucleic acid (DNA) techniques" (ICCBD 1994, 2). Five extensive paragraphs then review the "contribution of biotechnology to the objectives of the convention," including the way that "*in vitro* cell and tissue cultures can be used ... to increase biological diversity ... and may facilitate the selection of useful individuals resulting from natural or induced mutations." Also discussed are the ways in which biotechnology has "boosted the socio-economic development in industrialized countries by providing new opportunities for protecting the environment" (2). These statements are clearly attempts to reverse the claims made by the majority of Expert Panel IV and others that GEOs can lead to the reduction of biodiversity and negative socio-economic effects. This section is then followed by one paragraph listing "public concerns" and "most often mentioned potential threats related to biotechnology" (3). Although noting that a binding protocol is indeed one option, the interim secretariat concludes with suggestions for how international cooperation for biosafety can be enhanced by drawing on voluntary guidelines and codes of conduct such as those developed by the OECD, stating that these "could be considered for setting the basis at the international level for the safe transfer, handling and use of LMOs resulting from biotechnology" (4).

It is difficult to ascertain exactly how actors allied with the biotech bloc were able to influence this document. What is clear is that the equivalency discourse was still alive in the international arena. What is also apparent, from a review of the biosafety debate, is that this document is the last place where one can find the discourse of equivalency dominating discussions in these fora. In other words, the watershed Gupta suggests took place at Rio in 1992 did finally occur by late 1994, at COP-1. It was only at the time of these talks that the argument against any sort of new international instrument, based in the notion that the risks of biotechnology were negligible, had to be abandoned by the biotech bloc.

At COP-1, held in the Bahamas in November 1994, the interim secretariat's paper was presented as background on the topic of biosafety, and its precepts, rooted in the discourse of equivalency, were echoed in the position of the United States. The US position can be seen to reflect that country's domestic regulatory position as well as the immediate material interests of its biotech industry. In May 1994, the Flavr Savr had been deemed "substantially equivalent" to other tomatoes, paving the way for it to be planted in the United States as the first of the new GE crops. The year 1994 also saw Monsanto actively courting regulators in the United States, Canada, Argentina, and the EU for approval of Roundup Ready crops, which Monsanto hoped would enter the global grain trade after the next growing season.

Despite their best efforts, the secretariat's background report and the US position fell on deaf ears. The debate was not organized by equivalency, as the biotech bloc would have wanted, but was instead now dominated by the discourse of risk, and this discourse implied the need for some sort of management response. Spokespersons for countries of the global South argued that the report did not take seriously the potential risks of GEOs. And these countries were now joined in their criticism by allies from the North, including Denmark, the representative of which had co-chaired Expert Panel IV, and Norway, which, by this point, had adopted domestic regulations singling out GMOs for evaluation according to their contributions to sustainable development. The biggest challenge to the equivalency position, however, came from those the United States had hoped would be allies: the countries in Europe, such as the United Kingdom, the Netherlands, and Germany, that had growing biotech sectors. At COP-1, these countries, along with the rest of the EU, chose to adopt a very different strategy from that of the United States. Rather than fight a new international instrument, the Europeans came prepared to accept an instrument, as long as this instrument was voluntary, giving countries guidance only in how to undertake risk assessment and management for LMOs.

How are we to understand this position, given the material interests of certain European countries in a GE revolution in the early 1990s? In Gramscian terms, the European voluntary position was an attempt at accommodation of those calling for a biosafety protocol; it was a concession designed to have minimal impacts on the underlying material interests of the international biotech sector. Such a strategy had become, by this point, a fairly common neo-liberal response to cases in which it looked like a regulatory hammer was the only other likely option. This position also managed to bring divided interests in Europe together. Some European countries did have a vested interest in GE crops, and they wanted to see them approved for use in the EU. At the same time, the genetic engineering issue was starting to spark consumer concerns in Europe, and so other EU politicians wanted to be seen to be doing something in response. Austria, Denmark, and Sweden, for example, were particularly cautious about GE crops and calling for labelling of GE foods in intra-EU discussions (Charles 2001, 166).

Two people came forward as the organic intellectuals of what can be called the European arm of the biotech bloc to define this more accommodative, voluntary position. These were Helen Marquard, the head of international policy on biotechnology in the United Kingdom, and Dr. Piet van der Meer, a policy representative for the Dutch government who would eventually head the Dutch delegation to the Biosafety Protocol talks from 1996 and 1999. Starting in late 1994, Marquard and van der Meer, with the active support of their governments, spearheaded efforts through the UNEP (under the pretence of Agenda 21, Article 16 and not the CBD) to create voluntary international technical guidelines on biosafety. Officially, these guidelines were not meant to "prejudice an outcome to the CBD deliberations" on Article 19.3 (Chasek 1998). Nonetheless, they represented an escape route for European biotech interests that worried that a legally binding biosafety protocol could become an obstacle to the "sustainable development" of biotechnology.

In the context of the positions brought forward by the global South (which were still closely in line with Expert Panel IV) and the European move to push a voluntary biosafety instrument, the United States found itself out of step with the emerging common sense at COP-1. This shared view held that LMOs posed real risks that needed to be taken seriously in some form of international regulatory instrument. With little other choice, the United States acquiesced to the European position by the end of the meetings. As a result, it did not try to block the decision to set up an "open-ended ad hoc group of experts nominated by Governments without undue delay" to consider Article 19.3 of the CBD and then report back to COP-2 (with "open-ended" meaning that all countries could send experts) (CBD 1994, 32). This decision was understood to be setting the foundations for the development of a biosafety instrument.

By the end of 1994, then, a discourse of biosafety that accepted the need for a new international instrument was universally accepted among the participants in the biosafety talks. This acceptance was clearly dependent in part on the dynamics internal to international environmental institutions. In this case, the CBD created the context through which both the benefits and risks of biotechnology could be articulated in an international forum. Then, the precedent of the prior informed consent procedure in a parallel UN institution created the possibility for the horizontal adaptation of this procedure (rearticulated as AIA) to the field of biosafety. This acceptance was also dependent on the role of scientists, and the epistemic consensus emerging around the potential risks of biotechnology among ecologists in the early 1990s, as critical to the first steps in the formation of a biosafety regime.

This analysis demonstrates how the positions brought forward by nation-states in specific international contexts are the product of state and civil society complexes guided by particular class interests and informed by specific ideas and the organic intellectuals who shape them. In the case of the United States and biosafety, its allegiance was with the biotech bloc and the ideology of equivalency. Some countries, Malaysia and Ethiopia among them, brought forward positions strongly influenced by their cooperation with activist CSOs from both the North and South, and they shaped their positions in relation to the discourses of risk, sustainable development, and the idea of precaution. The EU, meanwhile, had mixed allegiances. This situation enabled certain organic intellectuals to forge an accommodative position that was still tied to the interests of the biotech bloc.

This analysis also demonstrates that ideas are not simply mobilized as resources by actors in international negotiations, as a simple reading of ideology would suggest. Here, ideas operated discursively, as sites of struggle that also set the boundaries for future debate. By 1994, the idea of biosafety had come to be embedded in the overarching discourses of risk and sustainable development that had gained currency in the context of contemporary ecopower. Within these discourses, proponents of GE could initially argue that the products of biotechnology represented minimal risks and were a tool for sustainable development that should not be slowed, but this argument became harder to make as the scientific discussion about the possible risks of GE continued. Eventually, by 1994, the risk discourse, in combination with the strategic move of the Europeans, enabled international agreement to emerge on the need for a biosafety instrument that would manage the risks of GE. While reigning in the biotech bloc, the risk discourse also had disciplinary effects on critics. For example, the majority of Expert Panel IV worked to articulate socio-economic concerns about genetic engineering in terms of possible risks to biodiversity in order to have these issues included in the debate. However, while these issues could be articulated in a way that tied them to the risk debate, the link between the socio-economic concerns and risk was seen as rather tenuous, and less than objective, to many of the biodiversity experts involved in the talks. This issue would remain a challenge in ensuing discussions for those who wanted to use the biosafety talks to address the full range of risks posed by GE in the developing world.

Of LMOs and GEOs
Given the number of years it took to achieve consensus on the need for a new biosafety instrument, the second and third discursive moments leading to the precautionary framing of GE in the biosafety talks came relatively quickly. Within six months of COP-1, it was accepted that a new instrument would focus on GEOs, in particular, rather than a wider class of novel organisms. Within another six months, the international community accepted that the Biosafety Protocol would be a legally binding instrument, rather than a voluntary one. How were these shifts made possible?

After the decision was made to establish an open-ended ad hoc group of experts nominated by governments at COP-1, it was also decided that another background document was needed before the first meeting of the ad hoc group of experts (CBD 1994). This was prepared by another new fifteen-member expert panel that met in Cairo in May 1995. It is in the report of this meeting, which came to be known as the Cairo report, that LMOs are defined as a narrow class of "all organisms produced through the use of recombinant DNA technology" (CBD 1995a, 25). Although a similar definition had been seen in the (officially silenced) Expert Panel IV majority report of a few years earlier, the Cairo report represents a discursive watershed because the committee that produced it included representatives from both the American and European arms of the biotech bloc (Terry Medley and Helen Marquard, respectively).

Why were these representatives of the biotech bloc now willing to accept a narrow focus on GEOs? For the Europeans, the European Council's regulatory directive 90/220 already made this move, in 1990, with its focus on "GMOs." The Americans appear to have been prepared to make this major concession because the report otherwise reflects their domestic position based on equivalency and science-based risk analysis. For example, the Cairo report states that "biotechnology does not differ from other technologies" and that it was "not necessary to apply risk management measures to all uses and releases of LMOs solely because the organisms were produced using modern biotechnological techniques" (CBD 1995a, 31). It also states that "risk assessment should be restricted to objective parameters" and not on criteria such as socio-economic aspects of biotechnology that "bring value judgements into the analysis" (23). Further, it agrees with the American view that "there would in many cases be sufficient information at the time of commercialization to allow the removal of any distinction between LMOs and organisms produced by traditional methods" (31).

On topics where this panel's position is less strident than an American one, these are elements that American delegates had already accepted as compromises in OECD discussions. For example, the report reaffirms the OECD's (1993) definition of familiarity as an important feature of risk assessment, but "not synonymous with safety" (CBD 1995a, 26). Most important for the United States and the European countries aligned within the biotech bloc was the moderate tone of the panel's report as a whole, which stressed the need for an international framework, such as the guidelines being developed through the UNEP, rather than a legally binding instrument.

By assenting to this report, the countries aligned with the biotech bloc signalled that they could accept a voluntary international instrument that would cover only GEOs. At the same time, this discursive moment was accompanied by a continued rejection, on behalf of the biotech bloc, of many of the elements of a protocol sought by the critics of GE that were found in the majority report of Expert Panel IV, such as the notion of risk assessments that would address socio-economic considerations, and a recognition that GEOs present unique and uncertain risks.

A Legally Binding Protocol

The Cairo report reveals the shared perspective that emerged between the European and American arms of the biotech bloc in the summer of 1995, and the ability of this moderate position to win the consensus of representatives of at least thirteen other countries. This report was the background document brought to the meeting of the open-ended ad hoc group of experts nominated by governments in Madrid in July 1995, which was convened to set recommendations for COP-2 later that year. Given its function, and the cooperation evident between the Americans and some Europeans in the Cairo report, this position could have represented a formidable ideological force in future biosafety discussions – a force that could have led to recommendations for a minimalist voluntary instrument not unlike the compromises being developed through the OECD. Despite this potential, however, a very different direction was taken at COP-2 of the CBD in Jakarta in November 1995. Decision II/5 of COP-2 states that the parties to the CBD will

> develop, in the field of the safe transfer, handling and use of living modified organisms, a protocol on biosafety, specifically focussing on transboundary movement, of any living modified organism resulting from modern biotechnology that may have adverse effect on the conservation and sustainable use of biological diversity, setting out for consideration, in particular, appropriate procedures for advance informed agreement. (CBD 1995b, 6)

This decision entrenched the second discursive moment, by stating that LMOs resulting from *modern* biotechnology (i.e., GEOs) were the main objects of concern. The decision is also indicative of the third discursive moment on the way to a precautionary Biosafety Protocol: it signals the arrival of a consensus to pursue a legally binding protocol. What happened within the relations of force of genetic engineering to enable this decision?

The first factor was the growing strength of a perspective critical of GE among states in the global South. This framing was reiterated in a report released under the auspices of the Third World Network (TWN) to respond to the Cairo report in the summer of 1995. The TWN report was written by key intellectuals of the anti-biotech movement, including two of the main critics of GE to serve on Expert Panel IV, Tewolde Egziabher and Vandana Shiva. It highlights the scientific uncertainties of genetic engineering, calls for a response rooted in the precautionary principle, and repeats the elements sought in a strong legally binding protocol that were first introduced in the majority report of UNEP Expert Panel IV. These elements include attention to the potentially "crippling" socio-economic impacts of genetic engineering in developing countries (Biosafety 1995, 21). What is noteworthy about the TWN report is that it is rooted *within* the discursive framework of risk. This gave the activists credibility in a largely technical discussion. As in other debates about risks in the context of ecopower, the authors of the TWN report appear to have been prepared to accept that it is the risks of LMOs that are the central issue because they believed that these risks were many, varied, complex, and uncertain. In other words, they were prepared to adopt the language of risk because they saw it as underpinning a call for risk avoidance, rather than risk management.

When the Open-Ended Ad Hoc Group of Experts on Biosafety met in Madrid in July 1995 to consider Article 19.3 of the CBD, delegates had in hand two conflicting technical documents on how to deal with the risks of LMOs: the Cairo report, largely influenced by the new European-American voluntary position, and the TWN report, co-authored by, among others, the lead spokesperson for the African Group of countries, Egziabher. In this polarized context, delegates from the South united around the call for a legally binding protocol, both at the Madrid meeting and at the subsequent COP-2 in Jakarta. This solidarity was the second critical factor that led to the decision reached at Jakarta.

Significantly, the solidarity among states from the global South did not occur because their representatives were all as critical of genetic engineering as Egziabher and his CSO allies. In fact, Argentina saw itself as a likely exporter of LMOs, and it did not want to jeopardize that position. What enabled these countries to unite was the discourse of risk within which they were operating, which included room for multiple readings of genetic engineering. As an illustration of this inclusiveness, Antonio G.M. LaViña of the Philippines, the spokesperson for the G-77/China at Jakarta, recalls that whenever he or others made official interventions on behalf of their negotiating group, they clearly referred to both the possible benefits and risks of LMOs (LaViña 2002, 36). The context of the CBD was important here too, since these countries saw the CBD as an instrument for enabling benefits from technology transfers as well as the conservation of biodiversity. It was thus a discursive context of risk, combined with the framing of the CBD, that enabled Southern solidarity on the call for a legally binding protocol.

The third and likely most important factor that led to the Jakarta decision was a further shift in the EU's position. At COP-2, the EU, in solidarity with the global South, agreed to the call for a legally binding protocol. This agreement was a move that went against the efforts of Marquard and van der Meer, along with other European allies of the biotech bloc, to push for a voluntary protocol.

It is important to note here that, in the realm of material capabilities, while biotech priorities were clearly important to governments in the United Kingdom, the Netherlands, Germany, and France in 1995, the industry's economic impact was still small. The United Kingdom's Zeneca had marketed the Flavr Savr tomato in 1995, but otherwise there were no biotech products on the market in Europe. Monsanto was seeking regulatory approval for its Roundup Ready crops, but biotech products themselves were not yet being produced in the EU. As key European Commission representatives note in hindsight, that "the EU did not have any strong biotech export interests gave additional weight to the development and environment angle in the negotiating position" (Bail, Decaestecker, and Jorgensen 2002, 167).

Meanwhile, in terms of ideas, the debate over GEOs was heating up. The summer of 1995 was the first season that a substantial acreage of GE crops was grown commercially in the United States, and GE soy and cotton were expected to be in production the following year. Commercialization meant that the debate over these crops, especially in regulatory circles, was growing on both sides of the Atlantic, and the notion that GEOs should be treated cautiously was gaining ground. This debate meant that the countries in the EU (notably Denmark, Norway, and Austria) that had aligned themselves with the critics of GEOs, and that argued that developing countries should be supported in their desire to build regulatory capacity, were given a stronger footing in intra-EU negotiations on a common biosafety position.

At this stage, however, the most important factor to influence the new EU position emerged in the organizational arena. The UNEP technical guidelines were close to completion by the time COP-2 was held, and the European countries allied with the bloc saw in these guidelines the prototype for a protocol they could live with, even if this were to become legally binding: the UNEP technical guidelines embedded the same US-EU biotech bloc compromise on how the risks of GEOs should be dealt with that we see in the Cairo report (UNEP 1995). This was not a surprise, given that both reports were spearheaded by Marquard of the United Kingdom, and neither report was authored by representatives from the full body of nations gathered under the umbrella of the CBD.[1]

The *UNEP International Technical Guidelines for Safety in Biotechnology* is a pragmatic document written by scientists. It discusses norms and practices for carrying out risk assessment of GMOs, and what needs to take place in developing countries to build regulatory capacity for this purpose. Like the Cairo report, the UNEP document holds a narrow focus on "organisms whose genetic make-up is unlikely to develop naturally, such as organisms produced by modern genetic modification techniques" (UNEP 1995, para. 21). However, the guidelines note that "it is generally considered that, in most cases, there will be low environmental risk from introducing into a similar environment ... well-known crop plants after they have been modified by altering or adding only one or a few genes, especially when compared with the risks of introducing entirely new or alien species" (8). This narrative of genetic engineering reflects the equivalency framing of GEOs, as embraced by the biotech bloc. It also makes no mention of how to deal with scientific uncertainty in risk decision making – a key concern of critics. In a useful example of the way that the risk discourse enables a separation of supposedly technical issues from normative ones, the document acknowledges that there may be relevant socio-economic issues related to GMOs, but it states that these are outside the "technical purview" of the committee that wrote the guidelines. This statement is then followed by what is clearly a non-technical, normative move: the document specifies that the guidelines are not intended to override obligations required under existing national, regional, and international legal systems such as the recently negotiated WTO SPS Agreement. This is significant because the WTO has a fairly narrow interpretation of risk assessment and the role of precaution in these assessments. This issue would resurface in the ensuing Biosafety Protocol talks.

Although officially not intended to prejudice deliberations on biosafety under the CBD, the UNEP technical guidelines were immensely important in this process. The document set the template for many aspects of the protocol when it was later negotiated. It also created clear demarcations for what would be considered properly technical concerns of the risks of GEOs. At COP-2 in Jakarta, the imminent arrival of these guidelines meant that European countries aligned with the biotech bloc were prepared to accept the call for a legally binding protocol. Negotiators for these countries reasoned that if they could influence the creation of a "reasonable" protocol, as they had done with the technical guidelines, this would not become a major hindrance to future trade opportunities (interview #9).

While far from an acceptance of the kind of protocol called for by Ethiopia and the activist CSOs, the EU's move to accept the call for a legally binding protocol along the lines of the UNEP guidelines was still a major schism with the North American arm of the biotech bloc in 1995, and this schism would only grow over the coming years. It was clearly related to increasing skepticism about GEOs in Europe, but this factor does not fully explain the schism. One further explanatory factor may be the different American and European attitudes about the role of the state. In the United States, neo-liberal values were solidly entrenched by the 1990s, resulting in moves to minimize regulatory intervention for GEOs. Europe, although under similar ideological pressures, has a much longer history of active state intervention in the economy. This may have been part of what was at play in the acceptance of a new layer of international bureaucratic intervention in the field of biosafety by the European arm of the biotech bloc. From an American perspective, the emerging EU position would be characterized as "benign imperialism" (Enright 2002, 96). One US negotiator commented: "We imperialize by exporting McDonald's, the Europeans imperialize by trying to export regulations" (interview #4).

In the face of this new European position and the unity shown by the global South, the United States reluctantly agreed to a legally binding protocol at COP-2 in Jakarta. This decision was known thereafter as the Jakarta Mandate. "Consensus" – albeit one that left the United States increasingly uncomfortable with the direction in which talks were moving – had once again been achieved.

Conclusion

Achieving agreement on the Jakarta Mandate was the third key discursive moment on the way to the embedding of a discourse of precaution in the Cartagena Protocol. This followed the creation of the field of biosafety as an area of policy, and then a narrowed focus on GEOs within this field. As with the creation of the field of biosafety, the second and third discursive moments can be seen to be the result of more than simply interstate and institutional dynamics: they were a result of shifting material, organizational, and ideational relations of force.

In terms of material capabilities, 1995 was the first year of commercial GEO production, and the United States made every effort to minimize the impact of the biosafety discussions on its export interests. The European allies of the biotech bloc, most specifically France, Germany, the Netherlands, and the United Kingdom, also made strong efforts in this direction. But the higher degree of skepticism in Europe towards GEOs, combined with a stronger proclivity towards active state intervention in risk issues, meant that the delegations from these countries were prepared to make accommodations before the Americans. This set the context for the United States – a non-party to the CBD that could block consensus only by encouraging other parties to hold a stance similar to its own – to accept concessions as well. Argentina, also a full participant in the biotech bloc, took a position along the same lines as that of the European countries. Argentina chose to see a future protocol as a tool that would eventually allow the benefits of GEOs to be realized while minimizing risks.

The Canadian delegation was notably silent on the biosafety issue at Jakarta because of at least three factors: First, and most important, Canada has a history of supporting multilateralism, and was a particularly active supporter of the CBD (interview #1). Second, the biosafety issue was still fairly new, and the Canadian delegation, headed by Environment Canada, had not developed an interdepartmental consensus on the issue (interview #8). Third, Canada had other priorities at Jakarta. COP-2 was the forum where the decision was to be made on where to house the CBD secretariat, and Montreal was the main candidate. Because of the importance of this decision for Canada, it may not have wanted to ruffle feathers at Jakarta. When attending COP-2, the Canadian delegation was specifically instructed by the federal cabinet not to intervene and to act solely as an observer on the biosafety issue. Internal government documents reveal that Canada had a difficult relationship with the United States and Australia (a supporter of the US position) at the meeting because of this position. Neither of these countries supported Canada's candidacy for a biosafety bureau, which would help set up and guide the biosafety talks, because they were uncertain of Canada's position on the biosafety issue (Herity 1996).

Organizational relations of force, and especially the activity of international institutions, were also clearly important in realizing the second and third discursive moments. I've discussed the formation of a shared position between the American and EU arms of the biotech bloc in the Cairo report, and the power of this consensus in influencing the UNEP technical guidelines. On the other side of the debate, there was close cooperation between representatives from the South who were critical of GEOs and their allies within European delegations. Significantly, however, not all of the organic intellectuals on this side of the debate were part of national delegations; many of the key voices actively engaged in framing the issue in relation to the South's interests came from CSOs.

I have already drawn attention to the rapidly produced TWN report of July 1995, and its importance in countering the Cairo report at Madrid and Jakarta. The World Wildlife Fund (WWF) also provided a series of seven case studies that played a similar galvanizing role among critics at COP-1 of the CBD (Tapper 2002, 269). The work of the TWN and WWF, along with other CSOs, including GRAIN, RAFI, CRG, and Greenpeace, all helped shape biosafety as a field fraught with uncertain risks that needed further study, and which definitely needed some kind of management instrument.

Institutional relations were also important in realizing these discursive shifts. Consider the EU's focus on GMOs in its 1990 regulations, rather than on a wider class of novel organisms. This precedent set the stage for the focus on GEOs in the UNEP technical guidelines and in the Cairo report. While the question of what kind of organisms would be included under the protocol was again revisited in the formal negotiations, the precedent of a narrow focus had been established in these institutional contexts, and it would be carried into the final protocol text.

Ideas were also clearly important, both as ideological resources and as discourses framing the debate. They were so important that, as Foucault would expect, identities and interests were shaped into the context of prevailing ideas, as opposed to being somehow formed prior to debate on the issues. In this case, actors coalesced around different ideological poles within the risk discourse. This move was evident in the solidarity achieved by countries from the global South on the risks and benefits of GEOs. On the other side of the debate, the ideology of equivalency became a rallying point for pro-biotech interests. That equivalency functioned in this way was apparent in a statement made by Japan, which had an emergent biotech sector.

At the meeting of the Open-Ended Ad Hoc Group of Experts on Biosafety in Madrid, the Japanese delegate stated, for the record:

> It has been widely considered that recombinant DNA techniques are an extension of conventional genetic procedures and that organisms produced by this technology present risks that are the same in kind as those posed by any other organism. (CBD 1995a, 9)

In response, the representative from India had noted in the meeting's record "that a number of representatives at the present meeting had expressed views that disagreed with that statement" (CBD 1995a, 9).

The statements by the Japanese and Indian representatives reveal that in 1995 there remained major differences among nation-states over what risk management of LMOs should mean, and that these differences were indeed tied to differing ontological understandings of the organisms in question. In an effort to narrow these differences, the Madrid meeting set out to establish a list of consensus items for potential inclusion in a future protocol. International consensus was reached on the idea that a regulatory framework should address issues such as procedures for AIA, risk assessment and risk management, capacity building, definitions of terms, and so on. Significantly, two issues were included on a separate list of issues that "though not enjoying consensus, were supported by many delegations," as they were accepted by most states other than the United States and the EU. These were the issues of "socio-economic considerations" as well as "liability and compensation" (CBD 1995a, 13). That there was still no consensus on the need to address these issues related to LMO imports suggests that critics had not yet been able to make the case for them. But there was more at play. The lack of consensus on the importance of these issues is an example of the irrelevance of "non-technical" issues within the particular reading of risk that had become the basis of negotiation. While the risk discourse did enable critics to make considerable headway in their efforts to place controls on the biotech industry through the biosafety talks, it continued to discipline their efforts in some of the areas where their concern was the greatest.

5
Staking Out Positions

In the lead-up to formal negotiations on a biosafety protocol, the idea that genetically engineered organisms were no riskier than non-GEOs was losing ground. At the same time, the idea of a more precautionary approach to GE was gaining ground. The precautionary principle had been explicitly referred to by the majority report of Expert Panel IV, and it was also central to the Third World Network report of 1995. But precaution was not the only, nor even the central, idea brought forward by critics of GEOs. Precaution was only one element of a broad critique that called for a comprehensive assessment of the potential risks and benefits, certain and uncertain, environmental and socio-economic, of GEOs, before their use in agriculture.

This chapter begins with a detailed examination of this "comprehensive" position, where it grounded, and how it, in combination with supportive material and organizational relations of force, enabled those activists leading the call for a biosafety protocol to gain adherents to their cause among states involved in the negotiations. I then consider a similar set of issues for the position the defenders of the biotech bloc brought to the formal negotiations. Having lost early rounds of the fight, so to speak, what strategies did allies of the biotech bloc employ to gain adherents to their vision of a minimalist protocol?

The third position I examine is that brought to the talks by the EU. Representing a complex compromise among distinct relations of force in EU member states in the late 1990s, this position is best characterized not in terms of its polarity within any particular ideological spectrum (at least initially) but by its shifting nature. Finally, because of the important role Canada played in the latter stages of the biosafety talks, the development of Canada's position, and then that of the Miami Group, which Canada chaired, is examined in detail. In this analysis of state-negotiating positions, I do not look at all aspects of the complex protocol negotiations. Instead, I am concerned with those parts most relevant to the question of how LMOs would be regulated in international trade, including, in particular, the parameters on which decisions to import LMOs would eventually be based.

Negotiations on a Biosafety Protocol began at the first meeting of the Open-Ended Ad Hoc Working Group on Biosafety (BSWG) of the Convention on Biological Diversity in Aarhus, Denmark, in July 1996 (BSWG-1). This meeting was followed by two meetings of the group in 1997 (BSWG-2 and BSWG-3) in Montreal – which became the home of the CBD secretariat in 1996 – and by another two meetings in Montreal in 1998 (BSWG-4 and BSWG-5). BSWG-6, intended to be the final meeting for the talks, took place in Cartagena, Colombia, in February 1999. The chair of the working group throughout this process was Veit Koester of Denmark, who had also been one of the co-chairs of Expert Panel IV. Between 90 and 130 national delegations participated in the various BSWG meetings, along with intergovernmental organizations and representatives from industry and environmental CSOs. Until the final meeting of the BSWG, very little text was actually negotiated at the meetings. This is because Koester focused the activities of delegates on presenting text on various elements of a protocol, and then on trying to reduce the number of options under each proposed article (displayed in documents as square-bracketed text), so that any disagreements between delegations could be clarified. When negotiations did begin to take place at BSWG-6, there were still more than 450 pairs of brackets (representing 80 percent of the text) in the draft protocol.

Proponents of a Comprehensive Protocol

Many of the same individuals who had led the process from Rio to the Jakarta Mandate continued to be the main proponents of a protocol once negotiations began. This group was spearheaded by Egziabher of Ethiopia, who represented his country in the talks and who was also the spokesperson for the African Group. Before Egziabher's participation, Ethiopia did not have a history as an outspoken leader in agricultural and environmental fora, nor could it be seen to have had a distinctive interest in the outcome of these talks (as compared with other countries of the global South) that might have catapulted it to the forefront of debate. Genetic engineering appears to have been an important issue for Egziabher personally (Egziabher also represented Ethiopia in the ongoing TRIPS talks), and he used his position in Ethiopia's government, combined with a discursive position that framed GE as an imminent threat to farmers and the environment in the South, to bring this issue forward on the global stage.

Egziabher's continued leadership on the biosafety issue was evident when the chair of the BSWG asked countries to submit draft articles of a text in the lead-up to BSWG-2. The African Group was the first with a complete text, released in October 1996. Several countries of the South followed, with texts raising similar concerns to those of Africa, including Malaysia, Sri Lanka, and Bolivia (CBD 1997). What these states had in common in material terms was an active interest in protecting diverse natural environments as well as agricultural systems based on small farms and diverse landraces, each of which they perceived to be threatened by the imminent arrival of GEOs. Again, however, it is important to note that these countries were not necessarily any more threatened than other countries of the global South. What really brought their representatives together was a shared perspective on the biosafety issue, and this perspective had been shaped by the debates that had taken place from 1990 to 1995. In the draft texts brought forward by these states, one can identify elements of the World Resources Institute's framing of biosafety from 1992, of the silenced Expert Panel IV majority report, and also of the Third World Network report of 1995.

That a report written by activists (the TWN report) appears to have been a significant influence on the texts proposed by Africa and others reveals the central role anti-biotech CSOs continued to play in the biosafety talks once the negotiations began. Among these CSOs were organizations that had already been part of the debates leading to the Jakarta Mandate, such as the TWN (based in Malaysia), the WWF, the Council for Responsible Genetics (US), the Washington Biotechnology Action Council (US), the Edmunds Institute (US), GenEthics Network (Australia), and Greenpeace International. Other important CSOs were the Institute for Agriculture and Trade Policy (IATP, US), the Canadian Environmental Law Association, and Ecoropa (Europe). For each of these groups, engagement with the biosafety issue was one part of larger campaigns against the commercialization of genetic engineering in agriculture, and they brought the same diversity of concerns about GE to this forum as they brought to domestic struggles over GEOs. They had concerns about the health and environmental implications of genetic engineering and about the effects of gene patenting, corporate concentration, the terminator technology, and the loss of family farms.

Together with national delegates such as Egziabher, representatives of these CSOs emerged as much more than just lobbyists in the biosafety talks. They were the organic intellectuals of the coalition pushing for a biosafety protocol. To identify these activists as organic intellectuals is unconventional, from a Gramscian perspective, given that the CSOs were not, in most cases, organic to (meaning emerging from within) the global South. However, these groups and individuals were organic to the transnational environmental movement that had emerged since the 1960s, and in the biosafety talks, this is exactly what mattered. It was within the discursive framework of ecopower, with its emphasis on risk management and sustainable development, that the biosafety debates were taking place. In this context, many delegates from the South turned to representatives of environmental CSOs as biosafety experts who could articulate their interests and concerns within the accepted language of risk. That the CSOs played this role in the meeting was recognized by an Austrian negotiator, whose recollections of the meetings note that the "Policy and Science Updates" compiled by a group of environmental CSOs were a key source of technical information for the delegates (Gaugitsch 2002). Nijar (2002, 267), a representative of the TWN, notes that the work of CSOs often extended to helping countries make proposals, and to drafting language that would then be brought forward on behalf of national delegations. Another activist stated: "I am sure that Val Giddings [of the Biotechnology Industry Organization] wrote lots of text that was suggested by delegates; my colleagues and I also wrote text that was suggested by delegates" (interview #2).

As noted above, the initial strategy the proponents of a biosafety protocol adopted was to establish the direction of the talks by bringing forward the first-draft texts. In the African text, we can see how the proponents of a protocol positioned themselves in relation to already existing discourses and ideologies. The African text called for a protocol that would cover all LMOs and products derived from them. LMOs were defined to include

> any living organism or part thereof which is capable of regenerating itself on its own or in the body or cell of another organism and whose genetic material has been modified by modern biotechnology in a way which does not occur naturally by mating or recombination, or any living organism or part thereof which has been a fossil but has been resuscitated through modern biotechnology. (CBD 1997, 9)

Modern biotechnology was then defined to specifically focus on genetic engineering as well as "new cell and tissue culture methods for specific purposes" (9).

There are three important things to note about these definitions. First, they continue to single out GEOs as representing a distinctive set of potential risks. This move demonstrates the allegiance between the African position and the position taken by CSO critics of GE that is rooted in complexity talk. Second, they adopt a literal reading of the term "living modified organism." LMO had come to be accepted as meaning any genetically engineered organism that is *alive*, instead of what was initially intended by the United States – that it be understood to be synonymous with a "novel" organism. While there were initial differences on this understanding, the strongest proponent of the equivalency framing, the United States, had already signalled a willingness to agree to a narrower focus on GEOs in the Cairo report of 1995, and this move enabled the new meaning of LMO to become the norm. Third, the African submission refers to LMOs resuscitated from fossilized DNA.

This latter idea, brought into the biosafety debate shortly after the film *Jurassic Park* was released, is a good example of the discursive politics of biosafety on several levels. It reveals the way that the African Group's position was influenced by some of the more extreme catastrophe narratives of genetic engineering circulating among environmentalists. In this case, the notion of resuscitated DNA had been suggested as a possibility in the GE debate but was considered more science fiction than science. That this idea made it into the African submission also reveals a lack of internal checks among the proponents of a protocol to ensure that they were making a credible argument, as well as a lack of technical expertise among the African officials in particular, who did not catch the spuriousness of the inclusion of resuscitated DNA. Lastly, this idea is a useful example of how a technical concept, when not supported by credible scientists (and thus lacking a discursive footing), can die a quick death in multilateral fora. The possibility of resuscitated fossil DNA was ridiculed by genetic engineering experts and never incorporated into any of the negotiated definitions of LMOs or modern biotechnology.

In its operational articles, the proposed African text states that LMOs, or products thereof, can be exported only once agreement for this has been obtained, in advance, from a state of import, based on information provided to it about the LMO and its possible risks and benefits (CBD 1997). This position is framed in relation to the principle of national sovereignty, as well as to the already existing institutional framework of prior informed consent (PIC), which works along the same lines. It is also clearly located within the risk discourse, and stakes out a particular ideological position within this discourse.

Risk assessment, according to the African document, is defined to include the identification of potential benefits and harms of an LMO or product thereof, including socio-economic impacts, as well as possible effects "which are contrary to the social, cultural, ethical and religious values of communities" (CBD 1997, 9, 85). In this reading, "risk assessment" refers to a comprehensive assessment of the benefits and risks of GEOs in relation to national priorities and the goal of sustainable development. It is not meant to be a technical exercise only. A number of reasons were put forward in biosafety fora for why such a science-based analysis might be inadequate, each of which can be seen to reverberate with the skepticism of official risk analysis identified by Beck (1992). Some activists argued that the funding required to do science meant that there could be "far more data on one side of the scale than on the other" when it came to assessing risks scientifically (Lewis 1998, 17). Others took the position that risk assessment is ultimately a political judgment. As one environmental CSO representative active in the biosafety talks stated:

We were in a conversation where it was very important to make the point that the issue of acceptable risk is fundamentally a question that science cannot answer. It is a value-laden question and fundamentally a political question. Therefore, societies, through their governments, are entitled to make their own decisions about what is an acceptable level of risk. The only discipline we felt was appropriate was national treatment. (Interview #5)

The belief that the import of an LMO represents a sovereign, democratic decision that should be based on all relevant parameters, whether technical, social, or ethical, was also reflected in the article of the African text calling for "public hearings in the process of approving the release, transfer or use, contained or otherwise, of such living modified organisms or products" (CBD 1997, 64).

In addition to its calls for a comprehensive assessment of the risks of LMOs, the African proposal calls for LMOs to be treated cautiously in cases of scientific uncertainty. Its preamble states:

In accordance with the precautionary principle, lack of full scientific certainty should not be used as a reason for postponing measures to avoid or minimize risk where such a risk is posed by LMOs resulting from biotechnology. (CBD 1997, 3)

This is basically an iteration of the precautionary principle from Principle 15 of the Rio Declaration. However, it has dropped the requirement that measures be cost-effective, as stated in Principle 15. The idea of cost-effectiveness had been introduced by the United States in the Rio talks and was seen as a constraint by, among others, the EU negotiators (Interview #9). The Africans and CSOs clearly felt this way too, removing the caveat from the definition of the precautionary principle put forward in the African text. The inclusion of the precautionary principle in the proposals of proponents of a protocol enabled the potential risks of LMOs to be related to other major transboundary environmental issues. This is exemplified in the following statement by a Greenpeace representative:

In view of the potentially irreversible harmful consequences of releasing GMOs into the environment and the food chain, the protocol had to allow parties to take preventative action in order to protect public health and the environment. Waiting for conclusive evidence of environmental damage can, in many cases, mean waiting until damage begins to occur. This can have not only irreversible but also costly consequences, as we have already seen, for example, with the depletion of the ozone layer, the impacts of climate change and the build-up of persistent organic pollutants in the food chain. (Gale 2002, 253)

When it came to socio-economic impacts of LMOs, the African text proposed that these be assessed not only in relation to import decision making but also in relation to their effects on exports from the global South. Specifically, the text contained an article dealing with the possibility that import substitution in the North might hurt traditional agricultural exporters, thereby indirectly leading to genetic erosion. This article states:

A Party that intends to produce, using a living modified organism, a hitherto imported commodity, shall notify the affected Party or the Party likely to be affected long enough, and in no case less than seven years in advance so as to enable them to diversify their production and to implement measures concerning the biodiversity that would be reduced following the disruption of production of the commodity in question. The Party substituting its import in such unnatural way shall, when the affected Party is a developing country, provide financial and technical assistance to the affected Party. (CBD 1997, 72)

As with the Expert Panel IV's majority report, this description of the possible socio-economic impacts of LMOs substituting imports is framed in terms of the need for long-term protection of biodiversity, the stated goal of the CBD.

The African draft text also stipulated that all products, particularly food products incorporating LMOs or products thereof, "should be clearly labelled" (CBD 1997, 62). This article was justified on the grounds that importers would have to know what a commodity is in order to know if it was approved for import. However, its inclusion reveals, once again, the close participation of activists from the North in the development of Africa's position. The call for GEO labelling had been gaining ground in industrialized countries, and activists from these countries wanted to use the Biosafety Protocol context to make further gains on the issue. Because of its importance in the GE debate in the North, the issue of labelling would prove to be one of the last to be resolved in the biosafety talks.

Finally, the African text includes strong language on liability and compensation to be held by the "States of Origin" should there be harms caused by LMOs (CBD 1997, 67). A Greenpeace representative summed up the rationale for this article: "Those who do not take action to prevent or reduce risks should be prepared to bear the full legal and financial responsibility for the consequences of their activities" (Gale 2002, 256).

Rooted in the principles of national sovereignty and the goals of CBD, as well as a strong interpretation of what precaution, risk assessment, and even liberalism (at least as regards labelling) should mean vis-à-vis unnatural GEOs, the African proposal was designed to give nation-states as many tools as possible to keep out LMOs they might see as undesirable, and to compensate them for any and all damages that could be caused by LMOs. Within this comprehensive framing of the biosafety issue, there is a precautionary reading of the risks of GEOs, but this is clearly not the only, nor the strongest, ideational current. One Canadian government representative recalls that, as a whole, the African proposal "set a kind of rude awakening for people ... because it was quite, quite strident" (interview #6).

After presenting their demands, the organic intellectuals leading the call for a comprehensive protocol worked to build consent among other countries in order to move their view of biosafety forward through the BSWG process. They began by garnering support among the G-77/China and among sympathetic European states, including, in particular, Norway, Denmark, Sweden, Austria, and Spain. Why were the biosafety activists able to get support from the global South as well as from the delegations representing certain European states?

In terms of material relations of force, what all of the states that backed the demand for a strong protocol had in common was that they would likely become, first and foremost, importers of LMOs. This meant they had little economic stake in a successful genetic engineering revolution and, if anything, saw potential economic losses (vis-à-vis import substitution). As in the debates that were taking place in North America and Europe, this consumer position lent itself to a particularly skeptical perspective when it came to weighing the potential risks and benefits of GEOs for human health and biodiversity protection.

In terms of institutional relations of force, what these states had in common was either an absence of regulatory systems for evaluating GEOs or systems that were based on similar principles to those called for by the Africans and the CSOs. The Norwegian proposed protocol text of 1997, for example, says that "production and use of living modified organisms should take place in an ethically and socially justifiable way, in accordance with the principle of sustainable development and without adverse effects on human health and the environment" (CBD 1997, 6). Echoing the African proposal, Norway suggests that one vehicle for incorporating such concerns is to carry out public hearings. In proposing these articles, Norway was the only OECD country in the biosafety talks to strongly support the view that "non-technical" criteria could be incorporated in decision making on the import of LMOs, "as long as these were non-discriminatory (applied equally to imports and domestic production)" (Ivars 2002, 193). The Norwegian position diverged from those of other OECD countries because of the institutional context in Norway. Unlike the European Council's regulatory directive 90/220 or the regulatory systems in place in other OECD countries, the Norwegian Gene Technology Act of 1990 incorporated such non-technical criteria, and this was the basis of Norway's submissions to the biosafety talks.

Organizational relations of force were also relevant for those representatives who came to the talks from countries lacking existing regulatory systems. The absence of regulatory systems gave these delegates the freedom to develop positions for their countries, and their own institutional roles would bias these positions. Specifically, in the context of the CBD, state representatives generally came from environmental departments of governments, and from biodiversity conservation branches in those departments. Because of this, these representatives appear to have exhibited a proclivity towards seeing the GE issue first as an environmental protection issue rather than, say, a trade issue (interview #1; Muller 2002, 140). The unity of purpose that came with the institutional (and ideational) positions held by most of the delegates of the South was a big reason why the G-77/China (eventually re-forming as the Like-Minded Group) managed to maintain cohesion throughout the biosafety talks.

As Gramsci would expect, the glue that bound state and CSO representatives who supported the call for a comprehensive protocol was ideational. This involved two key elements. Most important, the notion of precaution had by now consolidated into a distinct ideological position in the GE risk debate, and, by the late 1990s, several scientific and policy factors were allowing it to gain adherents. As one environmental activist said, "The science was shifting very strongly" to support the stance that GE critics had taken against genetic engineering starting in the 1980s (interview #5). In 1996, for example, research was published showing unintended gene transfer from transgenic organisms (Mikkelsen, Anderson, and Jorgensen 1996). This was followed by a study in 1997 on the transfer of injected DNA in the body (Schubbert et al. 1997). A 1998 paper then argued that there is a higher rate of outcrossing of genetic material from GE plants compared with non-GEOs (Bergleson, Purrington, and Wichmann 1998). Also in 1998, Pustzai, a Hungarian researcher based in Scotland, presented research on British television suggesting that transgenic lectin-expressing potatoes had harmed rats in laboratory studies. This study, in particular, led to a vociferous public debate in Europe over the impacts of GEOs that continued long after the results were published in the UK medical journal *The Lancet* in 1999 (Ewan and Pustzai 1999). Finally, in May 1999, Losey, Rayno, and Carter (1999) of Cornell University published a preliminary study showing that pollen from Bt corn could kill endangered monarch butterflies. Each of these papers, and the controversies they engendered, raised questions about whether the risks of GEOs were fully known and amenable to management, as proponents of biotech suggested (Gaugitsch 2002, 88).

Although the publication of these studies was part of the story, to attribute growing adherence to the ideology of precaution to new scientific knowledge alone is to oversimplify. The other ideational factor that shaped this was a growing suspicion of state regulatory systems in general to actually assess and manage risks effectively. This suspicion, which Beck (1992) identifies as a central feature of the contemporary risk society, ignited in the context of food issues in the late 1990s as a result of the British mad cow disease scandal.

In this controversy, years of official statements that there was no risk to consumers from eating beef from cows afflicted with bovine spongiform encephalopathy (BSE) had to be retracted by the British government, in the spring of 1996, after patients began showing signs of new variant Creutzfeldt-Jakob disease (CJD), a human form of spongiform encephalopathy. Evidence also came to light at that time showing that the UK Agricultural Ministry had suppressed evidence of risk and silenced its own scientists (Levidow 1999). This particular event shook British confidence in the government's ability and willingness to protect public safety from problems in the food system, and GE foods would also become subject to this mistrust. It also had wider effects across Europe and in the Biosafety Protocol talks. This is because the BSE scandal raised concerns about the very notion of science-based regulation.

The British government had actively followed the advice of the scientific community, advice based on the assumption that BSE could be transmitted only by a bacterium or virus. Any theories that held that BSE could be attributed to other factors were rejected out of hand because they did not fit long-accepted pathogenic models. It was only after incidences of CJD were observed that the mainstream scientific community took another look at a theory that stated that the disease could be carried by an incorrectly folded protein known as a prion. Prusiner, the scientist who had been ridiculed when he proposed this radical theory only a few years earlier, was awarded the Nobel Prize in biology in 1997 for discovering the prion. And, the BSE case quickly became a key justification behind the calls for a precautionary regulatory system for GEO trade: a system not based on the assumption that all the risks of a product are known before its use.

The growing strength of a precautionary framing of the risks of GEOs in the biosafety talks was one of the ideational elements that brought the G-77/China, as well as some delegations from Europe, to rally around the positions of Egziabher and the CSOs. Again, this solidarity would not have been achieved had the position been framed differently. Solidarity would not have emerged around calls for a moratorium or even a ban on all GEO introductions, for example. Although Egziabher and many of the CSOs would clearly have favoured a moratorium or a ban, the risk discourse within which they framed their position had as its implicit goal the management of the potential risks of GEOs in order to realize the benefits of those organisms. This framing meant that even Argentina, while closely tied to the biotech bloc, could go along with the protocol demands being made by the G-77/China through most of the BSWG meetings.

Despite the consent achieved on the position of Egziabher and his CSO colleagues among the G-77/China, in particular, by BSWG-6 (in early 1999), a division finally appeared within this negotiating group. Not surprisingly, this split was precipitated by Argentina, which, ultimately, had distinct material interests (it was, by this point, an exporter of GEOs), different institutional leanings (it was represented by a trade representative from its Geneva UN Permanent Mission) (Ling 1996), and different ideological priorities (it was particularly concerned with the apparent disregard for WTO trade rules put forth in some of the G-77/China positions, such as the desire for socio-economic assessments of GEOs). When Argentina left, it did so to join other countries that were also worried about the trade implications of a potentially restrictive protocol.[1] This group, which would become known as the Miami Group, is discussed below. Those countries of the G-77/China that remained – still more than one hundred – re-formed at that point as the Like-Minded Group. The name bespeaks the fact that these countries had, by this time, settled on an ideological position calling for a comprehensive, precautionary protocol. States were welcome to work within this framework, and those that could not accept the position were welcome to leave, as Argentina had, or to remain quiet.

Defenders of the Biotech Bloc

I have already demonstrated the close relationships that came to exist among states and industry in the biotech bloc in the early 1990s. In the Biosafety Protocol negotiations, this cooperation was manifest in virtually identical standpoints on most of the key issues among industry lobby groups and numerous specific states. This position can be characterized as the minimalist position, as it sought to minimize the implications of the Biosafety Protocol for international trade in GEOs, and, especially, trade in bulk commodities such as food and feed grain.

Initially, the United States was the strongest voice for a minimalist protocol among countries. Over time, similar perspectives also came forward from Australia, Japan, Canada, Argentina, Uruguay, and Chile, as well as from European countries, including, in particular, Germany and France. In terms of material capabilities, each of these countries was actively supporting genetic engineering research and had biotech industries at various stages of development, or had strong trade ties with countries that did. This material position meant that if these countries did not already have commercial

GEO exports (as did the United States), they saw themselves as imminent exporters, and they wished to avoid market restrictions on GEO products. More important, however, in terms of organization and ideas, most of these states already had regulatory systems or were in the process of developing such systems, which accepted the basic equivalency of GEOs and non-GEOs.[2] As a result, they shared with industry a common perspective on the types of risks involved with GE and how they could be managed, and they sought to universalize this perspective through the protocol negotiations.

Among industry CSOs advocating a pro-biotech position, key groups present in the talks included the Biotechnology Industry Organization (BIO, North America), the Japanese Bioindustry Association, the Senior Advisory Group on Biotechnology (US), and the Green Industry Biotechnology Platform (Europe). Companies such as Monsanto also sent delegates to many of the meetings. Notably, however, the industry presence was sparse and disorganized in the early stages of the process. At BSWG-1 in Aarhus, for example, participation by the private sector was limited to only a few participants from just three sectors (seeds, forestry, and aquaculture), and of these representatives, most had simply come to monitor discussions (Reifshneider 2002, 274). Over ensuing years, the industry presence grew, but there was still a lot of turnover. The result, according to the only person to have represented the private sector throughout the whole process, was "imperfect coordination" (274). One possible explanation for the biotech industry not devoting more resources to the biosafety talks, one which fits with a Gramscian understanding of state and civil society relations in hegemonic formations, is that industry interests were already being articulated through the positions of several states at the negotiating table.

Actors allied with the biotech bloc used two main sets of ideational strategies in their efforts to show that a protocol designed to minimally impact trade in GEOs was the only reasonable approach. The first strategy involved showing why it was scientifically appropriate to have a protocol with limited scope. The second strategy involved explaining why any other approach was simply wrong, by being scientifically unsound, going against global trade rules, or representing a threat (rather than a boon) to global food security.

As they had at the domestic level, forces allied with the biotech bloc focused on the basic equivalency of GEOs with non-GEOs to justify minimal regulatory intervention. Noting that the Jakarta Mandate was intended to address LMOs "that may have an adverse effect on the conservation and sustainable use of biological diversity," the United States and industry both argued at BSWG-1 that the AIA procedure being advocated by the G-77/China should apply only to a "positive list" of LMOs known to have adverse effects on the environment (CBD 1997, 26). In principle, this approach seemed reasonable because this was exactly the kind of list being developed for international trade in dangerous pesticides and other hazardous chemicals (the field that the AIA approach was modelled on). In practice, the defenders of the biotech bloc expected that a positive list would be limited, since most LMOs being developed could be considered equivalent to their parent organisms in terms of the risks they posed for the environment and would thus be exempt from inclusion on the list.

As with the position of those calling for a comprehensive protocol, the position allies of the biotech bloc were putting forward was also rooted in the risk discourse, but it embodied a very different take on the risks of LMOs. Rather than assuming all LMOs were inherently risky and taking scientific uncertainty as the norm, the allies of the biotech bloc assumed that LMOs and non-LMOs were substantially equivalent. In cases where they were not substantially equivalent, science-based evidence of harm would be necessary to put the organism on a list requiring AIA. This argument brought together the discourse of risk, that of equivalence, and a particular neo-liberal narrative of what acceptable, science-based trade restrictions look like – to back the exclusion of most GE commodities from an AIA procedure.

Despite the force of the comparison with other areas of international environmental law, within the first few meetings of the BSWG it became apparent that there was little sympathy for the position that only a small class of LMOs should be subject to AIA. This demonstrates that the precautionary position within the risk discourse was truly gaining adherents in this field for the reasons noted above, eventually forcing the biotech bloc to change its strategy. By the latter stages of negotiations, the focus of pro-biotech actors in terms of AIA was on exemptions rather than inclusions.

In addition to their arguments for what a biosafety protocol should look like, the defenders of the biotech bloc spent a considerable amount of energy deriding the proposals put forward by the African Group, CSOs, and others that advocated a comprehensive and precautionary protocol. As in earlier domestic struggles over GEOs, a central underlying current was the issue of the equivalency of GEOs and non-GEOs. This was taken as a scientifically established truth by many advocates of the minimalist position, and any view of GE that questioned this was painted as unscientific and misinformed. From interviews, it is clear that some technical experts allied with the biotech bloc were so convinced of this equivalency that they remained incredulous towards alternate framings throughout the process of developing the Biosafety Protocol (and afterwards), even as their political masters were forced to make concessions in order to win a deal (interviews #6, #7, #13).

A second set of arguments was shaped around the notion that the demands of the G-77/China, especially for precaution and non-scientific criteria for assessing LMO imports, ran contrary to international trade rules. Specifically, the WTO SPS Agreement was still taken as the norm by the biotech bloc, and it has very specific criteria for science-based risk assessment. For example, the SPS Agreement accepts that a state may bring in precautionary measures, including bans on substances, in cases of scientific uncertainty. However, in such cases, that state is expected to actively solicit further scientific research on the substance in question and to provide the evidence to back a ban, or else to relinquish it, within a "reasonable period of time" (WTO 1994a, Article 5.7). The G-77/China did not address how long precautionary measures could be in place. Their desire to curtail imports based on socio-economic considerations, and their call for early warnings about import substitution, also flew in the face of the WTO. These were all seen as threats to a neo-liberal trade regime, a regime that had long been touted by the biotech bloc, as well as by the larger transnational historical bloc, as the basis of future global prosperity.

In addition to being unscientific and contrary to the WTO, the demands of the global South were painted as simply unworkable and a threat to the global food supply. This position involved rehearsing the now familiar argument that GEOs represent an important tool for sustainable development that could be inhibited through a tough protocol. Industry groups and government representatives stressed that the global grain-trading system was built on the efficiencies gained by combining shipments from many farms, often from different countries. The system was not built on the segregation of streams of grain based on their variety, and certainly not based on separation by genotype. With this system, those LMO commodities requiring AIA could not be separated from non-LMO commodities. This situation would require importers to "take it or leave it" when it came to importing soy, or corn, or other products that could be GE, and "leaving it" was not really an option for many countries. According to one industry representative,

> We are talking staggering quantities: 240 million tonnes of grain. If you would remove the Canadian, US, Australian, Argentinean, and Brazil contributions from world food requirements, you would be removing 90 percent of the grain that is available in the world. (Interview #10)

This statement assumes that GEOs are essential to the global food supply, or that they could not be distinguished from non-GEOs. The first assumption is patently false. Even in Canada, where there had been rapid uptake of GEOs in agriculture, only 4 percent of Canada's exports were genetically engineered in 1998 (Smith 2000, para. 6.86). The second assumption, that GEOs could not be distinguished from non-GEOs, was true when the talks started, but it would prove to be a double-edged sword for the biotech bloc when molecular biological techniques for identifying GEOs were developed in 1999 (interview #8).[3] Then, as consumer resistance to GEOs grew, some members of the food-processing and distribution industry, including the Canadian Wheat Board (CWB), started to take stances in favour of segregating GE from non-GE varieties in order to meet consumer demand (CWB 2001). As one representatives from a Canadian environmental CSO stated,

> The Canada Grains Council [the industry representative on the Canadian delegation] were constantly saying "there is absolutely no way that you could segregate GE from non-GE in the system." This was the bedrock of their position: "Instituting a two-stream system is impossible" ... In the next breath you'd have the Wheat Board telling us that they did it all the time. You tell the CWB that you want certified organic of this variety of wheat [and], "No problem, we'll get it to you." (Interview #5)

The development of the genetic-identification technologies, along with the willingness of some food processors to accept the need for segregated commodity streams, led the proponents of GE to shift from an argument that segregation was unworkable to arguing that it was costly, adding between 10 and 30 percent to the price of grain (IGTC 2002, 8). The shift in this argument shows how the uncertain relationships between the gene giants and the grain traders discussed in Chapter 2 rose to the surface in the biosafety talks. The grain traders (such as the Canadian Wheat Board) remained aligned with the biotech bloc's efforts to get a minimalist protocol, but they were also prepared to go with whatever outcome best met consumer demands.

Organizationally, actors allied with the biotech bloc used several strategies for moving their position forward in the field of biosafety. As one delegate from the global South noted, states allied with the biotech bloc resorted to ministerial-level contacts, forum shopping (moving from one international fora to another to find the best setting), bilateral trade and investment incentives, and sponsoring workshops, as well as the use of influence to undermine the credibility of individual negotiators (Muller 2002, 142). Like their activist counterparts, industry CSOs were heavily involved in the negotiation process. They sometimes made direct interventions in plenaries and small group meetings, though they did most of their work in direct lobbying of national delegations (Fisher 2002, 124). These groups also undertook media outreach to try to build support for their position that agbiotech would benefit the world, and that a strong protocol could prevent those benefits from being realized. In an effort to build goodwill for their industry in the biosafety meetings, after the Jakarta Conference of the Parties, the biotechnology industry organizations even sponsored the production of a Braille version of the CBD "as a sign of their commitment to biodiversity" (Ling 1996, 1).

Despite their common goal, industry CSOs from different regions appear to have taken slightly different approaches in their efforts to influence the talks. US groups, comfortable with the highly confrontational lobbying tactics in Washington, brought these strategies to this international forum. European business groups, on the other hand, have a history of taking a more tactful, collaborative approach in trying to ensure that regulations are aligned with their interests. This difference was noted by European delegates, who stated that "meetings with European industry reps were always constructive, but they were more difficult with North American reps, who initially took a rather confrontational line without trying to understand the EU's position" (Bail, Decaestecker, and Jorgensen 2002, 173).

However, reflecting the larger trend in global environmental politics, these national differences in lobbying approaches shifted during the biosafety negotiations. (This trend is discussed in Levy and Newell 2000). By Cartagena in 1999, industry groups had organized an international coalition in an attempt to show that biotechnology was not simply a product of US-based multinationals. This was a strategy recently employed by industry in the Kyoto Protocol talks, with some success. In the biosafety context, the 2,200 companies from 130 countries that signed on to the Global Industry Coalition (GIC) came from across the food system, representing food processors, retailers, and seed companies (GIC 1999). What these companies had in common was that they benefited in one way or another from the globalized food system that had emerged since the 1970s. This group was thus quick to rally around the idea that a stringent protocol requiring segregated commodity streams would cause major upsets in the global grain-trading system.

While officially representing a diverse cross-section of actors in the global food system, the GIC actually had a much smaller base of active support. The organization had no formal presence outside the Biosafety Protocol negotiations: it was based in Canada and represented by spokespersons for the Canadian lobby group BIOTECanada, a group that had received more than $6 million from the Canadian government between 1994 and 1999 to promote biotechnology (Abley 2000).

The EU, the Compromise Group, and the CEE countries

From these descriptions of the proponents of a comprehensive Biosafety Protocol and the defenders of the biotech bloc, it is clear that European delegations fell on both sides of the debate. On the one side, Germany opposed a protocol in principle until 1998, and the Netherlands, Britain, and France all favoured an approach tailored along the lines of the UNEP technical guidelines, which reflected the equivalency framing of GEOs. On the other side of the European delegation, Austria, Norway, Denmark, Sweden, and Spain were sympathetic to the calls from the G-77/China for a strong, legally binding protocol for all LMOs based on AIA. This split represented a challenge for the EU in its efforts to come up with a common negotiating stance, as is its practice on international issues.

In the context of these differences, Marquard and van der Meer took the lead, once again, in defining a distinctive European position designed to appease both factions within the EU delegation (interview #9). The negotiation stance the EU brought to the initial meetings of the BSWG embodied what was defined as a "two-track approach" (CBD 1997, 5). The EU advocated the continued development and implementation of the UNEP guidelines, which the European allies of the biotech bloc saw as the prototype for a reasonable, science-based set of rules on trade in GEOs. The EU then also accepted the need to negotiate a protocol that would bring similar standards for risk assessment and management into a legally binding text.

This position can be seen to have been rooted in the discourse of risk, though it appears to have had a foot in both ideological camps when it came to the question of how serious the risks of LMOs were, and how they were best dealt with. To exemplify this, consider the following article from the preamble of the EU draft text, which is clearly located in complexity talk:

> The interaction between living modified organisms resulting from modern biotechnology and the environment, in particular in centres of origin and genetic diversity, is of a very complex nature not always fully elucidated by adequate scientific knowledge. (CBD 1997, 5)

This statement is followed by proposals on risk assessment and management that refer to the UNEP technical guidelines, guidelines that assume the manageability of most GE risks because of the equivalence of GEOs and non-GEOs.

The EU position was initially somewhat schizophrenic, but over the course of negotiations it evolved to become more consistent with many elements of the position of the G-77/China. For example, on the topic of labelling, initial EU proposals had no article on this subject at all. By the final round of negotiations, however, labelling of LMO shipments became a central demand of the EU.

Steffenhagen (2001, 52) identifies a similar shift on the question of whether the AIA called for by the South should require explicit or implicit consent. Explicit consent would mean that there could be no movement of LMOs without written authorization by an importing country. The G-77/China saw this as a crucial tool for ensuring that the North would finance regulatory capacity building through the protocol so that authorities would be able to respond to export requests. If implicit consent by importers would be enough for the export of LMOs, there would be no pressure on the North to help the South develop this capacity. Germany, alone among the Euro-peans, was concerned about the way that an explicit-consent procedure could block international trade in biotech commodities. Within the German gov-ernment, the Ministry of Economics and Technology, the Ministry of Health, and the Ministry of Education and Research had all argued (against the much weaker Ministry of Environment) for a simple notification procedure accompanying the transboundary movement of LMOs. When the rest of the EU would not agree to this, Germany accepted a role for AIA, but only on the grounds that it would be based on implicit consent. The line held by the ministries was that "we don't want to close the door on biotechnology in Germany" (Steffenhagen 2001, 49). Germany held this position, block-ing an EU consensus on AIA, until early 1999. By the Cartagena meetings in February of that year, however, Germany accepted the need for explicit con-sent, bringing the EU position in line with that of the G-77/China (by then known as the Like-Minded Group).

What enabled these types of shifts in the EU position between 1996 and 1999? At the level of material relations of force, EU countries were still not exporters of LMOs. This meant that an importer-consumer mindset contin-ued to exert a strong influence on the overall EU position. In terms of ideas, public skepticism towards GEOs was on the rise in Europe, and this was being translated into support for a precautionary approach. I agree with Thomas (2001, 337) when he argues that this rise cannot be attributed to any one actor or incident; instead, it was the result of "an artful dance of hard work, lucky coincidence, strategy and passion."

Hard work and passion were exercised by the activist CSOs that became involved with GE issues. In the fall of 1996, for example, Greenpeace tried to block the first shipments of GE soy into Hamburg. Shortly thereafter, a group of UK activists, attending a press conference held by US Secretary of Agriculture Dan Glickman at the World Food Summit in Rome, stripped off their clothes to reveal bodies painted with anti-GE slogans. Beginning in 1996, European activists also actively "decontaminated" experimental GE test plots throughout Europe, by pulling out GE sugar beets, soy, canola, and other crops (Thomas 2001, 338-40).

The coincidence that led to a mushrooming of concern was the emer-gence of the GE issue alongside the BSE scandal in the United Kingdom. On 20 March 1996, the then prime minister John Major announced that the deaths of at least ten people were linked to BSE. On 3 April only two weeks later, the EU's decision to approve the release of Roundup Ready soy was announced (Charles 2001, 170). The similarities between the BSE issue and the potential risks of GEOs, and the one announcement following so closely on the heels of the other, meant that when Greenpeace and other European activist groups raised questions and concerns about GEOs in 1996 and afterwards, the European press was ready to tell their story, and the public was ready to hear it and respond. As Charles (2001, 170) writes,

> It took almost two years for [CEO Robert] Shapiro and his fellow Monsanto executives to realize it, but from that moment the burden of proof, when it came to the safety of Europe's food, had firmly settled on the shoulders of the biotechnology industry.

The strategy that enabled an EU shift towards precaution included decisions of activist groups to focus on consumer issues. These issues, such as the demand that GE food products be labelled, and the demand for a precautionary regulatory system for GE foods, struck a chord with consumers in the wake of the BSE crisis. Drawing on the risk discourse and the idea of consumer choice, the labelling strategy resulted in major gains for activists. In 1997, the EU decided that all products containing GE ingredients would need to be labelled as such. Labelling opened up the possibility for another activist strategy: boycotts of GE foods (which were now identifiable in the grocery stores) and secondary boycotts of stores that sold GE products. Such boycotts, or simply the threat of them, led several grocery chains, including Marks & Spencer, to state publicly that they would remove GE products from their stores in 1998 and 1999.

In terms of organizational relations of force, numerous changes in Europe between 1996 and 1999 also led to a greater consistency between the position of the EU and that of the G-77/China. The EU labelling decision clearly influenced the EU's negotiating position on this issue in the protocol talks. Developments related to directive 90/220 also played a role. Even as EU-wide approvals were being granted under 90/220 for a growing list of GEOs, rising public and scientific concerns about the environmental implications of GEOs had led some states to invoke the safeguard clause (Article 16 of 90/220) to put in place provisional bans on specific GEOs. For example, in 1997, this article was used by Austria and Luxembourg to ban the import and cultivation of Bt corn (Mackenzie and Francescon 2000). France and the United Kingdom followed in 1998 with provisional bans on glufosinate-resistant canola (European Commission 2005).

Three separate elections at the level of domestic politics played an important role in shaping organizational relations of force in the EU. First, in the UK, Major's Conservative government was replaced by Tony Blair's Labour government in 1997, and the new government did not want to be seen to be blocking European consensus on biosafety (as Germany was doing) in the face of the brewing GEO controversy at home. Second, 1997 saw a socialist election victory in France for Lionel Jospin. While the previous French government was more closely aligned with the biotech bloc, Jospin's government (which also had greens in its coalition) was willing to take a more precautionary stand on GE issues. Finally, and most important given its role in the European delegation, in 1998 Germany also had an election in which a conservative government was replaced by a red-green coalition. This new government was much more interested in seeing strong laws on GEOs in place, though this stance took some time to filter down to the biosafety talks. As Steffenhagen (2001, 44-47) documents, the Germans continued to oppose explicit consent on AIA until just days before the Cartagena negotiations. This stance appears to have been dictated more by the civil servants in powerful ministries (allied with the biotech bloc) than by elected officials, and it changed only after intense scrutiny from German environmental CSOs and other European delegations.

These material, ideational, and organizational factors brought the European delegation as a whole to accept the need for a protocol that included an explicit-consent AIA procedure, labelling requirements, and numerous other demands put forward by the G-77/China and activist CSOs. Even so, by the time the final round of talks began in Cartagena, there were still major differences between the EU and the proponents of a comprehensive protocol. An important example was the European position on the precautionary principle.

As noted above, the Africans, along with Norway and others, called for language in the protocol that would allow countries to decide against the import of LMOs even when there was insufficient scientific knowledge, or a lack of scientific consensus. The Europeans had come to accept this stance, as it was in line with the discourse emerging in Europe. They had even included preambular text in their submission of 1997, which stated that "the provisions of this Protocol should contribute to protection in the field of biosafety, based on scientific risk assessment *and* the precautionary principle" (CBD 1997, 5, emphasis mine). Thus, it came as quite a shock to CSOs and countries from the South when the Europeans proposed, in a last-ditch effort at Cartagena, to offer as a compromise to the United States and other LMO-exporting countries the removal of the phrase from the protocol that would allow decisions to restrict imports "without full scientific certainty or consensus" (Gale 2002, 253; Chasek 1999, 11).

This proposal demonstrates that despite the growing sympathy for the precautionary principle in Europe, by early 1999, other priorities were still more important. The highest priority for the EU, according to a German government source quoted by Steffenhagen (2001, 55), was to "move towards a consensus with the Miami Group in order not to jeopardize a Protocol." However, this strategic approach was unsuccessful. Despite the EU's move on precaution and several other issues, the Miami Group did not agree to the EU's compromise text at Cartagena primarily because that text suggested putting off to a future date the question of how LMO commodities destined for food, feed, or processing should be dealt with. Rather than end the negotiations in this way, with one of its key issue not yet decided upon, the Miami Group proposed a suspension of the Cartagena talks.

By Cartagena, the Miami Group had crystallized around a position that was closely aligned with industry interests, as had the Like-Minded Group in a position in line with environmental CSOs. The Europeans had negotiated as a bloc all along and were seen as the third key negotiating group, holding positions somewhere in between the other two. The remaining countries of the world coalesced into two more negotiating groups, which I touch upon only briefly here.

The Central and Eastern European (CEE) countries, including the Russian federation, had interests in common with countries of the G-77/China. They saw the need for an international regulatory instrument on LMOs, since all but three countries also lacked regulatory structures for GE organisms (Nechay 2002, 212). Although this was not always the case, generally, the position of the CEE group was aligned with that of Europe, as many of the CEE states were hoping to become members of the EU, and these states were already in the process of harmonizing their regulatory systems with those of the EU. Most of the CEE countries sent only individual delegates to the talks – if they were present at all – and there was limited participation of CSOs from the CEE region, both on behalf of industry and on behalf of environmental organizations.

When it was decided that negotiations would proceed on the basis of interest groups, rather than in the usual UN format of regional groupings (I explain this decision in Chapter 6), a fifth group emerged "on the spur of the moment" among countries that were not full members of the EU nor of the Miami Group but that were members of the OECD (Nobs 2002, 186). Key participants in this group, which became known as the Compromise Group, were Switzerland, Japan, and Norway. Because it included countries with diverse positions on the issues at hand, the Compromise Group tended to focus its efforts on building a protocol that the three key groups could agree to, and they proved crucial in this task. Still, the Compromise Group's strongest relationships were with Europe, as was the case with the CEE group. This meant that the Compromise Group would likely follow Europe in major disputes with the Miami Group or the G-77/China. As Canadian minister of environment David Anderson noted, this was a "tactical advantage enjoyed by the EU throughout the negotiations" (Anderson 2002, 240).

This discussion of negotiating positions has brought to the fore the importance of both context and strategy. By "context" I mean the material, organizational, and ideational structures that provided the framework for positions taken in the Biosafety Protocol negotiations. "Strategy" refers to the actions policy actors take as they work within these structures to further their own interests. The contextual power of the idea of precaution, and the agency of activist CSOs in particular, deserve further examination.

While the BSWG meetings were taking place, the idea of a precautionary response to GEOs was clearly growing in strength. In the talks, precaution was initially presented as only one element within a larger comprehensive assessment position brought forward by the African Group and its CSO allies, but other elements of that position (such as the concern for socio-economic considerations) had not been well received by countries outside the G-77. Still, precaution itself became a distinct ideological position because it had discursive traction in a technical debate framed by risk. This traction depended on the growing scientific controversy over GEOs and the BSE fiasco in Britain.

Despite its growing acceptance among many participants in the talks, at Cartagena, a precautionary response to GEOs was not yet accepted as common sense. This was evident in the EU's decision to remove language enabling precautionary decision making from the final compromise it offered to the Miami Group, and the fact that this move was actually accepted, however begrudgingly, by all of the other negotiating groups at the meeting.

One EU negotiator noted that this action was taken because of the lack of intra-EU consensus on what precautionary decision making meant in practice (interview #9). Lack of consensus on this subject did indeed exist in the EU. Drawing on the idea that the genetic engineering process itself has unknowable implications, as emphasized by complexity talk, some European scientists and policy advisors argued that GEOs should be subject to an outright ban on the basis of precaution (e.g., Myhr and Traavik 1999). Others used the idea of precaution and the inherent uncertainties associated with genetic engineering to call for more stringent regulatory systems that carefully evaluated (and monitored) all of the anticipated and unanticipated effects of the engineering process on GEOs before their commercial release (e.g., von Shomberg 1998). These represent two very different precautionary policy responses. Nonetheless, to blame the EU's move at Cartagena on this lack of consensus does not fully explain the dynamics at work. I would suggest that at Cartagena, precaution was seen as one current within a conversation framed by the discourse of risk. By the time the Montreal meeting took place, a year later, when the EU would champion precautionary decision making at all cost, precaution itself had become the discursive framework within which the GE debate was taking place, and within which positions and interests were mobilized. I develop this argument in Chapter 6.

As agents, CSOs critical of biotechnology were particularly important in these interstate negotiations. Their level of participation was high even for an environmental treaty process, which already tends to have a more prominent role for CSOs than other international fora. CSOs were at times kept outside the meeting rooms – "when negotiations got sticky," as one activist put it – but these occasions were rare (interview #2). At the end of BSWG-5, one CSO representative stated in the plenary that "this has been the most open process in the international system for Civil Society participation" (quoted in Koester 2002, 55).

One activist interviewed believes that the prominence of CSOs resulted from the process initially being chaired by a Dane, Viet Koester, and the Danish tradition of strong public participation (interview #2). Koester, as seen in the majority report of UNEP Expert Panel IV (which he co-chaired), was an advocate, at least initially, for a strong protocol along the lines of those demanded by CSOs. Koester's recollections of the BSWG process indicate that he had to do considerable bridge building with some delegations to have CSOs accepted as valuable participants in the process (Koester 2002). In February 1999, the Biosafety Protocol negotiation process was passed on to a new chair, Juan Mayr, Colombia's minister of environment. Mayr had moved into politics from being an environmental activist in Colombia, and was thus also inclined to strong participation from environmental CSOs (interview #2).

Although the sympathy of the chairs of these meetings did likely play a role, the ideational authority of the CSOs was also important. Environmental activist groups are often presented in the regime literature as lobbyists, as actors who try to influence the position of nation-states (e.g., Betsill and Corell 2001). In this case, however, representatives of environmental CSOs were acting as much more than just lobbyists. In Gramscian terms, they functioned as organic intellectuals, especially within the coalition of states and organizations advocating the comprehensive position. Many state delegates turned to them for advice on environmental matters, and their expertise in this area gave these activists a discursive authority that delegates on the other side of the issue, as well as the BSWG chairs, had difficulty dismissing out of hand. Industry CSOs did not perform this role to the same extent because the biotech bloc already had organic intellectuals in the delegations of some of the most powerful countries at the negotiating table. The authority exhibited by activist CSOs in the biosafety talks is what Beck (1995) refers to when he states that the risk society produces increased opportunities for shaping contemporary society from below by social movements and counter-experts in particular.

The Canadian Position on a Biosafety Protocol: 1997-99

During the biosafety negotiations, industry representatives were fond of noting that while many of the delegates from the G-77/China appeared to come to the table with only environmental priorities, Canada was not driven by single interests. Instead, Canada represented a balanced perspective that came from the interdepartmental negotiation needed to develop a national position (interview #10). How balanced was this position, really?

In Canada, the Department of Foreign Affairs and International Trade (DFAIT) is responsible for managing and conducting all international negotiations and for signing agreements on behalf of the state. Although DFAIT is always a part of a negotiating delegation, the lead responsibility is often given to the federal department with expertise in the area of concern. In the environmental field, Environment Canada generally plays this role, as it did in negotiating the Convention on Biological Diversity in Rio. As a result, when the Biosafety Protocol negotiations started, it was John Herity of Environment Canada's Biodiversity Convention Office who chaired the process of developing a Canadian position.

Before each international meeting on biosafety, a Canadian position was formally determined through an interdepartmental consensus process involving a dozen (and sometimes more) federal agencies and departments. The key players were Environment Canada, Agriculture and Agri-Food Canada (AAFC), Industry Canada, DFAIT, Health Canada, and (after its creation in 1997) the Canadian Food Inspection Agency (CFIA). Formal consultations on the position involved the federal departments, provincial representatives, and stakeholders. Because Environment Canada was the lead agency, stakeholder groups included industry groups and environmental CSOs (coordinated through the Canadian Environmental Network, or CEN) and Aboriginal organizations (through the Assembly of First Nations).[4] Each of these groups participated on the Biosafety Protocol Advisory Group (BPAG), meeting regularly to give feedback on the government position. At least one member of each of the three stakeholder groups was also a participant in Canada's delegation to the talks, a practice that may be unique to Canada (interview #6). Before being finalized, a proposed position, along with "fallback" and "deep fallback" positions, would be reviewed by cabinet, which would have the final say on the position adopted. Final instructions to delegations would then be issued through DFAIT, based on cabinet directions when provided.

This description of the formal process through which Canada's position was developed gives little real insight into the forces that shaped it. This description also speaks only to formal mechanisms of influence by CSOs, while there were certainly informal ones as well. It is more useful to think about a country's position in the talks as an outcome of the balance of relations of force at any given moment in time. This approach accounts for material interests and capabilities, the participation of civil society, and organizational priorities established within institutions (domestic and international), as well as for ideational factors (such as prevailing domestic policy leanings, scientific developments, and so on). The simplicity of this proposed framework does not mean, however, that it is easy to predict how these factors will intersect. Consider Canada's position on biosafety up to the beginning of BSWG-2 in 1997.

By early 1997, Canada had firmly committed to the biotech bloc. It had a growing commercial biotech industry, and the discourse of equivalency was embedded in its domestic regulations. And, as one might expect, Canada's position on biosafety was organized by this ideological perspective. This is exemplified by the definitions of LMOs and biotechnology that Canada proposed at BSWG-2, which were intended to be consistent with its domestic emphasis on novelty and novel traits, and not on any specific method through which the trait was introduced (CBD 1997, 11). Embedded in the confidence of gene talk, the Canadian position was based on the assumption that GEO exporters such as Canada had the answers to the questions that many other countries were asking. As one Canadian delegate noted:

> We did not have the apprehension towards these products that much of the rest of the world had, because we had capacity to deal with it ... because we were ahead of much of the world in the science. (Interview #8)

As another Canadian official recalls, "It was a perspective driven by the view that we had the best regulatory system in the world" (interview #6). Given this confidence, it may seem surprising that Canada took a less aggressive stance than the United States up to the beginning of BSWG-2, as well as in the pre-BSWG phase. Before the BSWG process started, Canada's position could be characterized as that of a participant observer. At the time, for example, Canada even took a backseat on the issue of whether there should be a protocol (interviews #2, #8). In some of its initial positions in the BSWG process, Canada had an active interest in trying to understand and meet the concerns of the G-77/China. As Herity (1997) stated in a speech to the Canada Grains Council in 1997, "We are trying in this to avoid having Canada appear to be too much of a bulldog in pushing the concepts that we believe to be important for Canadian interests." To this end, Canada accepted that the protocol should be a means for guidance and technical support for capacity building in developing countries. On other issues of importance to countries of the South, Canada did not completely shut the door – at first. On the issue of liability, Canada's submissions to BSWG-2 states: "Canada believes that at this juncture the protocol should not have an article on liability and compensation, but will consider proposals from other delegations" (CBD 1997, 69). Similarly, on socio-economic considerations, Canada stated that it "reserves further comment on this issue until the meaning and significance of these considerations to other delegations are understood" (73). Furthermore, while Canada stated that the question of whether AIA should apply to all LMOs or just to a subset of them should be decided "early in the process," it did not actually take a position on this issue (22). The absence of a position was in stark contrast to the stance of the United States, which came to BSWG-2 vehemently opposed to an AIA procedure that would include all LMOs in its ambit.

Several factors may have contributed to Canada's flexible approach before BSWG-2, despite its deep position within the biotech bloc. In terms of material capabilities, when the biosafety talks began in 1996, most delegates from the North did not think that the prospective agreement would have a significant impact on trade in agricultural commodities (Falkner 2002). While the inclusion of such commodities may have been the intention of the African proposal, countries such as Canada began the negotiations under the assumption that the protocol would likely cover only LMOs intended to be introduced into the environment (such as seeds), and not LMOs intended for food or feed. Then, in the realm of ideas, multilateralism continued to hold a central position in Canada's foreign policy (interviews #1, #8). In the absence of any other specific policy directives, this approach guided Canada's responses to the biosafety proposals of other countries. Finally, in terms of organizational relations of force, Canada took its domestic regulatory system seriously. On the question of AIA, for example, all novel organisms were subject to notification and review under Canada's domestic regulations. This meant, as one BPAG participant from the CEN put it, that it would be "difficult to imagine how Canada could enter these negotiations proposing anything *less* than its own domestic framework" (CEN Biotechnology Caucus 1997, 6, emphasis in original). It would also appear that Environment Canada's lead on the biosafety file made a difference in the early stages of the BSWG process. During this period, Environment Canada, which had built a strong domestic and international reputation of being supportive of the CBD and other multilateral environmental agreements, took on a steering role in establishing a "flexible" position (interviews #11, #5).

By contrast with the lead-up to that meeting, beginning at BSWG-2, Canada's position became more closely aligned with that of industry groups and the United States. This occurred over two stages, each of which was related to shifts among the forces that determined Canada's position. The first stage took place in 1997, while the second coincided with Canada's participation in the Miami Group, beginning in 1998.

At BSWG-2, in early 1997, Canada's position was already shifting from what it had proposed before the meeting. One key example was the new Canadian position on AIA. At BSWG-2, Canada now argued, along with the United States, that AIA should be limited to situations where LMOs are being introduced into "centres of origin and diversity, are known to be infective, invasive, or pathogenic, or [where] there was insufficient information available to make an assessment of the organisms' likely effects" (Winfield 1997, 1). This position was based on the premise that it is only these LMOs that "may have an adverse effect on the conservation and sustainable use of biodiversity," the LMOs of concern under Article 19.3 of the CBD. It was, according to one activist, "the most restrictive proposal presented at the meeting, with the possible exception of the US" (1).

In an attempt to support this new stance, one Canadian background document notes that Canada could justify a limitation on the scope of AIA for LMO imports "if they are considered 'substantially equivalent' to an organism previously assessed" (Lewis 1997b, 19). Herity (1997, 5) suggested this approach in his Grains Council speech when he said that "where you have the same novel trait introduced into 30 or 40 different varieties, each one need not be individually assessed." This line of argument would have been compatible with the Canadian interpretation of equivalency in domestic regulations, but it was never pursued in the biosafety talks because it implied that the first of a set of similar varieties of grains, for example, would have to undergo AIA and a concomitant risk assessment. Canada now rejected this idea.

Instead, Canada adopted the same line of argument that the United States and industry were advancing. This argument held that only a restricted "positive list" of organisms should be subject to AIA, just like the lists of hazardous chemicals that were governed by the PIC procedure. This stance made sense in the context of international precedents, but it lacked consistency with the requirements of Canada's domestic regulations. Those regulations subjected all novel organisms to notification and review before they were approved for environmental introduction, food, or feed. It also showed a lack of willingness to take seriously many of the South's concerns about the possible impacts of LMOs. What had changed to create this new Canadian position?

In terms of material relations of force, 1997 finally saw growing recognition of the potential trade implications of a biosafety protocol, along with recognition that it would have implications on more than just GE crops. I do not have the figures for 1997, but in 1998, total Canadian agricultural exports were valued at $22.6 billion. GEOs were only about $840 million, or 4 percent of that total. But because GEOs were not segregated from non-GEOs of the same crop, about $2.8 billion of that year's exports would have been affected by trade restrictions on LMOs (Smith 2000, para. 6.86). These kinds of figures brought Canada, the United States, and Argentina, in particular, to reassess their positions in light of commercial stakes involved (Falkner 2002, 5).

In the organizational arena, a growing recognition of trade implications meant that pressure was mounting on Environment Canada from industry groups and departments aligned with industry in the Canadian government, such as the CFIA, to reorganize the process through which the Canadian position was developed. (The CFIA is Canada's chief regulator of the environmental impacts of novel organisms, but it also has a mandate to promote trade in Canadian food products). Environment Canada's first response, in 1997, was a decision to set up, alongside the BPAG of balanced stakeholder interests, a new "significantly sized body of expertise in ... agribusiness only to inform our process on an ongoing basis" (Herity 1997). Recognizing that Canada was in a unique position in these talks, as "probably the only developed country, in the near term, [likely] to become a party to the protocol which was actually carrying out a significant trade in LMOs," the Canadian delegation drew on this new group to help organize and present a round table on the realities of the global bulk-commodity distribution chain and how this could be affected by onerous biosafety regulations at BSWG-3 in the fall of 1997 (Anderson 2002, 237).

Another institutional change that occurred in 1997 was that one of the main Environment Canada participants involved in the process, Des Mahon, left the interdepartmental group when he was seconded to the CBD secretariat to help organize the biosafety negotiations. Mahon had been one of three central figures back in the early 1990s to design Canada's regulatory framework for novel organisms, so he was a powerful voice within the bureaucracy on these issues. One participant in the BPAG recalls that

> when [Mahon] was in the room, he was a pound-the-table kind of guy who would say, "Hey, Environment Canada is the one with the regulatory authority in this area, we are the national focal point [for the CBD]. We are here. This is our file, not DFAIT's, not CFIA's." (Interview #5)

Eventually, Mahon would become a key assistant to Koester, chair of the BSWG. In that position, he wielded considerable influence on the negotiations, including taking part in the writing of Koester's chair's text in Cartagena. Still, this took Mahon away from his role as a participant in the development of Canada's national position. As the BPAG participant quoted above noted, "Without Des there ... the environment and biodiversity interests in the development of the Canadian position were extremely weak" (interview #5). These interests were still represented by the CEN participants in the BPAG, but they were weakly represented within the government itself.

While the question of who actually championed an environmental perspective – and the ability of those individuals to influence the Canadian position on biosafety – is enormously relevant, it is important to recognize that discursive factors were also at work that made it increasingly difficult for an environmental voice to be heard within the context of the Canadian delegation in 1997. Most important, I would suggest, was the growing ability of the discourse of equivalency to discipline the GEO discussion in Canada. In the case of Canada's domestic regulatory system developed in the late 1980s and early 1990s, equivalency was clearly influential, but this was also balanced with real attention to the concerns of ecologists, as noted in

Chapter 3. By 1997, the equivalency of GEOs with non-GEOs, along with the notion that assessment needed to be based on sound science, trumped all other ways of framing the issue. One example of the discipline of these discourses is the way some Canadian officials came to view the question of socio-economic considerations in AIA decision making. Although trade concerns were among the main reason that Canada did not want to enter the discussion about the socio-economic impacts of LMOs, the argument put forward on this issue was usually based on the "facts" that LMOs are not unique or different in any way and that socio-economic impacts cannot be measured scientifically. This dynamic comes to the fore in the following example.

By early 1997, CEN representatives on the BPAG had presented a strong case that decision making based on socio-economic considerations could be supported as a fundamental tenet of democratic self-determination, and that this position could be construed as consistent with the principles of the CBD and with the GATT Agreement, Article XX, which states that products may be legitimately refused by a country based on *l'ordre public* (CEN Biotechnology Caucus 1997, 14-18). CEN representatives drew attention to the fact that both Canadian parliamentary committees that had studied the issue of biotechnology in the mid-1990s reached the conclusion that there was greater need for consideration of the social, economic, and ethical issues in decision making (e.g., Standing Committee 1996). These concerns had registered with key officials on the Canadian delegation, such as Herity, who explained them in some detail in his speech to the Canada Grains Council in 1997 (Herity 1997, 4). Nonetheless, the official line was the following:

> The problem that ... the Canadian position had with [socio-economic considerations] ... is that there is no way of measuring, there is no scientific truth associated with that ... Predictions of socio-economic changes were seen to be more controllable by the nation involved rather than uncontrollable, and therefore why speculate about what might happen if it is controllable? (Interview #8)

A moment later, the same official acknowledged:

> If you look at it from the perspective of developing countries, they knew that they could not control it. They knew that forces, or simply lack of capacity, would mean that these LMOs would be in their country, would be seen perhaps as attractive to farmers who lived there. They may or may not completely disrupt the routine of farming practice – particularly by peasant farmers who save seeds – who, under the constraints with multinational seed companies, wouldn't be able to save their own seeds. (Interview #8)

When pressed on why these concerns, which this official clearly understood, were not seen as legitimate in the Canadian position, the interviewee replied: "Well, the belief among those who develop GMOs is that *corn is corn*" (interview #8, emphasis mine).

Because "corn is corn," other concerns about LMOs, no matter how reasonable, can be put aside. This is the best example to arise from my interviews of the power of the discourse of equivalency in shaping Canada's position. This position was constituted within the learned codes of what had become common sense in the extended state and civil society complex of the biotech bloc. It functioned to place limits on agency, even when some of the state actors involved were clearly able to sympathize with those whose concerns would be deemed irrelevant by the way the problem was constructed. The discipline of this discourse meant that those members of Canada's delegation who took seriously the concerns of the South about the unique socio-economic implications of LMOs were effectively silenced in the formation of Canada's official position. Particularly troublesome in this example is that equivalency is raised in the context of socio-economic concerns. This move is remarkable given that it was actually in relation to questions of intellectual property rights and technology use agreements (both socio-economic issues) that the biotech bloc had previously trumpeted the uniqueness, rather than the equivalency, of GEOs.

By 1997, the power of the discourse of equivalency in the Canadian biosafety position enabled Canada to cooperate with other countries in pushing for a regime that was no different from those negotiated for other traded commodities through the WTO. Canada's cooperation in the formation of the Miami Group, beginning in 1998, was the second stage in the shift away from its "flexible" position of early 1997.

Canada and the Miami Group

The Miami Group had its inaugural meeting in July 1998 in Miami. The group's specific goal was to "bring potential impacts on trade to bear on the negotiations" (Enright 2002, 97). Its core participants were countries that had already made a strong commitment to biotechnology (the United States, Canada, and Argentina) or that expected to do so in the near future (Australia). In a clear example of the biotech bloc's successful hegemony at the domestic level, these states saw their national interests as being tied to the future export of GEOs, and they wanted minimal interference in such exports.

In an effort to show that a trade-oriented position did not necessarily come from GEO exporters only, the initial four partners in the Miami Group worked to build allies for their position, courting Uruguay and Chile (Argentina's closest trading partners) in particular. The establishment of the alliances with Uruguay and Chile is an instructive example of the organizational

efforts of the biotech bloc. Canadian biotech experts, including nutritionists initially trained in the art of selling the safety of agbiotech by the
Monsanto-sponsored Dietitians Network, were sent to Chile and Uruguay
to run information sessions for their government representatives (L. Stewart
2002b; BIOTECanada 1999). Their costs on these trips were covered by Canada's International Development Research Council and the Canadian International Development Agency, until these agencies realized they could be
criticized for supporting such blatant efforts to promote biotechnology in
the name of international development (Abley 2000). New Zealand, Brazil,
Thailand, and Mexico were each also invited to join the Miami Group. While
some of these countries did sit in on meetings, none actually joined. Mexico
had recently put a moratorium in place on GE corn because of worries that
it could affect indigenous varieties of teosinte, so it wanted a protocol that
would support this action, which it saw as vulnerable to a trade dispute
under the North American Free Trade Agreement (NAFTA) or the WTO
(Galvez 2002, 207-9; interview #14).

The six Miami Group countries began working from a common position
at BSWG-5 in August 1998, though they identified themselves as a formal
negotiating group only when the format for negotiations shifted at Cartagena
in February 1999 (Ballhorn 2002). From its inception, Canada chaired the
group and Canada's spokesperson became the Miami Group's chief spokesperson when formal negotiations started at Cartagena.

In the protocol negotiations, the Miami Group, along with industry organizations, argued for the least trade-restrictive measures possible. The
group's underlying position was that the protocol should be fully consistent with other international instruments and with WTO agreements in particular. Although formally operating by consensus, this procedure had its
limits when it came to trade disciplines. As the Canadian chair of the group
noted, "The consensus method of operating resulted both in moving group
positions towards the positions of the other negotiating groups on some
issues, and in *holding the line on issues of fundamental importance*" (Ballhorn
2002, 110, emphasis mine). These issues of fundamental importance were
the trade and commodity-related articles. In this area, the Miami Group
position contained three paramount demands: (1) a "savings clause," which
would ensure that the protocol would not supersede other international
agreements; (2) LMO decision making based on sound science, not the precautionary principle and/or socio-economic considerations; and (3) the exclusion of LMO commodities from the AIA procedure.

The Miami Group's first demand, that the operative part of the text include a savings clause, became a key issue at BSWG-5. This was seen as im-
perative because Article 30.3 of the Vienna Convention on the Law of Treaties
states that more recent treaties on a given subject matter take precedence

over previous treaties, and the "earlier treaty applies only to the extent that its provisions are compatible with those of the latter treaty" (UN 1969, Article 30.3). As a result, the Cartagena Protocol would take priority over WTO agreements relevant to commodity trade, labelling, and so on.[5] Rather than accept such a situation, the Miami Group sought an article that would, as one Canadian official put it, "record the understanding of negotiators that what they did here is not intended to change rights and obligations under trade and other regimes" (interview #12).

From the perspective of global environmental politics, this demand signifies an important development on two levels: it illustrates the growing strength of neo-liberals inside the state and civil society complexes of the Miami Group to dictate the terms of an international environmental agreement; it also shows how simply conformity with the international trade regime could be achieved. A savings clause could effectively annul any decision made in the biosafety talks that would run counter to the WTO, regardless of whether or not such a move could result in threats to the environment or public health.

The Miami Group's second key demand was that there be no operative language in the protocol allowing countries to restrict imports of LMOs based on the precautionary principle. For Canada, this stance was a significant departure from earlier positions. One Canadian official noted that Canada had started talking about precaution in the 1970s and 1980s vis-à-vis the United States and acid rain (interview #8). A CSO participant in the international negotiations reiterated this point, stating that around the world, "precaution is the oldest public health approach. That is how you can justify quarantine" (interview #2). Because of its history with the precautionary principle, in January 1997, Canada had proposed that the precautionary principle, as found in the CBD, be placed in an operative, "principles" section of the protocol. Such a section, the lawyer for Environment Canada on the Canadian delegation wrote, is a "fairly new development in environmental conventions/treaties" in that it has features of a preamble, by consisting of general statements of obligation, yet, unlike preambular text (which is only referred to for interpretation), this section is "part of the operative portion of a treaty" (Daniel 1997, 1).

Given this willingness to consider the precautionary principle in the "operative portion of the treaty" in early 1997, Canada had shifted considerably by the Cartagena meeting of February 1999. At that meeting, Canada, on behalf of the Miami Group, was arguing vociferously that precaution should be understood only as an "approach" and not a "principle" of international law, and that precaution had no place in operative sections of the Biosafety Protocol (Chasek 1999, 3). The Miami Group did not argue against the need for *any* precaution vis-à-vis LMOs, but, in an internalization of this discourse, the group argued that the protocol itself, with its basis in science-based risk assessment, embodied the precautionary approach called for in Principle 15 of the Rio Declaration (Chasek 1999).

Why this shift against an operative precautionary principle, especially in the Canadian position? One Canadian government source believed that "some of the difficulties lay in ... not having a unified position [among government departments] ... at that particular time, on the precautionary principle" (interview #6). While this may be true, it is also true that no efforts were made within the Canadian delegation to develop a common perspective on the uses of precaution in Canadian regulations and legislation (until after the protocol negotiations were completed), nor were such efforts made in the Miami Group. More important, I would suggest, were two events that were external to the biosafety talks. The first was the growing resistance to the introduction of new GEOs in Europe in 1998 and 1999. This caused the United States and Canada to worry that Europe might try to justify a moratorium on GEOs on the basis of precaution, despite a lack of specific scientific evidence that these organisms posed a significant health or environmental risk. The second and closely related event was the WTO dispute between Europe, on the one hand, and Canada and the United States on the other, over the use of growth hormones in beef production.

The EU had justified a ban on hormones in beef imports since 1981, on the basis of their potential risks to human health. The United States and Canada took action on this issue in the mid-1990s. In 1995, the United States worked through the Codex Alimentarius Commission to have international standards adopted that accept residual levels of certain hormones in meat. (The Codex Alimentarius Commission is a joint body of the UN Food and Agricultural Organization and the World Health Organization that develops international standards for food safety and facilitates intergovernmental coordination of food-related policies. It is recognized as the international authority in this sphere by the WTO.) Then, in 1996, following heavy lobbying efforts from Monsanto, the United States lodged a complaint with the WTO over the hormones issue. Canada lodged a complaint the same year.

A WTO dispute resolution panel ruled against the EU in August 1997, stating that the EU's ban was not WTO-compliant because it was not based on international standards, as the Agreement on Technical Barriers to Trade expects (referring to the new Codex standards), nor was it in line with the SPS Agreement, which expects concrete evidence of harm in order to justify an import ban. The panel agreed that the EU could have higher standards than the international standard of Codex, but that such standards have to be scientifically justifiable (WTO 1998a). It noted that although the EU provided studies demonstrating the health risks of the hormones themselves, these studies were inadmissible, as the issue was about the specific risks of hormone *residues* in foods. Based on the evidence provided, the panel ruled that the EU ban was not based on sufficient scientific evidence.

The EU appealed this decision in late 1997, referring to the precautionary principle, which it argued was an accepted general customary rule of international law applying not only to the management of risks but also to their assessment. The EU position stated that provisions of the SPS Agreement dealing with precaution, including Article 5.7 on provisional measures, "do not exhaust the applicability of the principle." Article 5.7 allows countries to adopt SPS measures "in cases where relevant scientific evidence is insufficient," provided that they "seek to obtain the additional information necessary for a more objective assessment of risk and review the measure accordingly, within a reasonable period of time" (WTO 1994a, 17).

EU lawyers argued that "these provisions do not prevent Members from being cautious when setting health standards in the face of conflicting scientific information and uncertainty" (WTO 1998a, 17). Discursively, this position frames precaution as a political judgment. It emphasizes the notion that, in cases of scientific uncertainty or a lack of scientific consensus, politicians must have the freedom to take measures that they deem suitable to minimize unacceptable harms, because it is ultimately they, rather than scientists, who are accountable to citizens for risk decisions.

The EU lost its appeal, and invocation of the precautionary principle did not help its case. While language reflecting the precautionary principle is found in the SPS Agreement, it is not specifically mentioned by name. In its ruling, the Appellate Body stated:

> The status of the precautionary principle in international law continues to be the subject of debate ... We consider it is unnecessary, and probably imprudent, for the Appellate Body in this appeal to take a position on this important, but abstract question. (WTO 1998a, 17)

Despite being ruled against twice, the EU did not agree to let hormone-tainted beef onto its territory. Instead, it said it would undertake further studies to demonstrate the risks of hormones in a way that would satisfy the SPS Agreement. The US and Canadian response was to impose trade sanctions (FOEI 2001).

That this case occurred at the time when negotiations were starting on the Biosafety Protocol was of enormous significance to the biosafety talks. The outcome of the hormone dispute case, in 1998, was a debate about sound science versus precaution in the context of a discourse focused on human health risks. The US State Department and Canada's Department of Foreign Affairs and International Trade in particular were determined to prevent an international agreement from containing language that would enable countries to take decisions against the import of any product without strong scientific evidence of harm. Conversely, this same case would lead the Europeans, by late 1999, to make the establishment of a precedent allowing for precaution central to their negotiating stance. These poles within the risk discourse laid the foundation for the latter stages of the overall biosafety debate.

Aside from shaping the discussion on precautionary decision making (and on the savings clause, as I discuss in the next chapter), the sound science–precaution debate functioned to continue to restrict discussion on socio-economic considerations. On this subject, as one environmental CSO delegate noted, "there was just a position of total rejection ... There was no receptivity [within the Miami Group]" (interview #5). Significantly, this rejection came from the Miami Group *and* the EU, because both sides in the sound science–precaution debate accepted the basic tenet that decision making should be based on technical risk parameters, whether actual or potential. In this context, non-technical considerations had no bearing. The strength of a perceived division between technical and non-technical concerns is revealed in the following statement by a Canadian delegate on the issue of socio-economic considerations:

> When the protocol negotiations were launched, it was the only game in town. So when there were concerns about intellectual property, impacts on social structures in developing countries, lack of [regulatory] capacity, consumer right-to-know issues, etc. ... there were attempts to import them – in our view, inappropriately – to overload the protocol with issues and objectives that go well beyond the protection of biodiversity. (Interview #13)

This official's statement makes sense if one separates the environmental and health risk issues from the many other concerns raised by the adoption of LMOs in agriculture. It is important to recall, however, that the CBD was conceptualized as having three interrelated goals: conservation, sustainable use, *and* equitable sharing of benefits. A focus on risk issues, narrowly understood, does not necessarily address the equity dimension, and, as a result, would not necessarily meet the goals of conservation and sustainable use either. This broader conceptualization of the issues was lost, however, in a debate that was now fully defined by a sound science–precaution polarity.

The Miami Group's final key demand was the total exclusion of bulk shipments of LMOs intended for food use (and thus not environmental introduction) from the AIA procedure. By the end of the Cartagena round of negotiations, these bulk shipments, which represented the vast majority of

traded LMOs, became known as LMOs intended for food, feed, or processing, or LMO-FFPs (the term was first used in a version of Koester's chair's
text at BSWG-6 [CBD 2003, 41]). In a related move, the Miami Group also
sought to exclude products of LMOs (such as corn oil) and LMOs destined
for contained use (for research purposes or field trials). This position was
clearly driven by the material interests of the biotech bloc to avoid a protocol that could disrupt bulk-commodity shipments. By 1998, these shipments
already included a small but growing percentage of LMO-FFPs.

Despite the Miami Group's adherence to sound science on other fronts,
the effort to exclude bulk commodities was not as easy to substantiate scientifically, and Miami Group delegates were aware of these limitations. A
Canadian background document from 1997 drew attention to a comprehensive 1993 report from the US Office of Technology Assessment on harmful
non-indigenous species in the United States. This report

> indicates that unintentional introduction of non-indigenous species (non-
> modified) such as contaminants in packing materials, *bulk commodities*, and
> ships' ballast have had very significant economic and environmental im-
> pacts in the U.S. To date, these unintentional introductions have generally
> caused more harm than *intentional introductions* of non-indigenous species.
> (Edge and Forsyth 1997, emphasis mine)

Another Canadian document from 1998 notes that exempting LMO commodities from AIA "misses out LMOs that are inadvertently released (spill-
age, escapes, stolen material, i.e. low probability of exposure to the
environment) but could still be detrimental to biodiversity" (Edge 1998, 8).
These documents show that Miami Group delegates were aware that the
group's call to exclude bulk commodities was made on trade grounds and
was scientifically questionable.

Canada's decision to participate in the Miami Group was clearly made at
the highest political levels, and it was based on trade concerns that extended well beyond the protocol itself. How did this impact on the formal
process in place to determine Canada's position? The decision to get involved with the Miami Group resulted in a considerable increase in participation within the interdepartmental working group from DFAIT, and the
participation of AAFC and CFIA trade lawyers, beginning in early 1998.
One BPAG member recalls how, in January 1998, "suddenly there were a
whole bunch of trade lawyers in the room" (interview #11). Another BPAG
participant states:

> The way that they hijacked the Canadian position was really quite some-
> thing ... The trade people came in and said: "Look, the Europeans are trying

to escape WTO disciplines, they are trying to roll back the SPS Agreement."
(Interview #5)

By linking the protocol to larger trans-Atlantic trade disputes, these trade
lawyers were able to dominate the Canadian position starting in 1998.

In addition to an increased physical presence in the interdepartmental
committee of trade interests within the Canadian government, 1998 also saw
Richard Ballhorn of DFAIT become the co-chair of the Canadian interdepartmental working group (along with Herity of Environment Canada). At
intergovernmental meetings, it was Ballhorn who became the spokesperson
for Canada, and, later, for the Miami Group of LMO-exporting countries. As
co-chair, DFAIT took a very different approach from Environment Canada.
When Environment Canada chaired the interdepartmental group, according
to one CSO participant, it "actually went in as intra-governmental broker
... as a neutral forum for the discussions" (interview #5). DFAIT, on the
other hand, strongly pushed its trade agenda. The result of this different
style, according to this same participant, was that "the environmental and
biodiversity interests in the development of Canada's position were very,
very weakly represented within the federal government" (interview #5). This
opinion was echoed by a government official, who believed that because
Environment Canada co-chaired the group, "its voice perhaps wasn't as
strong as it might be" (interview #6).

Eventually, beginning at BSWG-4, Herity of Environment Canada stopped
participating in delegation deliberations at the BSWG meetings in order to
take on the role of chairing a working group for the Secretariat of the
Convention on Biological Diversity (interviews #11, #8). The absence of
Mahon, and then Herity, left environmental CSOs to bring forward the environmental position on the BPAG. CEN representatives such as Michelle
Swenarchuk of the Canadian Environmental Law Association, Mark Winfield
of the Canadian Institute for Environmental Law and Policy, and Raphaël
Thierrin of the Canadian Organic Growers took this role seriously, but they
were constrained by a climate in which the ideas of equivalency and sound
science were hegemonic and trade considerations dominated.

In the Canadian delegation, the differences between a flexible, more environmentally oriented position and one dominated by trade concerns led
to significant tensions (interview #6). An Office of the Auditor General (OAG)
report, based on interviews done in 1999 and 2000, states that "participants
[from the federal agencies involved] described the interdepartmental process as long, difficult, and tense" (Smith 2000, para. 6.93). Observers from
other countries also noted the "tensions and divisions" within the Canadian delegation (interview #9). The OAG interviews reveal that one key problem with the interdepartmental process was that there was little incentive
to adopt a broader, corporate perspective on the issues at hand, as opposed

to holding onto narrow departmental views (Smith 2000, para. 6.93). Interdepartmental conflicts became significant enough that "central agencies facilitated the final agreements that were reached before the Cartagena and Montreal meetings, in particular by involving senior political levels" (Smith 2000, para. 6.95). This is significant given that the central agencies (the Prime Minister's Office, Privy Council Office, and Federal Treasury in particular), as well as the biotech coordination committee at the deputy ministerial level, had been the major bodies to push the biotech agenda within the Canadian government in the early 1990s (Kuyek 2002). Was it the central agencies that called for a "no compromise" position at Cartagena? Or did they facilitate the more accommodative position that the Canadian delegation took to Montreal in 2000? It is difficult to answer these questions conclusively, but it is possible to answer yes to each of these questions – despite their apparent contradictions – given the implications for protocol negotiations of the larger political struggles over biotechnology at that time, as the next chapter examines.

Conclusion

Before 1997, Canada's position was established in relation to its domestic regulatory structures and with a view to being open to the concerns of the global South. However, beginning at BSWG-2, and then through the formation of the Miami Group in 1998, a narrow perspective on Canada's trade interests clearly gained the upper hand in shaping Canada's agenda in the biosafety talks. One environmental activist called this a "stupid interpretation of Canada's interests as an exporter ... a block-headed version of that ... [because] it assumed that exporters can bludgeon their way into [foreign] markets using trade disciplines" (interview #5). The discursive authority of science was important to have onside (and this was true for all positions in these debates), but the Canadian position on biosafety was not constrained by the need to be scientifically grounded. When the science may have been in conflict with a $3-billion commodity trade, as it appears to have been in the determination of a position on LMO-FFPs, bulk-commodity trade interests set the agenda. The reality of material capabilities clearly played an important role here. Canada did not have a segregated stream for GEOs. This meant, in 1998, that it was the entire soy, corn, and canola commodity streams that were at stake and not just the portions of these streams that were GE. Had Canada been forced to adopt a labelling policy, and concomitant segregation (as was done in Europe), the politics of Canada's biosafety position would have played out differently.

Several Canadian delegates (within government as well as CSO representatives; particularly interviews #5 and #6) suggested that interdepartmental organization, and the actions of some of the key actors involved, may have made a difference to policy outcomes in Canada. The evidence presented here suggests, however, that the larger discursive structures of neo-liberalism, equivalency, and now sound science, and the immediacy of the beef hormone dispute and the EU's growing reluctance to accept GEO imports, were much more important. Because of these factors, Canada and the Miami Group were inflexible on many of the key issues of the Like-Minded Group and the Europeans at Cartagena in 1999, and inflexibility on the commodities issue, in particular, was largely responsible for the collapse of the talks in Cartagena.

6
A Precautionary Protocol

Given the divergences among positions on biosafety just described, it is difficult to see how a single text could emerge from the biosafety protocol talks. Nonetheless, by the early morning hours of 29 January 2000, negotiations on a protocol were complete, and the outcome was seen as a surprising victory for many of the environmental CSOs involved (Dawkins 2000a; interview #2). From the perspective of these organizations, the protocol takes progressive steps in several areas. It accords nations the right to give explicit consent on LMOs intended for introduction into the environment (such as seeds), and it affirms their right to reject those imports, as well as bulk commodities containing LMOs (LMO-FFPs), on the basis of their risks or on the basis of precaution when those risks are uncertain. What made these outcomes possible in Montreal in 2000, given the Miami Group's intransigence on many of these issues in Cartagena a year earlier?

Numerous material, organizational, and ideational factors (including the three discursive moments introduced in Chapter 4) came together in 1999 to allow one final shift in the biosafety debate: precaution came to supplant risk as the overarching discourse guiding the biosafety discussion and its outcomes in several critical areas. In Gramscian terms, we might say that the ideology of precaution had become hegemonic: precaution was now the common sense guiding the talks. But this formulation does not go far enough in describing the productive effects of the precautionary perspective in the talks, which is why I deliberately employ the notion of discourse here. In Foucauldian terms, precaution moved from being a discourse of resistance, in opposition to the equivalency–sound science position within the framework of risk, to becoming the discourse of normalization in the field of biosafety, establishing the boundaries of truth in the final stages of debate. As a discourse of normalization, precaution pulled the rug from under the equivalency argument, leading those who held this position to frame their interests within the new context. The result was a debate that was very different from conversations that had taken place in Cartagena, one that would establish clear rules of precautionary decision making. As a discourse of normalization, the Cartagena discourse of precaution also disciplined those who brought forward the notion of precaution in the first place (as part of the comprehensive assessment position) but whose broader concerns still did not fit into the new discursive framework.

What do I mean by this shift to a discourse of precaution, and what was its significance? Just like the discourse of risk had supplanted that of hazard in the early 1980s in the United States and Canada, in the final round of the biosafety talks, actors were now shaping their arguments in the context of the reasonableness of a precautionary response to GEOs. They were also revisiting their interests in the new context. Within the risk discourse, a claim that LMO imports could be stopped at the border without conclusive scientific evidence of harm was seen as an unwarranted trade restriction, out of proportion with the minimal risks of GE. Once precaution came to define the discursive terrain, a position calling for such restraint made perfect sense as long as specific conditions were met.

It is important to emphasize that the precautionary discourse emerged in the context of the discourse of risk, as previous chapters have demonstrated. As such, a precautionary framing of GEOs is not necessarily antagonistic to the risk framing. In fact, it is critical to see that these two ways of constructing the GEO problem are closely linked, with the latter providing a foothold for the former. Still, as the discursive lens of precaution gained in strength, we can identify major shifts in how policy issues were perceived by the main actors at the negotiating table, foreclosing certain options while enabling others.

In many ways, a conversation organized through the discourse of precaution brought the GEO discussion back to some of the assumptions of the hazard discourse that guided the field of GE safety in the mid-1970s: the implicit cost-benefit calculation had been reversed once again; living GE plants were recognized as inherently risky, rather than presumed to be at least as safe as conventionally bred cultivars. Furthermore, while the potential harms of these GEOs could be calculated in some instances, it was again accepted that there were complexities involved that meant this calculation was no certain exercise. It was also now accepted that risk management would not necessarily deal with all the risks of GEOs, leaving open the possibility that a better strategy might be risk avoidance and the adoption of alternatives. Finally, the discursive framework of precaution once again prioritized consumer risks and benefits over those that befall producers of GEOs.

As Litfin (1995, 255) demonstrates in her analysis of the international debate over the ozone layer that had taken place a decade earlier, the shift to a precautionary discourse in the Biosafety Protocol talks also came largely from the outside. And, as in the case of the ozone talks, the shift to precaution was not based on conclusive scientific evidence. In many ways, this shift was a response to critics of GE raising concerns about potential dangers and uncertainties of, and grounds for suspicion for, the products of GE. The real determinant for the uptake of a discourse is not its objective truth value. Rather, such a shift is a function of the willingness of a growing network of actors to accept, internalize, and reproduce the discourse in conversation and day-to-day practices *as* truth. In the case of the discourse of precaution, this network of actors initially arose in the context of developments in Europe. The network comprised anti-GE activists, distrustful consumers, skeptical scientists, a sensationalist media, and European politicians and civil servants wary of the possibility of another trans-Atlantic trade dispute. In the context of the biosafety talks, this network first involved activist CSOs and their allies. Eventually, however, those working within the precautionary discourse included actors from all the negotiating groups, including delegates from the Miami Group bent on achieving a compromise between LMO exporters and importers that they could live with.

The focus of this chapter is on the factors that enable a discursive shift to occur in international negotiations, and on how this changes the field of struggle within which actors mobilize their interests. That the ascendancy of the discourse of precaution in the final stages of the biosafety talks took place can be seen first by zeroing in on the signs that pointed to the possibility of this shift at Cartagena, and then by looking at evidence from the final agreement reached in Montreal. Reinforcing the point that a discursive shift is never only discursive, I then discuss the material, institutional, and ideational factors that made this final shift possible, and analyse some of its productive effects in the context of the negotiations. Let's begin, though, with a brief chronological overview of the final stages of biosafety negotiations.

The Process: BSWG-6 to ExCOP

True negotiations on a biosafety protocol text never really got underway until BSWG-6 in Cartagena, Colombia. When they began on Sunday, 14 February 1999, negotiations were meant to be finalized within a week. This was to be followed by a short Extraordinary Meeting of the Conference of the Parties to the CBD (ExCOP) to adopt the Biosafety Protocol.

BSWG-6 began with a plenary, involving more trhan six hundred participants from 138 national delegations, as well as CSOs, intergovernmental organizations, and companies (Chasek 1999, 1). The plenary was followed by intense concurrent negotiations within two sub-working groups, two contact groups, and a legal drafting group. These groups were asked to arrive at a consensus on key issues by the Wednesday, but this goal was not achieved. At that point, the chair of the meeting, Viet Koester from Denmark, after extensive consultation with the chairs of each of the concurrent negotiations as well as with a Friends of the Chair group (represented by one person nominated from each regional group), produced a chair's text that attempted to find compromises among the stances of the major negotiating blocs (Chasek 1999, 2-3). Koester tabled this draft protocol text at a plenary on the Thursday, stating that any changes to it could be made only by consensus. At that point, further negotiations continued among only a very small group, the Friends of the Minister, convened at the initiative of the Colombian minister of environment Juan Mayr (who was to become chair of the ExCOP).

The convening of the Friends of the Minister was the first instance in these talks when nations negotiated through interest groups rather than through regional groups. The Friends of the Minister included two representatives from each of the five negotiating groups that had emerged (the Miami Group, the Like-Minded Group, the EU, the Central and Eastern Europeans [CEE], and the Compromise Group) (Samper 2002, 64-65). It met steadily over the final days of the BSWG-6 to try to reach a final compromise but was again unable to reach agreement on whether products of LMOs ("products thereof") and bulk commodities intended for food, feed, or processing (LMO-FFPs) were to be included in the protocol, the issue of socio-economic considerations, the role of the precautionary principle, and the savings clause.

At the close of BSWG-6, on Monday, 21 February, Koester decided to gavel his draft through – a decision he regrets in hindsight for its lack of diplomatic tact (Koester 2002, 58). This prompted more than fifty interventions from states and negotiating groups listing problems they had with the draft protocol. These statements also drew attention to the lack of transparency in the negotiation process that had taken place over the latter part of the week and the weekend. As a result, when the ExCOP was convened on the Monday afternoon, there was still no consensus on the protocol text. Still, the Koester text, as I refer to it, served as the basis for all ensuing negotiations.

Taking into account the concerns of many countries about the negotiation process during BSWG-6, ExCOP chair Juan Mayr reorganized the format to ensure greater transparency. Negotiations would continue among a group of ten representatives, one each from the EU, the CEE countries, and the Compromise Group, two from the Miami Group (one from the North and one from the South), and five representatives from the Like-Minded Group. These negotiations did not occur behind closed doors. Instead, they took place in a large room where all other delegations could be present to observe the proceedings, though CSO representatives were not privy to the talks. Early on the Wednesday morning, after another thirty-six hours of laborious negotiations, the EU proposed a new protocol text in an effort to reach a last-ditch compromise that the Miami Group would accept. This was the text that, among other changes, took out the references to a "lack of full scientific certainty or consensus" as a sufficient reason for prohibiting the import of LMOs (Chasek 1999, 11).

Despite that most of the groups still had concerns with the EU proposal, one by one, the spokespersons for negotiating groups stated that they were willing to accept the compromise proposed. As one negotiator said, "For the first time we thought there was a chance of reaching an agreement that night, and the tension in the room and pressure on the Miami Group to accept the proposal grew with every minute" (Samper 2002, 67-68). After four other groups, representing a total of 140 countries, indicated their willingness to accept the EU proposal, the Miami Group stated that it could not accept it (Falkner 2002, 17-18). Although it had a list of problems with the proposal, the Miami Group was particularly concerned with the EU's suggestion that the question of whether LMO-FFPs should be subject to a full AIA procedure be put off until the first Conference of the Parties serving as the Meeting of the Parties to the Protocol (COP-MOP-1). Following its rejection of the European proposal, Canada made a motion to suspend the ExCOP. At this point, as one American negotiator recalls, Canada "was jeered by representatives of well over one hundred countries" (Enright 2002, 101). Still, recognizing that no further progress could be made in Cartagena, it was agreed by consensus that the plenary would be suspended to reconvene no later than COP-5, in May 2000.

Following Cartagena, Mayr convened three exploratory meetings to try to break through the Cartagena impasse. These were not formal negotiations; they were intended, first, to gauge the political will of nations to continue the negotiation process, and, second, to work informally to bridge the remaining differences among the groups. The initial meeting took place in Montreal in July 1999, the second in Vienna in September, and a third immediately before negotiations resumed in Montreal in January 2000. In Vienna, Mayr instituted a new process designed to increase the transparency of the discussion format. The "Vienna setting" brought two representatives from each of the five groups to the table, along with the chair. These meetings were open to all delegations, as well as CSO observers and the media, thanks to considerable lobbying on the part of environmental CSOs after the closed-door Montreal meetings. In addition to serving as a format for these informal discussions, the Vienna setting would become the negotiating format for the resumed ExCOP in January.

Another informal but important event took place in September 1999. On the invitation of the chair of the Like-Minded Group, Egziabher of Ethiopia, several Miami Group delegates visited his country. African biosafety delegates used the opportunity to demonstrate to representatives of LMO-exporting countries that most farmers in Africa rely on traditional landraces for their subsistence, and to impress upon them how LMOs, whether intended as seeds or as commodities, could disrupt traditional agricultural practices. Several key Miami Group delegates would later comment that this trip had been an important moment in the building of understanding between the Like-Minded Group and the Miami Group (interview #8; Enright 2002; Ballhorn 2002). In the conclusion to his reflections on the negotiation process, John Herity of Environment Canada thanked "Ethiopia's Tewolde [Egziabher] for showing and teaching so many of us what it was really all about in the first place" (Herity 2002, 350).

When negotiations resumed in Montreal on Monday, 24 January 2000, significant progress had been made on key issues. In particular, a Canadian suggestion to have LMO-FFPs dealt with through a procedure separate from AIA had been floated before the first informal consultations and then formally proposed at Vienna (Winfield 1999). The other negotiating blocs were beginning to converge around a compromise that would accept this proposal as long as states would still be able to prohibit the import of these LMO commodities. This compromise, discussed below, was critical to the Biosafety Protocol's conclusion. However, there were still significant differences among groups on many issues, including the use of precaution to block imports, the relationship between the protocol and trade agreements (the savings clause), and the question of the documentation requirements for LMOs. Negotiations took place in working groups and contact groups on these issues through the week, but by the time the ExCOP was to end, on the Friday, they were not yet resolved.

This impasse was broken by deft diplomacy on the part of Mayr. He threatened to table a new chair's text on the savings clause, precaution, and the commodities issue, which he would then take to a vote in plenary, if no agreement could be found among the negotiating groups themselves (Enright 2002, 104). This approach was apparently taken on the advice of the group of ten European ministers of the environment and the European environmental commissioner who were present at the meeting (Bail, Decaestecker, and Jorgensen 2002, 184). This threat increased the pressure on the Miami Group in particular to agree to a compromise in informal talks on the Friday afternoon and evening. The Miami Group had concerns with many aspects of the text that Mayr was developing, but it decided to focus its efforts on what it saw to be the biggest problem: the issue of the documentation that would accompany shipments of LMO-FFPs. Although seemingly minor, the question of documentation was critical to the Miami Group and the wider biotech bloc because it was closely related to wider debates about the labelling of biotech products, and because of the segregated commodity streams that stringent documentation requirements could result in.

A compromise on the documentation issue was finally reached around three in the morning of Saturday, 29 January. Canada, according to reports from participants in these ministerial-level backroom talks, was the last country to assent to an EU-Miami Group compromise on this issue (interview #9; Bodegard 2002). With no time to revisit other concerns the Miami Group (or others) had with the protocol text, it was brought before a plenary and adopted unanimously within the hour (interview #12).

Early Signs of Precaution

This overview of the final stages of the biosafety talks places a high degree of importance on the process and the persons involved. Several people did have a considerable impact. Without the Vienna setting, Egziabher's invitation for the Miami Group representatives to come to Africa, or Mayr's threat to table a chair's text, progressive outcomes may not have been realized within one year of the Cartagena collapse. However, I believe the critical factor that led to the Miami Group's compromises on several of its key issues in Montreal was a shift by early 2000 in the discursive context of the biosafety talks from the framework of risk to that of precaution. The first sign that the logic of precaution was indeed gaining adherents in the biosafety talks was the entrenchment, in the final protocol text, of the three discursive moments discussed in Chapter 4. These moments – the acceptance of biosafety as a distinct regulatory field; a focus on GEOs rather than on a wider class of novel organisms; and acceptance of the need for legally binding rules, rather than voluntary guidelines – laid the groundwork for the larger discursive turn discussed here, though they still did not make it inevitable. The two examples from the final protocol text that best demonstrate that this groundwork had been laid are the rigorous AIA procedure and the definition of LMOs.

Advance Informed Agreement

The concept of an advance informed agreement procedure already existed in Article 19.3 of the CBD. Still, what it would look like in practice was far from decided. During negotiations, the G-77/China (later the Like-Minded Group), along with their activist CSO allies, sought an AIA procedure that would cover all classes of LMOs, and they had managed to gain the sympathy of many European countries on this issue. On the other side of the debate, the Miami Group wanted to restrict the scope of AIA, while Germany argued that the procedure should be based on implicit consent. In the end, the Like-Minded Group basically won this debate. Once Germany acquiesced to the intra-EU consensus to support the Like-Minded Group's demand for *explicit* consent, the Miami Group was brought around at Cartagena to accepting the principle that there would be no movement of LMOs subject to AIA without consent.

Strategically, the Miami Group appears to have been willing to make this concession in exchange for having the EU and others recognize that the largest class of traded LMOs (by volume), bulk-commodity shipments destined for food, feed, or processing, would be beholden to a set of import procedures separate from AIA. Separate procedures for LMO-FFPs, the Miami Group continued to argue, should not require explicit consent (Steffenhagen 2001, 55; Winfield 1999).[1] Still, that this group agreed to AIA for all other LMOs was already a major concession in itself. This concession can be explained only by the little weight that the minimal risk position – the ideology that saw LMOs as basically equivalent to non-LMOs, presenting no inherent new risks – carried in the field of biosafety by 1999. The demise of this position is glaringly apparent if we consider the differences between AIA, as found in the protocol, and previous iterations of the prior informed consent (PIC) procedure in international environmental law.

The PIC procedure is applicable only to products already known to present a significant risk. To be subject to this procedure under the PIC treaty of 1998, a chemical or pesticide must be banned or severely restricted for health or environmental reasons in at least two countries in different regions. Furthermore, information must be compiled on specific incidents of damage in developing countries, the adverse effects, and the way in which the formulation was used (PANNA 1998). These criteria mean that the list of substances subject to the PIC procedure is limited. Under the Cartagena Protocol, all LMOs intended for direct introduction into the environment are subject to AIA. This reveals a major change in the internationally accepted understanding of when the AIA/PIC procedure is appropriate, demonstrating a shift towards a presumption of potential harm, rather than a presumption of harmlessness.

One would expect this presumption to be a major concern for LMO exporters, and it certainly was in the early stages of the talks. By Cartagena, however, the presumption that LMOs could not be assumed to be harmless had become the norm. The chair of the working group that negotiated AIA at Cartagena admits that "negotiation of most aspects of the AIA procedures themselves were not especially contentious" (Schoonejans 2002, 304). These negotiations were not contentious because, even by the Cartagena round, the Miami Group was no longer willing to fight on this issue. In the end, while the LMO-FFP debate was not resolved at Cartagena, negotiations on the main procedures for AIA were. They were included in Koester's text and remained relatively unchanged (except for different procedures developed for LMO-FFPs discussed below) in the final text of the protocol. By Cartagena, then, a presumption that LMOs intended for introduction into the environment were potentially harmful had become common sense. This presumption was a far cry from the equivalency assumptions that underpinned the Miami Group's initial position. Final outcomes on the issue of definitions also suggest that the biosafety conversation was already starting to shift before 2000.

Definitions

In Chapter 4, I noted that the United States, through its assent to the Cairo report of 1995, indicated a willingness to accept an instrument focused narrowly on GEOs. Despite the symbolic importance of that move, the United States, along with Canada and Mexico, entered formal negotiations proposing definitions of LMOs consistent with their domestic emphasis on novelty. These definitions would include all organisms with novel traits, regardless of whether the organisms were created through the use of recombinant DNA techniques (Galvez 2002; interview #7). Because of these proposals, the debate about definitions had to take place all over again. Nonetheless, the tide on this issue had already turned. As one Canadian official states, "We lost that argument in the first year" (interview #7).

The final text of the Cartagena Protocol defines LMOs as "any living organism that possesses a *novel* combination of genetic material obtained through the use of modern biotechnology" (CBD 2000a, Article 3(g); emphasis mine). Modern biotechnology is then defined to mean the application of:

a) In vitro nucleic acid techniques, including recombinant deoxyribonucleic acid (DNA) and direct injection of nucleic acid into cells or organelles, or

b) fusion of cells beyond the taxonomic family that overcome natural physiological reproductive or recombination barriers and that are not techniques used in traditional breeding and selection. (CBD 2000a, Article 3(i))

From these definitions we can see that LMOs are considered "novel," but this term no longer has its North American meaning. As in the UNEP technical guidelines, novelty was reinterpreted to refer to organisms that are specifically produced using rDNA and related techniques, even though the North American usage was intended to do the exact opposite.

At least one European official believes that GEOs became the focus of the protocol as a matter of political expediency: No country really wanted to create rules that would impact on all organisms that were changed in any way, including the methods of traditional breeding (interview #9). In my view, this reading misses the larger significance of these definitions. These definitions provide further evidence that the language of substantial equivalence, novelty, and familiarity, as used in the US, Canada, and Argentina, and the discourse of equivalency that these terms had embedded in regulation, were losing their standing in this international forum. Among their productive effects, the final definitions of LMO and modern biotechnology created the space for a discourse of precaution – a discourse informed by complexity talk and focused solely on GE as a unique, risky, and uncertain enterprise – to gain an upper hand, should the balance of forces move the talks in this direction (as they eventually did by early 2000). The discursive implications of this move come through in this statement of a Like-Minded Group spokesperson:

> When we started negotiations, the industry view [was,] "What's wrong with genetic engineering? You mix genes. You marry a woman, you mix genes. Big deal! They are the same." That has essentially died now ... The claim that genetic engineering does nothing that would not happen naturally is *not relevant anymore.* (Interview #14, emphasis mine)

Montreal Outcomes

The final articles on AIA and the definitional issues discussed here, each already finalized in the Cartagena round, point to the growing influence of a precautionary framing of biosafety in the biosafety talks at that time. And evidence from the final round of negotiations in Montreal, in January 2000, shows that precaution had indeed moved from being one ideological current in a debate framed by risk to becoming the underlying framework within which previously unresolved issues were concluded in the final text. This evidence can be found in the final compromises on the LMO-FFP issue, the precautionary principle, and the savings clause.

Living Modified Organisms for Food, Feed, or Processing

Early in the negotiations, the Miami Group tried to limit the requirement of AIA to those LMOs that had demonstrated negative impacts on the environment. When the Like-Minded Group would not relent in its demand to subject all LMOs to risk assessment and precautionary import decisions, the Miami Group refocused its arguments on the idea that, at the very least, bulk commodities (LMO-FFPs) should be exempt from AIA. This position was based on the view that an AIA procedure could cripple the global grain-trading system. However, the argument was also framed in terms of the notion that a regulatory measure should be proportionate to risk. As one Canadian official stated,

> If we tried to regulate FFPs in the same way as deliberate release, I just think the whole trade system would shut down ... That was out of proportion to any risk posed to the environment. (Interview #12)

For its part, the EU supported the Like-Minded Group on the inclusion of commodities within the scope of AIA. The environmental commissioner of the EU states that this position was a way of supporting the interests of developing countries (Wallström 2002, 248). The EU's 1997 decision to require labels on GE foods also appears to have influenced its position.

Despite opposition from both the Like-Minded Group and the EU, the centrality of the LMO-FFP issue to the Miami Group was not lost on Koester. In his compromise text, Koester proposed that LMO-FFPs be completely excluded from the AIA procedure. In making this concession towards the Miami Group, he hoped that it would accept concessions that he had made towards other groups. The Miami Group welcomed Koester's proposal on the LMO-FFP issue, but the Like-Minded Group and the EU rejected it. In ensuing backroom Friends of the Chair discussions at BSWG-6, the EU suggested a proposal that would have allowed importing countries to loop back into the AIA for LMO commodities (Bail, Decaestecker, and Jorgensen 2002, 176). Significantly, this proposal was quite similar to the final compromise on LMO-FFPs concluded a year later in Montreal. But at Cartagena, it became apparent to the EU that the Miami Group was simply not open to considering any proposals that could link LMO-FFPs with AIA. The Miami Group appeared to have come to Cartagena with demands on a take-it-or-leave-it basis (Bail, Decaestecker, and Jorgensen, 2002, 177; Winfield 1999).

Following the collapse of the Cartagena talks, the LMO-FFP issue became one of the main issues for further informal discussions. While one Miami Group negotiator writes that "the Cartagena experience ... served to make clear to other negotiation groups that the Miami Group was not prepared to accept a text that did not meet its needs," evidence shows that, after Cartagena, it was actually the Miami Group negotiators who began to show greater flexibility in order to come up with a compromise with the EU and the Like-Minded Group, for reasons that I examine below (Ballhorn 2002, 110). Notes from a June 1999 meeting of the Canadian BPAG reveal that at the first informal consultations following Cartagena, Canada suggested "putting information on crops approved for growing in Canada on a website, so that a party of import could see what was likely to come and make decisions about what it would accept." At the time, the EU response was reportedly that this idea was "impractical [and] too far removed from reality" (Winfield 1999, 2). Nonetheless, a Compromise Group proposal on an alternative procedure for commodities along these lines eventually formed the basis for a solution on this issue (CBD 1999b). Both the EU and the Like-Minded Group came to accept a web-based process (through a new biosafety clearing-house) that would provide information about LMO-FFPs approved in countries of export, as long as countries were still able to make import decisions on these LMO-FFPs before their presence in international trade (CBD 2000a, Article 11).

Even though all sides made some concessions on the LMO-FFP issue, the Miami Group in particular moved significantly from its initial ideological position on this issue. By the time it arrived at the Montreal ExCOP, the Miami Group was prepared to have all LMOs, including LMO-FFPs, subject to procedures that could see countries refuse to import them in accordance with the Cartagena Protocol. This move signifies that the ideational relations of force had truly shifted between Cartagena, where the Miami Group held its ground on this issue, and Montreal. It no longer made sense to posit that LMO-FFPs should not fall under the protocol simply because of their intended uses as food and feed; instead, all LMOs could be expected to be subject to some degree of scrutiny by importers. In this context, the best that the Miami Group could do was try to minimize any hassles for the world grain trade that would result from the new norm. This effort saw the Miami Group focus on the exclusion of LMO-FFPs from formal AIA procedures and then on the question of the documentation requirements for LMO-FFPs at the Montreal ExCOP, rather than on the complete exclusion of LMO-FFPs from precautionary decision making it had sought earlier. In Gramscian terms, this move can be seen as an accommodation of resistance designed to minimize immediate economic impacts on the biotech bloc's still growing, but now highly contested, hegemony.

Precautionary Decision Making

Beginning at BSWG-2 (and even as early as the UNEP Expert Panel IV majority report), organic intellectuals among the G-77/China and activist CSOs sought rules that would allow countries to bar LMOs on the basis of the precautionary principle. By 1999, this demand had been recognized as reasonable by several European countries and by the majority of members of the G-77/China.

In Koester's text, the strength of the idea of precautionary decision making was acknowledged in four places. First, the preamble to the protocol drew attention to the precautionary "approach" (rather than "principle," a concession to the Miami Group) (CBD 1999a, 18). Second, the article that articulated the objective of the protocol referenced the precautionary approach (18). Third, the general principles of risk assessment note that a lack of scientific knowledge or consensus should not be interpreted as indicating "an absence of risk" (39). Finally, the article on decision procedures for AIA states:

> Lack of full scientific certainty or scientific consensus regarding the potential adverse effects of a living modified organism shall not prevent the party of import from prohibiting the import of the living modified organism in question. (23)

Initially, the EU, along with the Like-Minded Group, supported each of these inclusions of precaution in the Koester text. Nonetheless, in its bid to get a compromise with the Miami Group, the EU's final proposal at Cartagena removed the precautionary language from the operative decision-making clauses, restricting it to the non-operational preamble and risk assessment annex only.

The willingness of each of the delegations, even the Like-Minded Group, to accept this EU compromise should have weakened the chances of the issue of precautionary decision making being revived in the final negotiating session in Montreal (Graff 2002, 415). The exact opposite actually took place. Rather than come to Montreal prepared to sacrifice its stance on precaution once again, during 1999 the European Commission had worked to develop a common position on precaution among EU member states (European Commission 2000), and it came with the intention of seeing this position embedded in the protocol. Among the reasons for this EU move was the recently adjudicated WTO beef hormone dispute.

On the other side of the debate, 1999 saw industry and member countries of the Miami Group, especially the United States and Canada, continue their opposition to precautionary language in decision-making articles of the protocol text. They argued that there was little clarity on what precaution meant in practice. One industry representative commented: "If you ask any three Europeans what the precautionary principle is about, you'll get five answers" (interview #3). Miami Group representatives rejected proposals for articles that allowed for precautionary decisions because they offered no "detail" of the kinds of information gaps or level of scientific disagreement that could trigger a precautionary decision, nor any limits on how importers would be expected to act (to avoid potential harm) in the context of gaps or disagreement (Bowdens Media Monitoring 2000, 5).

Industry representatives and the Miami Group also saw the precautionary principle as a veiled attempt to build a barrier to trade that could not be substantiated (interview #13). And judging from its experience with the beef hormone dispute, it appears that the Miami Group did have grounds for concern. In the appeal to that case launched in September 1997, the EU had argued that the ban was a precautionary move, and that the precautionary principle applies not only to the management of risks but also to their assessment. The appeal saw precaution as a principle that could be called upon to restrict a product even when risk assessment research had not clearly demonstrated that significant harm could be caused by the product, simply because harm was theoretically possible (WTO 1998a). From the perspective of the Miami Group, it appeared as if the EU wanted the precautionary principle in order to "step outside the framework of trade law," and to "no longer be ... subject to those disciplines" (interview #13).

When the status of the precautionary language within the operational text of the protocol was finally discussed in the Vienna setting, towards the middle of the Montreal ExCOP, the basis of discussion was the Koester text from 1999. The Like-Minded Group, Compromise Group, EU, and CEE group all supported the articulation of precaution as presented in that text. As its opening stance, the Miami Group stated that it was now willing to support references to precaution in the objective, in addition to its use in the preamble and in the annex on risk assessment, but not in the decision-making article (Chasek 2000, 6). The issue was then sent to a contact group for further negotiation. In the new setting, the Compromise Group, on the initiative of Japan (Akasaka 2002), proposed operational text based on the language of Article 5.7 of the SPS Agreement (the article that outlines criteria for the establishment of provisional regulatory measures in cases of scientific uncertainty) (WTO 1994a).

In hindsight, it may be difficult to distinguish the Compromise Group's position from that of the Miami Group. Indeed, Japan did have strong sympathies with the Miami Group on this issue (Akasaka 2002), and the SPS-consistent framing of precaution appears to have simply added some of the detail that the Miami Group felt was lacking from earlier formulations. The possibility that the Cartagena Protocol could contain language on decision making in the context of scientific uncertainty that differed from that of the SPS Agreement had been one of the Miami Group's major concerns, so on this level the Compromise Group's proposals also appear to have been in keeping with the Miami Group's position. The discursive significance of the Compromise Group's proposed text is that it started to meet key Miami Group concerns in a way that the latter group's members had not been able to achieve among themselves. Miami Group members had not yet reached an agreement, after almost two years of working together, on treaty language that would guide import decisions in cases of scientific uncertainty (Miami Group 2000). It was the Compromise Group's proposal that allowed Miami Group members to see their interests reflected in operational language, thereby providing the United States and its allies a point of entry into the emergent discourse of precaution on LMOs.

In reaction to the Compromise Group's proposal, the EU accepted that an article written along the lines of SPS 5.7 provided greater definition to the notion of a "lack of scientific certainty or scientific consensus" found in the Koester text. However, it did not accept the provisional nature of the proposed language because that language implied, as a representative from Norway put it, that the "precautionary principle could be referred to only temporarily" (Ivars 2002, 199). On this issue, the EU supported the case activist CSOs had been presenting in the hallways of the Montreal conference centre where the ExCOP was being held. Friends of the Earth International, for example, was handing out leaflets to delegates stating that precaution should not be understood as simply about "temporary measures," since "uncertainty may not be resolved by just a few more tests, it may be a permanent characteristic of the technique or situation" (FOEI 2000, 1). The Like-Minded Group also had problems with the Compromise Group's proposal on these grounds.

In a clever negotiating move, the Europeans responded by introducing a revision of the Koester text as a "compromise" between the original chair's text and the Compromise Group's proposal. In fact, however, the new EU text was precisely the text that it had brought to Montreal, coming out of the internal EU discussions (interview #9). It was drafted to reflect the policy document prepared (but not yet released) by the European Commission on the precautionary principle (interview #9; European Commission 2000). In response, the Miami Group representative in the contact group made minor modifications to the European proposal. Still, the main elements of this European proposal survived in the final protocol text. The result of these was Article 10.6 of the protocol (with Article 11.8, which deals with LMO-FFPs, reiterating the same language):

> Lack of scientific certainty due to insufficient relevant scientific information and knowledge regarding the extent of the potential adverse effects of a living modified organism on the conservation and sustainable use of biological diversity in the Party of import, taking also into account risks to human health, shall not prevent the Party from taking a decision, as appropriate, with regard to the import of the LMO in question ... in order to avoid or minimize such potential adverse effects. (CBD 2000a)

Technical experts on the Canadian delegation saw this article as an improvement over the article in the Koester text because it was clearer in its definition of scientific uncertainty (interview #1). Still, the text was not formally accepted by the Miami Group in the contact group meeting. It would be accepted only later as part of a package that also included language on the savings clause and LMO-FFPs. Participants noted that in the closed-door ministerial-level negotiations leading to this final package, the EU was unwilling to budge any further on the precautionary language in the decision-making article, unlike a year before in Cartagena. On the other side of the table, it was the United States that held out the longest on the precautionary language (Meacher 2002, 225), though it ultimately accepted the precautionary articles.

That the idea of precautionary decision making was revived at the Montreal ExCOP by both the Europeans and the Like-Minded Group, that these groups were unwilling to "sacrifice" this demand again, and that the Miami Group, and the United States in particular, was willing to compromise on this issue can be explained only by the normalization of the precautionary discourse in the biosafety negotiations by early 2000. Despite their public stance against the precautionary principle as a decision-making tool, even the Miami Group came to Montreal prepared to accept some type of precautionary decision-making language (interview #6). Both the negotiation dynamics and the public statements of Miami Group representatives at the

Montreal meetings help substantiate this. This evidence shows that over the course of the Montreal negotiations, Miami Group representatives were more concerned with the details of precautionary decision making than about rejecting it outright.

This new stance came about because the Miami Group's mantra of sound science as a sufficient basis for import decision making – though well rehearsed over the preceding two years – was now falling on deaf ears. The only choice left to the Miami Group was to take up a position within the precautionary discourse itself. One important consequence of this move is that it allowed the biotech industry's strongest state allies to help shape what a precautionary response to LMOs would look like in an import decision-making context. Most important for the Miami Group in this regard was its desire to ensure that precaution could not become an excuse for blanket bans on GEOs under the protocol (interview #1).

The shift to a discourse of precaution had not taken place only in the negotiations over import decision making. As a matter of fact, negotiations on this issue were limited. One Canadian official recalled: "We never really got to precaution at Cartagena ... and even in Montreal, we never got to it until the last few days." At Cartagena, "the only debate was around the language: Do you reference Rio 15, do you bring Rio in as a new article, or do you put [in] language?" (interview #1). When delegations pointed out to the chair of ExCOP, Juan Mayr, that the precautionary decision-making articles found in the Koester text had never been formally debated, this was a surprise to him (interview #12). Mayr was surprised because a precautionary stance towards LMOs *had* been vigorously debated, but not in the formal negotiations as such. I return to the broader dynamics of the discourse of precaution below.

The Savings Clause

The question of whether the protocol would have a savings clause became increasingly important to LMO-exporting countries as the biosafety talks progressed, especially as GE commodities entered international markets beginning in September 1996. The neo-liberal justification for a savings clause, as articulated by a delegate from Brazil, was that

> the WTO agreements have a contractual nature, in which parties have exchanged very specific trade concessions among themselves to reach a final overall agreement ... Should these concessions be unilaterally disavowed [through new rules pertaining to only one set of commodities], the whole construction would be in jeopardy. (Nogeira 2002, 134-35)

Most countries from the South were not swayed by these arguments. When confronted by a group of trade lawyers from LMO-exporting nations, Egziabher of Ethiopia is reported to have stated:

International law is not unmodifiable, in fact it is far more modifiable than the average living organism; and if current international law does not cater for our concerns then we must work to change it; and please do not be surprised if we are not overly committed to international law as it currently stands, as much of it was drawn up while our countries were still colonies. (Egziabher, quoted in Nevill 2002, 147)

When Koester and his aides prepared his draft text at Cartagena in 1999, a savings clause was included, based on (unfinished) discussions in one of the working groups. This article was a direct replication of the relevant text of the CBD (Article 22 on relationship with other conventions), in the hope that this precedent would ensure its eventual acceptance by all delegations. This article states that the protocol's provisions will not affect a party's rights and obligations under any existing international agreement to which it is also a party (UNEP 1992). In the chair's text, this clause was qualified with the phrase "except where exercise of those rights and obligations would cause serious damage to biological diversity" (Chasek 1999, 8).

Despite the hopes of those who drafted it, the savings clause in the Koester text was not acceptable to any of the negotiating groups. Miami Group representatives argued that the qualifier would isolate the protocol from the context of international law, so they wanted it deleted (according to Egziabher 2002, 118). Many countries from the South and from Europe saw the qualifier as placing a high burden of proof for the placing of import restrictions on LMOs – higher than that suggested by the precautionary principle, which they supported. "Who is to judge the seriousness of the damage, thereby triggering the later-in-time treaty?" they asked (Afonso 2002, 427).

The Miami Group's push for an unqualified savings clause strengthened through 1999, largely because of the European market's growing rejection of GE foods, combined with the EU's push to have the precautionary principle enshrined in the protocol. The United States and Canada worried that the EU would try to use the Cartagena Protocol, especially if it included the precautionary principle, to justify a moratorium on GEO approvals that was not science-based. LMO exporters wanted to be sure that the EU would continue to be bound by the norms of international trade agreements (such as the principles of non-discrimination). In any conflict between the protocol and the WTO, they wanted the WTO agreements to be accepted as the higher authority.

By the summer of 1999, the European rejection of an unequivocal savings clause was solidly entrenched, as I describe below. The EU supported the notion that trade agreements should be respected, but it believed that regarding LMO imports and potential environmental effects, the Cartagena Protocol should be the main instrument for regulating any trade dispute.

The Europeans agreed with the Miami Group that the protocol should be as consistent as possible with the intent of other relevant agreements. But the EU interpreted this as meaning that decisions should be based on science-based risk assessment *and* precaution in the absence of clear scientific evidence. On questions of non-discrimination, the EU had been arguing for an article in the protocol to enshrine the principles of most-favoured nation and national treatment (Afonso 2002, 424). It saw such an article as replacing the need for a savings clause. Notably, the Koester text included the non-discrimination clause and the qualified savings clause, but this was a compromise that the EU felt would lead only to further confusion.

In the lead-up to the September 1999 Vienna meetings, the gulf between the EU and the Miami Group on the issue of the savings clause was narrowed to some extent by the efforts of Mayr. During the Vienna round of consultations, Mayr produced a "non-paper" on the relationship issue in which he listed four underlying concepts for which he elicited agreement from all of the negotiating groups. These were:

1. The main purpose of this protocol is biosafety;
2. We recognize that there are other international agreements relevant to sustainable development with rights and obligations;
3. The Protocol and other international agreements are of equal status; and
4. Trade and environment agreements and policies should be mutually supportive. (Afonso 2002, 431-32)

Once it was clear that all the parties could agree to these four basic principles, Mayr believed that a compromise on the issue of the savings clause was within reach, even though an articulation of these concepts in legal terms would still have to be decided upon (Mayr 2002, 225). Once again, outside factors were important in determining the eventual solution to this issue.

The most important factor was the resolution of a very similar debate in the talks on the Rotterdam Convention on the Prior Informed Consent (PIC) Procedure for Certain Hazardous Chemicals and Pesticides in International Trade. Although the PIC solution, as some came to call it, was arrived at in September 1998 and had been discussed in the corridors of the biosafety talks, it did not formally surface in the negotiations until halfway through the Montreal ExCOP, when the chair of the working group dealing with the relationship issue proposed it as a way through the impasse between Europe and the Miami Group (Chasek 2000, 5; Afonso 2002, 433). In the PIC treaty, there was no savings clause in the body of the text. Instead, this clause appeared as part of three preambular paragraphs:

Recognizing that trade and environmental policies should be mutually supportive with a view to achieving sustainable development,

Emphasizing that nothing in this Convention shall be interpreted as implying in any way a change in the rights and obligations of a Party under any existing international agreement applying to chemicals in international trade or to environmental protection,
Understanding that the above recital is not intended to create a hierarchy between this Convention and other international agreements. (UNEP 1998, preamble)

Given the acceptance of Mayr's non-paper, which had clearly been informed by these clauses from the PIC treaty, it would appear that all sides in the savings clause debate could agree to these three points in principle. Furthermore, even though many delegations raised questions about how this language would be interpreted, the recent precedent of these clauses in international environmental law made it difficult to reject them outright (Ivars 2002, 199). As a result, the PIC text became part of the final biosafety protocol text, with only minor changes in language. The second paragraph was made more general by deleting everything after the word "agreement," and the phrase "in any way" was also removed on the request of the EU (Afonso 2002, 435). In the third paragraph, the EU also sought the replacement of "create a hierarchy between this Convention and" with "subordinate this protocol to." These changes were crucial for the EU, since it interpreted the revised text to mean that, in the case of conflict, the protocol would not simply be beholden to the WTO, despite the apparent savings clause found in the second paragraph (Afonso 2002, 435-37).

Although the issue of the savings clause is the least clear of the victories for critics of GEOs (the implications of which I discuss in Chapter 7), the final outcome was certainly not the text that the Miami Group had hoped for. The Cartagena Protocol, along with the PIC treaty, set a dangerous precedent from the perspective of some neo-liberals. Kerr (2002) argues, for example, that those who negotiated these two multilateral environmental agreements on LMOs and hazardous chemicals appear to have been accorded the authority to establish international trade law. That the Miami Group agreed to this compromise reveals how little real negotiating power it had, vis-à-vis the EU, at the Montreal ExCOP. How can we account for this situation?

I believe that the answer lies, once again, in the discursive shift that had taken place during the negotiations. A discourse that empowered a precautionary stance towards LMOs had become the norm, both inside the biosafety talks and in the wider public sphere. This discourse prioritizes precautionary environmental protection over trade interests, whereas the WTO's SPS Agreement (reflecting a narrower reading of risk) does not go nearly as far in this direction. This new context left the Miami Group scrambling to find a compromise, and the result appears to be text that puts the protocol and the WTO agreements on an equal footing. This outcome is far from a "savings" clause.

The examples of the negotiated outcomes in the issues of the LMO-FFPs, precautionary decision making, and the savings clause each show how the biosafety conversation had moved, by early 2000, from a discussion about how to manage the risks of LMOs without compromising international trade to a conversation on how to treat LMOs in a precautionary light, even if that action might differ from already accepted rules of trade. What were the factors that made this discursive shift conclusive in January 2000, when it was clearly not yet realized in February 1999?

An Analysis of Relations of Force in 1999-2000

To understand the discursive shift from risk to precaution in the biosafety negotiations, we need to closely examine the relations of force in this field in 1999 and early 2000, both outside and inside the talks. In terms of material relations of force, the international trade in GEOs was continuing, but only a handful of nations were (or, more important, saw themselves as) LMO exporters. The majority of the world's nations perceived themselves as importers of GEOs, and this situation was unlikely to change in the immediate future. In Europe, the region (outside of the Americas) with the strongest biotech industry, a 1999 survey of ninety-nine European seed companies showed that they had earmarked only 38 percent of their R&D budgets to developing crops using genetic engineering techniques (Arundel, Hocke, and Tait 2000). So European companies were not yet placing their bets on GE. That most of the world's nations saw themselves as importers and consumers continued to have an impact on how national delegations approached the biosafety issue.

Within the other two sets of relations of force, the organizational and ideational, three specific factors were critical to the shift that enabled the outcomes discussed above. The first was the institutional precedent of the PIC treaty on the international trade in certain hazardous chemicals and pesticides. As we saw in Chapter 4, the PIC treaty had already inspired the AIA procedure now found in the Cartagena Protocol, even though the scope of the finalized procedure is very different. The PIC talks also provided the solution on the issue of a savings clause. Mayr's work may have led the Cartagena talks towards a similar outcome, but such a result was not certain. The horizontal application of the PIC solution allowed this issue to be resolved in a way that did not subordinate the precautionary Cartagena outcomes to related WTO agreements.

The second major factor that enabled a discursive shift towards precaution was the set of developments in the EU that led this group of countries to stand firm on the precautionary clauses, as well as the repercussions of these developments in North America. The third factor was the balance of organizational relations of force in the negotiations that saw proponents of a protocol create the conditions for a precautionary, albeit workable (from the point of view of LMO exporters), agreement.

GEO Politics in Europe

The year 1999 saw a major official change in the EU's approach to GEOs. I have already mentioned the importance of the WTO beef hormone dispute in informing the EU position on biosafety in January 2000. This case had actually concluded in 1998. Why was the beef hormone dispute apparently more important to the EU in early 2000 than in early 1999? The answer lies in the June 1999 imposition by the EU of another de facto temporary precautionary ban. On 24 June, the EU effectively introduced a moratorium on the introduction of new GEOs (Bail, Decaestecker, and Jorgensen 2002, 180).

The EU moratorium was the result of several factors. In civil society, objections to GEOs had continued to grow through late 1998 and into 1999. As in earlier waves of opposition to GEOs, this was still a multi-pronged attack. Activist CSOs and even major public figures such as Prince Charles criticized genetic engineering for a host of reasons, from its potential impact on wildlife and human health to the implications of the terminator technology on farmers in the global South. Prince Charles's critique was articulated in religious terms, echoing Rifkin in the 1970s. In a June 1998 op-ed in London's *Daily Telegraph,* Prince Charles wrote: "Genetic engineering takes mankind into realms that belong to God, and God alone" (quoted in Charles 2001, 222). Buoyed by such high-level moral support, activists used their 1997 labelling victory to lobby supermarkets to remove GE products, while continuing the by-then widespread practice of field trial "decontamination."

In European policy circles, public concern combined with the ongoing scientific discussions that raised doubts about the safety and predictability of GE led to a growing emphasis on the precautionary principle as a reasonable regulatory response to GEOs in agriculture. Rather than seeing this principle as antagonistic to sustainable development, as it was framed by industry and the Miami Group in the biosafety talks, most European intellectuals, policy makers, and regulators saw the precautionary principle as complementary with their understanding of ecological modernization (Cohen 2001). In practical terms, greater emphasis on precaution meant that regulators were able to ask, and demand, answers to more probing questions about the possible impacts of GEOs on the environment and human health than the biotech companies had been asked in the United States, Canada, or Argentina.

The idea of a precautionary response led to a rethinking of the basic framework for assessing GEOs. EU regulators began demanding answers to the kind of questions that environmentalists had been asking for years: How can we effectively monitor the long-term ecological effects of this GEO on the environment? And how can we trace a GEO, once released, should it lead to unanticipated effects that require it to be completely removed from the food system (Traxler et al. 2001)? Earlier debates about the adequacy of concepts such as familiarity for assessing the ecological effects of GE crops were also being revisited (Torgerson 1996). And in the context of the EU's immediate regulatory environment, a debate emerged about the fast-track notification system that the European Novel Food Regulations of 1997 had established in the same regulations that required labelling of GEOs. That regulation allowed products of GEOs that could be deemed "substantially equivalent" to non-GE counterparts to bypass detailed risk assessments (European Parliament 1997a). This approach had been welcomed by the biotechnology industry as progressive, but it appeared to fly in the face of the growing tide of interest in a more precautionary regulatory approach. The raising of these kinds of questions and concerns, in parallel with a growing list of "safeguard" bans of specific GEOs by Austria, Luxembourg, France, and others, meant that by June 1999, it had already been nine months since the last GEO was approved for commercial release in the EU (USDA 2003a; European Commission 2005).

The biotech bloc's choice of response to the challenges facing its products in Europe must also be seen as part of the reason for the EU moratorium of 1999. In 1998 and 1999, in the face of a European regulatory slowdown, corporate, CSO, and North American government arms of the biotech bloc redoubled their efforts to build support for the biotech project among European governments. For example, a working group of the TransAtlantic Business Dialogue – an organization of European and North American businesses that informally develops trade policy proposals for streamlining regulatory hurdles in the United States and Europe – put forward a proposal to set up a European counterpart to the US FDA to speed up and harmonize the approval of new biotech products (TABD 2002; Levy and Newell 2000, 18). An EU-US biotechnology group was also set up under the umbrella of the Transatlantic Economic Partnership (TEP) in 1998.[2]

The strategy the biotech bloc had adopted, that of working with European governments, was chosen because government support had clearly been so important to its North American successes. In Europe, this strategy was the bloc's downfall by the summer of 1999. Although government regulators had indeed slowed down in their approvals of GEOs, they were not the primary source of the backlash to GEOs. The backlash was rooted in civil society. It came first from consumers and small-scale farmers, among others. These groups could identify the GEOs in fields and stores, and, through the guidance of well-organized civil society activists, they rejected them en masse. (For more detailed accounts of anti-biotech activism in Europe in the late 1990s, see Purdue 2000; Charles 2001; Schweiger 2001; and Schurman 2004). Unfortunately, from the perspective of the biotech bloc, the proponents of GEOs had devoted scant resources to bringing civil society onside in the ten years or more that they had been active in Europe courting politicians and regulators.

Almost as an afterthought, Monsanto initiated a US$5 million European advertising campaign in June 1998 to promote biotechnology as, among other things, the solution to future world hunger problems (Charles 2001, 215; Schurman 2004). But this campaign was too little, too late. Monsanto ads were quickly swept away by Prince Charles's comments in the *Daily Telegraph*, which appeared in the same month. The European experience with GEOs in 1998 and early 1999 bears out Gramsci's insight that "a social group can, and indeed must, already exercise 'leadership' before winning governmental power" (Gramsci 1971, 249). The biotech bloc did not exercise leadership in European civil society, the sphere of political activity that Gramsci argues matters the most. As a result, the bloc was not able to maintain its influence over how GEOs would be governed on that continent.

Public concern, boycotts of GEOs, field trial "decontaminations," ongoing scientific debates, and regulatory slowdowns all occurred without a formal EU-wide policy response until June 1999. At that time, the slowdown became a full-fledged de facto moratorium when environment ministers from five countries (France under its new left-green coalition, Italy, Denmark, Greece, and Luxembourg) declared at a meeting of European environment ministers that, pending the adoption of stricter regulations on environmental releases and labelling and traceability of GEOs, they would suspend further authorizations for the growing or marketing of GEOs and any new field trials (European Council and Commission 1999). In fact, concern in civil society was so strong that other EU governments also needed to be seen to be doing something. In addition to the suspension announced by those five countries, this same meeting saw Germany, Belgium, the Netherlands, Sweden, and Finland declare that they would be unlikely to approve new GEOs for marketing, though they would continue field trials (Schweiger personal communication, 3 July 1999).

From the outset, the European regulatory slowdown was framed as a precautionary measure and not as a ban on GEOs (Schweiger personal communication, 3 July 1999). It was presented as one step along the way to having a more detailed regulatory directive for giving GEO approvals (the primary objective of which would be precaution). At the June 1999 environment ministers meeting, a draft of this directive was approved. This draft, if enacted, would give (time-limited) approvals for GEOs only after applicants addressed an extensive set of concerns about these organisms, including ethical issues, labelling and traceability requirements, and risk assessments designed to take into account long-term and indirect effects of GEOs on the environment and consumer health, in addition to the short-term, direct effects already being assessed.

It was EU developments on GEO regulation, together with the outcome of the WTO beef hormone dispute, that led the EU's negotiators to adopt their strong positions on precaution and the savings clause at the Montreal ExCOP in January 2000. As three EU negotiators stated, "what was becoming clear was that the battlefield was far broader than just a protocol under the Convention on Biological Diversity" (Bail, Decaestecker, and Jorgensen 2002, 179). The de facto moratorium made the EU vulnerable to a trade dispute under the auspices of the WTO, and the beef hormone dispute had shown them that the language of the SPS Agreement was not necessarily strong enough to justify a moratorium in cases of scientific uncertainty. As a result, by the time they came to Montreal, their "main objective was to have the principle recognized as a principle of international law" (interview #9). This also meant that the EU would fight as hard as it could against any Miami Group attempt to subordinate the protocol to the WTO through a savings clause. Should these two goals be realized, "it was felt that the Protocol might bolster the EU's defences in the event of a WTO challenge to its regulatory framework for safety in biotech" (Bail, Decaestecker, and Jorgensen 2002, 167).

While the EU moratorium explains the gusto with which the EU defended the precautionary clauses in Montreal after abandoning them in Cartagena, it does not explain why the Miami Group was willing to give in to the EU on these issues in early 2000 when it had been prepared to let the talks collapse a year earlier. Importantly, although the EU negotiators came to Montreal with a harder line on precaution and the savings clause, their bottom line was still the same as it had been in Cartagena a year earlier. Their political leaders instructed them to negotiate a text with the LMO exporters rather than agree to a deal with other groups that the exporters could not accept (Bail, Decaestecker, and Jorgensen 2002, 180). Unlike in 1999, however, in 2000 this bottom line allowed a deal to be reached. The EU's new internal situation helps explain this outcome, but there is more to the picture. We must also consider the way that developments in Europe contributed to reshaping the global discursive dynamics of the GE issue, and the pressure that these dynamics exerted on the Miami Group's position.

By the end of 1999, the GEO issue was constantly in the EU media, and news of the EU moratorium had brought the issue back to the front pages of newspapers in the United States and Canada. Meanwhile, CSO involvement in spurring on a precautionary moratorium in the EU had rekindled the spirits of anti-GE forces on this side of the Atlantic. In Canada, for example, September 1999 saw Greenpeace, the Council of Canadians, the Sierra Club of Canada, and others relaunch their anti-GE campaigns. In addition to the EU moratorium, these activists were buoyed by the global finance industry's increasingly cold feet when it came to supporting GE research. To exemplify this shift, in a report that circulated on the internet in May 1999, entitled "GMOs are Dead," Deutsche Bank analysts argued that GE seeds were becoming a liability for farmers and that the large food manufacturers were turning away from GE ingredients (Ramey, Wimmer, and Rocker 1999). They urged investors to sell their stocks in Pioneer Hi-Bred. This report was followed by a second, released in July, that made a similar case for selling off shares in the other major biotech seed companies (Mitsch and Mitchell 1999). Activists also felt vindicated by emerging scientific developments, especially the paper on the detrimental effects of pollen from Bt corn on monarch butterflies that Losey, Rayno, and Carter (1999) had published in *Nature* in May of that year.

The aim of these North American activists was to achieve the same gains that had been realized in Europe: a moratorium on the introduction of new GE crops and laws requiring the labelling of all GE foods as well as stricter, more precautionary, assessments of their potential risks. It was this rejuvenated campaign that triggered the public relations efforts undertaken by the Canadian government and representatives of the biotech industry, discussed in Chapter 2.

The overall atmosphere of the biosafety talks in Montreal in January 2000 was itself infused with the renewed attention the GEO issue was receiving. This attention focused in part on the precautionary principle as common sense. Evidence for this can be found in the way that the proponents of the biotech bloc tried to engage in the biosafety debate in the media. In January 2000, two Canadian national newspapers (the *National Post* and the *Globe and Mail*) devoted editorials to attacking the precautionary principle during the fourth week of January 2000. The *Globe and Mail* editorial stated that the principle, if it had been implemented in the past, would have prevented the development of electricity, antibiotics, and agriculture. Nonetheless, the same editorial acknowledged that anyone condemning the principle today appeared to be adopting the viewpoint that it is "better to be sorry than safe" (SCP 2000, 3).

In seeking to explain the outcome of the biosafety talks, one Canadian official pointed to the importance of the scientific developments that had taken place between 1996 and 2000. The year 1996 was

> before debates arose about effects on monarch butterflies ... before those sorts of controversies ... In the overall evolution of the debate about science underlying risk assessment for LMOs ... I think there was a lot of confidence and not a lot of acknowledgment of the scientific uncertainties that were associated with that risk assessment. (Interview #6)

This statement exemplifies the discursive shift I am drawing attention to here. This shift was not simply informed by new science, or scientific uncertainty; it was also informed by a different value judgment on how a society,

·collectively, ought to respond to the science and uncertainty. This combination of facts, values, and a way of expressing them in policy terms is what I have identified as the precautionary discourse. Precaution was a departure from the risk discourse that had shaped the international regulatory field of GE for over two decades, and it came to maturity only after ten years of consideration and debate in the field of biosafety. A Swiss negotiator summed up the way that this discourse, as it surfaced in discussions of GE controversies in the media, had restricted the Miami Group's options coming into the Montreal ExCOP:

> In hindsight, the media coverage in summer 1999 had closed the door once and for all on the political feasibility of all groups, in particular the Miami Group, agreeing on a non-solution. (Nobs 2002, 189)

Aside from the institutional precedent of the PIC treaty, the shifts in the EU, and the North American response to these shifts, there is a final set of factors that need to be considered in order for the discursive shift to precaution in the biosafety talks to make sense. These factors were the organizational relations of force in the talks themselves and, in particular, the work of the EU environment ministers, the chair of the ExCOP (Juan Mayr), and the activist CSOs.

Organizational Dynamics in the Biosafety Negotiations

At the level of organization, EU environment ministers were critical in the final year of the biosafety talks. Their active intervention in the Montreal negotiations of 2000 was important, but I would suggest that the key task undertaken by this group did not actually occur inside the talks themselves; instead, it was their defence of the Biosafety Protocol talks against attempts by the United States, Canada, and Japan to shift the issue of GEO safety to the WTO that proved most significant.

Sensing growing international resistance to GEOs, the American delegation to the Seattle WTO talks in November 1999 brought a proposal to initiate new international negotiations on the regulation of GEOs in the context of the next round of WTO talks dealing with agriculture. At the same meeting, the Canadian and Japanese delegations proposed to initiate a "broad, fact-finding ... approach" that would lead to a dialogue on the biotech issue within the WTO (interview #13, see also Gear 1999, and Falkner 2000). Critics of the biotech bloc saw these moves as veiled attempts to shift the biosafety debate to a forum more amenable to neo-liberal values, a forum that would be less transparent and less easily influenced by CSOs and countries from the South. Braithwaite and Drahos (2000, 28) call this political strategy "forum shifting," and historically it has been executed successfully in numerous fields, especially by the United States. Were new negotiations, or even a dialogue, successfully initiated under the WTO, it would take the pressure off the Miami Group to reach a solution in the context of the Biosafety Protocol.

The response to these proposals was strong from many corners, but particularly among EU environment ministers. At one point, the European trade commissioner had agreed to the idea of new talks under the WTO. The very next day, fourteen EU ministers of the environment flew to Seattle to publicly contradict the position of their trade commissioner. They rejected the attempt to start new biotech talks and reiterated their support for the Biosafety Protocol process. This move successfully killed the North American and Japanese proposals (Winfield 2000). It is notable that this event took place only two weeks before the Environment Council of the European Parliament met. At the meeting of the EU Council of Ministers on 13 December 1999, new negotiation directives were issued for the upcoming ExCOP in Montreal. These directives reaffirmed that trade and environment agreements should be mutually supportive and stressed the importance of the protocol having legal status equal with that of other international agreements, rather than being subordinate to such agreements (Afonso 2002, 432). Having concluded their directives, a majority of these EU ministers also resolved to attend the upcoming biosafety talks in Montreal to ensure a conclusion (Bail, Decaestecker, and Jorgensen, 2002, 180). Recognizing that the EU would be firm on this issue, four days later, Mayr initiated his dialogue on the inclusion of preambular text emphasizing the equal status of environment and trade agreements, rather than the subordination of one to the other (Afonso 2002, 432, 433).

Mayr's skill as chair deserves mention. His creation of the Vienna setting, a transparent negotiation process that allowed negotiations to proceed by interest groups (rather than regional groups), will likely set a new standard in the negotiation of multilateral environmental agreements. His non-paper on trade-related issues was instrumental in resolving the savings clause issue. And finally, his decision to threaten to bring a chair's text to the plenary in the final stages of the ExCOP – despite his awareness that a compromise without the participation of the LMO exporters was virtually meaningless – illustrates a keen ability to read and harness the broader political currents shaping the biosafety debate in early 2000: He recognized that the Miami Group was no longer in a position to walk away from the talks.

As at the beginning of the biosafety negotiations in 1996, organizational efforts by the initial proponents of a strong biosafety protocol were enormously important to the outcome of these talks. The solidarity achieved among members of the Like-Minded Group, under the leadership of Egziabher, was remarkable considering the number of countries in that group and the divergent perspectives they brought to the table. In the final stages of the talks, the economic and political stakes were such that the main debates on trade-related issues were between the United States, Canada, and Argentina on the one hand, and the EU on the other. Still, these issues would never have come out into the open in the context of the CBD had the countries that made up the Like-Minded Group, together with activist CSOs, not worked to bring this forum into being in the first place.

It is difficult to overstate the role of CSOs critical of GEOs in the biosafety talks, and in the politics of agbiotech more broadly. These groups harnessed a public skepticism towards the new genetic technologies, born of people's concerns as both consumers and citizens, and channelled it into very specific proposals on how GEOs ought to be governed. They then worked both inside and outside formal political institutions to see these governance strategies adopted (although, as we have seen, the proposals often changed through negotiation). The lobbying of these groups also ensured that the Vienna setting, the new negotiation format established in the final stages of the Cartagena talks that will likely be copied in future multilateral contexts, was open to public and media scrutiny.

In the final round of biosafety negotiations in Montreal, CSO critics again had a major impact. Canadian environmental groups, for example, started a campaign early in the week calling for the presence of the Canadian minister of environment, David Anderson. They sought Anderson's participation because the CEN representatives on the Canadian delegation were fully aware of that delegation's negotiating position. While not allowed to divulge that position, these representatives recognized that it would not go nearly far enough to meet the demands of the Europeans and the Like-Minded Group, and that without a political representative present who was able to override Canada's mandate (especially on documentation requirements for LMO-FFPs, hindsight reveals), no deal would be reached (interview #5). The chair of the talks, Mayr, had also recognized the importance of Anderson's presence and had personally invited him to participate on two separate occasions, but in each case Anderson had declined (Samper 2002, 72).

The civil society campaign to bring Anderson to the talks included the posting of "wanted" posters around the conference centre, even in toilet stalls, with Anderson's face and name on it. It also used the media to get the message out that Canada was not sending its own environment minister to host the talks, while many other countries had sent ministers to participate. Present were environment ministers from Portugal, Denmark, Sweden, the Netherlands, Germany, Italy, Spain, France, the United Kingdom, and Greece, along with the Environmental Commissioner Margot Wallström from the EU, among others (Meacher 2002, 231). Anderson had planned to undertake other business that week, but the unrelenting call for his participation did eventually pay off. Anderson came to the ExCOP at mid-week, and he proved critical during the final discussions in shifting Canada's position in a way that enabled a compromise with the Europeans (interviews #8, #9; Falkner 2002).

Activist CSOs were very active in Montreal during the ExCOP. In the conference centre, their media room was said to be running

> a continuous parade of press conferences ... Farmers from Europe, from the South ... Press conferences in Spanish, French, English ... They could give a sound bite in the language of your choice. The NGOs were running rings around [the industry representatives] in terms of communications and media strategy. (Interview #2)

Outside the conference centre, Biotech Action Montreal, Greenpeace, the Council of Canadians, and other organizations had organized a series of events, including a biotechnology teach-in, street protests, and a round-the-clock vigil at the "biodiversity camp" (Gale 2002, 260). This vigil was noted for its dramatic impact by a number of people I interviewed. One French negotiator is reported as saying:

> Inside, we could hear the chants [from the protests,] and we knew your environment minister didn't have much negotiating room because he didn't have the support of the people. (Quoted in Landsberg 2000, A2)

Aside from their efforts to inform the public about their stance on the biosafety talks through the media and protests, the more experienced CSO critics also retained their roles as organic intellectuals for those groups trying to establish a strong protocol. By this point, some of these activists had been a part of the biosafety discussion for a longer period than the vast majority of the diplomats in the negotiations. Phil Bereano of the Washington Biotechnology Action Council is a good example of a seasoned activist engaged in the Montreal ExCOP. He had been involved in these discussions as early as the Rio Earth Summit of 1992, where he spoke alongside Vandana Shiva on behalf of the Biotechnology Working Group in the first international workshop on biotechnology issues, and he remained involved in subsequent meetings (Diamond 1992). Activists like Bereano continued to nurture their relationships with delegates, effectively serving as consultants for some state delegations as they established negotiating positions on emerging issues (interview #2; Gale 2002, 259).

In contrast with the activities of the CSOs critical of GEOs in Montreal, industry CSOs were not nearly as effective in having their voices heard and, more important, believed. For example, whereas activists brought in farmers from around the world to denounce GE crops, the Global Industry Coalition flew in four farmers from the American Midwest and the Canadian prairies to promote GE. Journalists could not help notice the difference between the press conference held by these four "white, earnest, middle-aged men in glasses and suits" and the "much more colourful, multi-lingual events" held by Greenpeace and others (L. Stewart 2000, C4). While activist press conferences decried "genetic pollution," a Manitoba grain farmer, speaking at an industry press conference, was reported to have made the following argument: "I get up in the morning and use margarine on my toast, so there's no reason not to use Round-Up-Ready canola in my fields" (C4). While the outward appearance of industry press conferences mattered, discursive traction was once again an important factor. The precautionary discourse had become the authoritative framing of the biosafety issue. By contrast, the discourses that the industry apologists had adopted simply found no footing. This is clear from this statement by a Canadian delegate:

> They were making business pitches to the wrong audience. This is Monsanto sitting there saying it is here to feed the world. But everyone in the room knows perfectly well, including the person speaking, that this is not the truth. They're here to make money ... In the end, *nobody believed them*. (Interview #5, emphasis mine)

Another organizational factor that deserves mention is the dynamic that took place within the Miami Group. The Miami Group functioned as a single unit throughout the talks. However, it appears that there were tensions at the final stages among the six countries making up the group. These tensions were likely because, as one Canadian official noted, "the lowest common denominator, from my perspective, came from two countries that had no particular interest in the protocol" (interview #1). Another commentator stated that, in the final day of the talks, Chile and Uruguay were distancing themselves from the Miami Group's official positions (Winfield 2000). Three insiders in the final set of backroom talks have also written that each of the developing-country members of the Miami Group appeared to be prepared to accept documentation requirements for LMO-FFPs when the United States and Canada were not (Bail, Decaestecker, and Jorgensen 2002). And at that point, Canada and the United States also apparently had differing bottom lines. For Canada, the key issue was documentation for LMO-FFPs, while for the United States, it was the precautionary principle and the savings clause (Anderson 2002; Bodegard 2002; Meacher 2002). The emergence of these internal fissures led the Miami Group to agree to a deal, as a group, that neither of its two lead members was fully satisfied with on its own. This suggests that the formation of the Miami Group, although a useful strategic move on the part of Canada, the United States, and Argentina for moving trade issues forward in 1998, may have actually forced the final accommodations from these nations most strongly allied with the biotech bloc.

Although the Miami Group was often painted by anti-GE activists as the villain in these talks, the evidence presented here shows that much of the activity that enabled solutions to be reached in Montreal actually came from within the Miami Group, as its representatives developed responses within the new discursive field of precaution. One Canadian official admits that, in Cartagena, "we were not functioning in the way that Canada is perceived" (interview #1). This appears to have led more accommodative forces in the Canadian government to work hard to reach a deal that the rest of the world would accept while, at the same time, ensuring that it was one the Canadian minister of environment, David Anderson, could defend to his cabinet and Canadian industry. This was no small task.

For his role in orchestrating participation among Miami Group delegates to the visit to Ethiopia in September 1999, John Herity should be recognized, because this event appears to have been an important step in getting the Miami Group to accept the need for the compromise on LMO-FFPs (interview #14). It was also another Canadian civil servant who developed the proposal for a web-based biosafety clearing-house mechanism for LMO-FFPs, as an alternative model to AIA, that could meet the needs of importer countries for information and the ability to make precautionary decisions while causing minimal impacts on the global grain-trading system. The emergence and acceptance of the clearing-house mechanism made the difference between the Cartagena debacle and a conclusion at Montreal that the Canadian and US governments could sign on to. The Biosafety Clearing-House may not have been an optimal arrangement from the perspective of the biotech industry, but it was an accommodation to critics of GE that could be lived with without causing major economic upsets. By January 2000, such a solution was the best they could hope for.

Disciplinary Effects of the Cartagena Discourse of Precaution
Through the final stages of the Cartagena Protocol negotiations, an emergent discourse of precaution shaped the discussion and outcomes in the areas discussed here. This was clearly a blow to the biotech bloc and a major concession to critics in the context of the global struggle over agbiotech. However, the particular Cartagena discourse of precaution that was being shaped through these negotiations also had disciplinary effects on those who first brought the idea of a precautionary response to GEOs forward in the talks. Consider the results in the debates over "socio-economic considerations" and "products thereof," both issues brought forward by the Like-Minded Group and its CSO allies.

Socio-Economic Considerations

The idea of an article on socio-economic considerations emerged from the call by activist CSOs and some delegates from the global South for assessments of the broader societal impacts that could result from the introduction of GEOs. This call was strongly resisted by the Miami Group and the Europeans because it challenged the neo-liberal assumption, embedded in the WTO agreements, that potentially negative social and economic impacts cannot be used (in all but a very few circumstances) to justify trade restrictions. As an example of how the WTO treats this issue, a 1984 GATT ruling had rejected Japanese trade restrictions on US leather goods that the Japanese government argued were in place to protect the modest standards of living of a highly disadvantaged minority population (the Dowa) (GATT 1984, paras. 43-44, cited in Mackenzie et al. 2003, 238).

The potential for direct conflict with the WTO was a major strike against the inclusion of an article on socio-economic considerations in the protocol. However, as we saw with the precautionary clauses (which are also not accepted by the WTO in the way they are iterated in the protocol), there was room in these negotiations to institute new norms in the global trading system if the political will was there to push them through. Furthermore, that some countries, Norway and India among them, already evaluated the contribution of GEOs to "sustainable development" – defined to include the economic, environmental, and social dimensions of sustainability – meant that precedents existed for the inclusion of socio-economic impact assessment as part of a risk analysis process.

Once they recognized that an inclusion of socio-economic assessment on its own terms was unlikely, the organic intellectuals of the comprehensive assessment position repeatedly tried to frame these concerns in the discursive terms of the biosafety debate: socio-economic implications of LMOs were presented as risks to biodiversity. For example, proponents of an article on socio-economic considerations argued that social and economic impacts of GEOs could have downstream effects on the conservation and sustainable use of biodiversity. As a result, assessing and taking precautions against damaging social and economic impacts of LMOs could be an important part of environmental and health protection. Consider the example of coffee put forward by Stabinsky (2000, 263): Coffee is now a smallholder crop in most areas of the world. The adoption of (hypothetical) genetically engineered coffee varieties could potentially lead to a shift from small- to large-scale coffee production. This economic effect could, in turn, have important implications for ecosystem integrity in the regions where coffee is produced.

As we saw in Chapter 5, framing socio-economic impacts of GEOs as potential risks to biodiversity made few converts among negotiators outside the Like-Minded Group. The final round of negotiations, framed through a precautionary lens, *could* have revitalized the debate on socio-economic considerations. After all, the precautionary principle and socio-economic considerations had been part of the same comprehensive assessment position initially brought to the talks by the African Group, CSOs critical of GEOs, and their allies. However, precaution had shifted from its position as a discourse of resistance to one of normalization, first adopted by the EU in its moratorium and eventually even accepted by the Miami Group at the Montreal ExCOP. This shift meant that the initial proponents of comprehensive assessment would also become subject to the discipline of the precautionary discourse as its full implications became clear.

As a discourse of normalization, precaution, like the risk discourse in which it has its roots, maintains a dichotomy between technical and non-technical concerns. The acceptance of this basic dichotomy in the biosafety discussions began early in the formal negotiations. Socio-economic considerations were initially debated (at BSWG-2) in the context of risk assessment (CBD 2003, 80). But as Stabinsky (2000) points out, in subsequent negotiations, the issue of socio-economic considerations came to be dealt with in a completely separate working group from the group that developed the risk assessment procedures. Even though it was in the context of risk assessment procedures that an evaluation of socio-economic impacts could have been accepted as part of the study of LMO impacts, the group dealing with risk assessment procedures (consisting mostly of scientific representatives) rejected any attempts to raise socio-economic issues as being outside its technical purview. The outcome of these moves was that the question of socio-economic considerations came to be framed as separate from other impacts of LMOs on biodiversity, and the issue was relegated to an independent article. This situation stacked the discussion against those who saw the "technical" (e.g., environmental and health) and "non-technical" (e.g., social, cultural, and economic) implications of GEOs as intrinsically interrelated.

In the eventual debate about language on this issue in Cartagena, the only way that Egziabher and his colleagues could justify including an article on socio-economic considerations at all was to refer back to the parent treaty, the CBD. Given that the CBD includes text that appears to contravene trade rules for the sake of protecting biodiversity and indigenous communities (on behalf of "sustainable development"), this looked like a fruitful direction to take the debate in.

This direction was indeed fruitful: An article on socio-economic considerations was eventually included in the Koester text, and this survived in the final protocol text. But the proponents of this article had to make a major concession to the Miami Group and the EU. Article 26 of the Cartagena Protocol states that parties can take socio-economic considerations arising from the impact of LMOs on the conservation and use of biodiversity into account in reaching decisions on whether or not to allow imports of LMOs.

However, such decisions and measures must be "consistent with international obligations." This condition ensures that parties still live up to any obligations they have under the WTO trade regime or any other multilateral trade agreements they are party to. What this condition could mean in practice is discussed in Chapter 7.

The debate and final outcomes of the socio-economic considerations issue demonstrate the disciplinary power of ideational relations of force in the biosafety negotiations. On the one hand, the ideology of precaution, brought to the talks by a small group of organic intellectuals critical of GE in the early 1990s, had flourished to the point that it was received as common sense by the final round of negotiations. On the other hand, the broader comprehensive assessment position, of which precaution was just one element, had not become accepted as the norm in international civil and political society. In the end, international acceptance of the Cartagena discourse of precaution – a framing that was narrowly focused on "objective" environmental risks – helped sideline debate over socio-economic issues related to LMO imports.

Evidence that the balance of ideational relations of force was not on the side of a more comprehensive position can even be found in the positions of countries in the Like-Minded Group itself. Consider the way that the representative of China, in retrospect, framed that country's three key concerns about LMOs. These concerns were (1) potentially harmful impacts on the environment and agriculture, which could harm biodiversity and thus local communities; (2) potentially harmful impacts on human health, including possible long-term, but currently unknown, effects; and (3) uncertainty about the manageability of the risks of LMOs (Lijie 2002). Of these three points, only the first relates to possible social impacts of LMOs but frames these as the indirect result of environmental damage caused by the transgenic organism. Although the latter two points suggest the need for precautionary LMO decision making, the Chinese position does not articulate any direct socio-economic implications of LMOs, nor does it mention the potential for harm to biodiversity resulting from the socio-economic implications of LMOs. This example shows how even among members of the Like-Minded Group, some (and possibly many) countries did not recognize the issue of socio-economic implications of LMOs as a distinct category of impacts warranting attention in the protocol. This internalization of a narrow interpretation of risk (and precaution) in the very negotiating group that ostensibly fought most strongly for comprehensive assessments in the biosafety negotiations (Kwaja 2002), helps explain the lack of stronger articles on socio-economic considerations in the final text.

In academic circles, there have been attempts to raise and define the social and economic implications of new technologies for decades, going back, for example, to Schumacher's *Small Is Beautiful* of 1973. Furthermore, national legislation in several countries, including the National Environmental Policy Act of 1970 in the United States, mandates social impact assessment (SIA) as part of broader principles of environmental impact assessment. (For an overview of the methodologies of SIA, see Stabinsky 2000). Nonetheless, the Biosafety Protocol negotiation history demonstrates that a specific policy discourse encompassing a multi-dimensional response to the social, cultural, economic, health, and environmental risks associated with technology adoption had not yet achieved the resonance of either risk or precaution. To some extent, the concept of "sustainability" has been used by many groups to try to incorporate these various dimensions in, for example, the idea of sustainable agriculture. However, the meanings of sustainability are so varied and contested that this term appears to have lost its prescriptive value in policy making.

This lack of a galvanizing framing presenting the "comprehensive assessment" of LMOs as common sense became evident at the Montreal ExCOP when the Like-Minded Group was asked to articulate how exactly it wanted to deal with socio-economic considerations in the protocol. At that point, the proponents of this approach were at a loss on how to present this in practical terms:

> When given a chance, neither the G-77 nor NGOs did a really good job of defining what it means ... The people who could have done it ... India, Ethiopia, or others, just did not bring it out in a way that was compelling or even realistic, from a regulatory design perspective. There are methodologies for technology assessment, but somehow these did not come out on the table. The notion of alternatives assessment, for example, is more recent and wasn't on the table at that stage. (Interview #5; on alternatives assessment, see O'Brien 2000)

This moment cost the Like-Minded Group and its CSO allies dearly. It speaks to the work that remains to be done in articulating, and then gaining adherents across civil society and among state representatives to, the idea of a more comprehensive process for evaluating new technologies such as GEOs.

Significantly, while the negotiations on the Cartagena Protocol were completed in 2000, there remain many opportunities for the furtherance of this dialogue within and outside the institutional context of the protocol. Furthermore, that the final Cartagena consensus on precaution actually helped discipline conversations about socio-economic considerations in the protocol talks does not mean this will always remain the case. As Foucault would expect, the adoption of precaution as a discourse of normalization also enables a number of new points of resistance, some of which continue to challenge the separation between technical and non-technical risk issues in the precautionary discourse. The ongoing debate about the inclusion of socio-economic issues in decision making is taken up again in Chapter 7.

In the end, there is an important irony in the Biosafety Protocol's distinction between socio-economic considerations, as called for by the Like-Minded Group, and "technical" risk issues. Vandana Shiva is reported to have noted, in an early biosafety negotiation session, that the very separation of LMO-FFPs from other LMOs, as demanded by the Miami Group, was itself based on socio-economic considerations: the desire to keep goods moving in the international grain trade (Stabinsky 2000). This distinction would not hold water solely on scientific grounds, yet it survived a set of negotiations guided, according to most of its participants, by "scientific" consideration for biodiversity. When I drew the attention of one of the Canadian officials to this point, he had to agree:

> You've got the Miami Group wanting commodities out, from a trade perspective, and you have Europeans wanting pharmaceuticals out, purely from a trade perspective. To hell with biodiversity! Those negotiations were not fun. There was so much at stake. (Interview #7)

This statement points to the fact that discursive power, and particularly the power of a scientific discourse of LMOs, was far from the sole determinant of outcomes in the biosafety talks. I return to this idea below.

Products Thereof

There is another example that shows the disciplinary effects of discourse in the biosafety talks. The eventual removal of the phrase "products thereof" from the protocol text, after much debate, reveals the way that terminology, once widely accepted, can be productive in ways that would not have been anticipated when it was initially put to use. In this case, the designation of "living" in the term "living modified organism" carried a very different meaning at the end of the biosafety talks from what it meant at the beginning, with unexpected implications.

Countries from the Like-Minded Group wanted to include all LMOs and products thereof within the scope of the protocol, on the understanding that these products could also have both biological and socio-economic impacts (Egziabher 2002, 119). Other delegations, especially those representing Miami Group countries but also among the Europeans, argued that this inclusion made no sense within the scope of the Jakarta Mandate, which referred only to *living* modified organisms. As one Canadian official put it, "This is a convention on biological diversity, which is a matter of *life*, so that things that are not affecting biological diversity aren't relevant to the

Convention" (interview #8, emphasis mine). In my interviews, not one informant recalled that originally the term LMO was introduced by the United States to expand the focus away from *genetically* modified organisms. Instead, they saw LMOs as referring to a subset of GEOs that are alive. This latter framing of LMO was the interpretation of Malaysia and several other countries from the South at Rio, but it was not yet the norm at that time. Once this reading became the norm in ensuing protocol talks, it effectively silenced debate on "products thereof."

This shift in the understanding of the import of "living" reveals the power of discourse in international negotiations. Still, it is important not to overstate the productive effects of terminology. Were the scientists in the Biosafety Protocol talks convinced that "products thereof" could cause the same damage to biodiversity as LMOs themselves, they could very well have decided to include them under the scope of the protocol, even if those products were not alive. As it turned out, the arguments by Egziabher and others, that "DNA fragments in GMO products can find their way into other organisms through the natural bacteria-mediated processes of horizontal gene transfer," simply did not resonate as a relevant risk with the majority of technical experts in these talks (Egziabher 2002, 119). In the end, the notion that products thereof should be excluded from the protocol's scope because they were not alive served economic interests (of canola oil exporters, for example) *and* it reflected the common sense view of the majority of the scientists involved in the talks. The outcome may have been otherwise if Egziabher and his allies had been able to provide stronger evidence to substantiate the view that products of LMOs posed specific risks to other organisms through horizontal gene transfer.

The Limits of Discursive Power
The aspects of the final protocol discussed in this chapter so far all point to the power of ideas as discourses. Political positions are framed, and defended, in relation to specific ideas. Sometimes, as we see with precaution, these ideas take on a life of their own, overwhelming the ability of those who first proposed them to control their impact. Ideas, however, are only one sphere of political activity in our theoretical framework. This suggests that the power of discourse may be limited, especially if it directly challenges the material interests of a group with hegemonic ambitions, such as the biotech bloc. These limits were evident in the compromise eventually reached on LMO-FFPs.

Although the final article on LMO-FFPs means commodities may be subjected to the same import decisions as LMOs intended for direct introduction into the environment, the distinct procedure established for these commodities (the use of a web-based clearing-house mechanism rather than an AIA procedure) was a major concession for proponents of a strong proto-col. This must be seen as a compromise because during negotiations it was widely accepted that much of the grain sold as food and feed can just as easily be used as seed, thereby still presenting potential risks to biodiversity. The lack of AIA for LMO-FFPs means that these risks may not be assessed and managed with the same degree of care as are the risks of organisms that do go through the AIA procedure. The exemption of bulk commodities from AIA was due to nothing less than the economic clout of the Miami Group, in conjunction with the desire of the EU to reach an agreement that LMO exporters would accept.

Pharmaceuticals

There are two further examples that point to the limits of discursive power in the biosafety talks: the issues of pharmaceuticals and documentation. Many of the delegates to the biosafety talks recognized that LMO pharma-ceuticals, once administered, are released to the environment and thus also have the potential to cause damage to biodiversity. Governments around the world are paying increasing attention to the way that antibiotics and fertility drugs, for example, after having moved through the human body and being flushed into rivers and streams, can cause a variety of down-stream impacts to human health and natural ecosystems. Since it was the release of a genetically engineered rabies vaccine in Argentina that sparked the calls from countries of the South for a biosafety protocol in the first place, the Like-Minded Group pushed hard for the inclusion of pharmaceu-ticals under the protocol, and it had discursive power on its side. Within the discourse of risk or that of precaution, it is difficult to argue that phar-maceuticals should be exempt from the same standards of risk assessment that other LMOs are subject to. That the inclusion of pharmaceuticals was common sense to many of the scientists involved in the biosafety talks is revealed in this statement by a chief negotiator for the Central and Eastern European (CEE) group:

> If LMOs can be dangerous, they all – including material that are intended to and can produce LMOs – should be handled with caution ... In this respect I was for the most part satisfied with the extension of the scope of the Protocol to include pharmaceuticals as well. (Nechay 2002, 215)

A Miami Group scientist agreed that any exemption for pharmaceuticals in the protocol "doesn't make any sense to me, as a scientist" (interview #7).

Despite the logic of including pharmaceuticals, the Miami Group, the EU, and the Compromise Group all fought against their inclusion under the protocol. This position can be explained only by looking at the eco-nomic importance of the pharmaceutical industries for countries in these groups. Switzerland, for instance, a member of the Compromise Group, is a pharmaceutical powerhouse. EU negotiators acknowledge that pharmaceutical companies were also influential on the EU position in a way that its agbiotech companies were not (interview #9). In the face of a united front among OECD countries, the Like-Minded Group did not stand a chance, even though the demand to include pharmaceuticals under a precautionary protocol was reasonable.

The result? Article 5 of the final protocol states that the protocol shall not apply to the transboundary movement of LMOs "which are pharmaceuticals for humans that are addressed by other relevant international agreements or organisations." Most analysts of the protocol understand this article to mean that pharmaceuticals are exempt from the protocol (e.g., Marquard 2002, 295), and this is how I interpret it here. Egziabher and some others, however, argue that the impacts of pharmaceuticals on biodiversity are covered by the protocol, since no international agreement dealing specifically with pharmaceuticals actually mandates risk assessments for potential impacts on biodiversity (Egziabher 2002, 121).

Documentation

The final issue to be decided upon at the Montreal ExCOP was the question of documentation to accompany LMOs. This issue also demonstrates the limited power of the precautionary discourse when it was positioned directly against the perceived interests of the global grain-trading system. For LMOs to be covered by AIA, there was little controversy. Article 18 states that documentation accompanying LMOs

> intended for intentional introduction into the environment of the Party of import and any other living modified organisms within the scope of the Protocol, clearly identifies them as living modified organisms; specifies the identity and relevant traits and/or characteristics, any requirements for the safe handling, storage, transport and use, the contact point for further information and, as appropriate, the name and address of the importer and exporter; and contains a declaration that the movement is in conformity with the requirements of this Protocol applicable to the exporter. (CBD 2000a, Article 18.2b)

When it came to LMO-FFPs, however, the Miami Group strongly resisted the notion that these be accompanied by documentation at the same level of detail.

There were two motivations for this position. First, exporter countries recognized that documentation would mean segregation of food grains by genotype from the field to the port of import in another country. Given that the global food system segregates grain by variety only on rare occasions, there could be considerable costs associated with such segregation. In fact, the imposition of those costs could take LMOs out of the supply chain entirely as grain traders would not want to incur the expense of running parallel systems for LMOs and non-LMOs. The biotech seed industry and the large grain traders alike did not like the sound of either option. The second motivation for resistance to stringent documentation requirements for LMO-FFPs had to do with the labelling debates that raged around the world in the late 1990s. In countries as divergent as Australia, Japan, and Norway, laws were being drafted that would require GE food to be labelled as such. The biotech bloc fought these proposals in each and every case, since, as one industry insider said, labelling could be the "kiss of death" for GE foods (interview #10). GEO food labelling was a practice that the United States and Canada were considering fighting at the WTO as an unfair labelling practice (under the Agreement on Technical Barriers to Trade) because, in their view, it unfairly discriminated between "like" commodities. The last thing they wanted was an international agreement requiring GEOs to be labelled as such in transit (interview #13).

From the perspective of the importers of LMOs, which each of the other negotiating groups in the protocol process saw themselves as representing, the idea of detailed labelling of LMO-FFPs seemed logical:

> As biosafety was the aim of the Protocol, it made good sense to enable parties to identify and trace GMOs at any time so that they could take adequate measures to control or prevent harmful impacts of GMOs on ecosystems and food chains. (Gale 2002, 257)

As in the adoption of any discursive strategy, however, the position calling for detailed documentation for LMO-FFPs was not taken only because of its apparent reasonableness. Even though all sides in the biosafety debate formally accepted that the issue of documentation requirements had nothing whatsoever to do with end-product labelling, consumer demands did help shape the EU's position. The then minister of the environment in the UK, Michael Meacher (2002, 235), writes:

> Although the documentation requirements in the Protocol are not concerned with consumer information, consumers' insistence on choice ... can only be met if there is adequate international provisions at least to ensure that traded GMOs are accompanied by documentation that specifies their identity.

The European mandate coming to the Montreal ExCOP was therefore to get LMO-FFPs included in the treaty and to have some means of identifying shipped commodities as LMOs.

The enormous importance of the LMO-FFP documentation issue for both the Miami Group and the EU led it to receive the most attention in the final days of the Montreal ExCOP. When Mayr threatened to submit his own compromise text to the plenary for a vote, the Miami Group still had issues with how certain topics were dealt with in that compromise, including precaution and the savings clause. However, its most immediate concern was with documentation. The text Mayr was developing is said to have included strict documentation requirements for all LMOs subject to AIA as well as LMO-FFPs and, as Canadian environment minister Anderson writes, "the Miami Group knew that ... if we did not sort [the documentation issue] out with the EU ministers, we would leave Montreal without an agreement" (Anderson 2002, 240).

This disagreement culminated in bilateral ministerial-level meetings that lasted into the early morning of 29 January, delaying the plenary and another potentially divisive vote. In those meetings, the United States and Canada were most strongly against any documentation requirements for LMO-FFPs (Meacher 2002, 232). At some point that morning, it was the United States that came forward with a proposal (originally suggested by the Japanese delegation earlier in the negotiations) for language stating that LMO-FFP grain containers be accompanied by documentation that "clearly identifies that they 'may contain' living modified organisms and are not intended for intentional introduction into the environment, as well as a contact point for further information"(CBD 2000a, Article 18.2a). The EU was prepared to accept this compromise, as long as the article included a sentence stating that

> the Conference of the Parties serving as the meeting for the Parties to this Protocol [COP-MOP-1] shall take a decision on the detailed requirements for this purpose, including specification of their identity and any unique identification, no later than two years after the date of entry into force of this Protocol. (CBD 2000a, Article 18.2(a))

Given what was at stake, five of the six countries in the Miami Group were willing to accept this compromise. Initially, Canada did not accept it, as it went beyond the mandate that Anderson had negotiated with the federal cabinet (interviews #8, #9; Bodegard 2002). Still, when the plenary was called, Canada – not willing to be the only holdout on a protocol – assented to the documentation language and to the remaining final compromises on precaution and the savings clause.

The final outcomes in both pharmaceuticals and documentation show that discourses are clearly not the only form of power capable of determining the course of international negotiations. Recalling our theoretical framework, ideas are only one of three sets of relations of force. At times, developments in the realm of ideas do reconfigure material and organizational relations of force. Such changes are evident in many of the examples discussed in this chapter. In other cases, straightforward economic clout (as we see in the case of the pharmaceutical question and the Canada-US position on documentation) and organizational power (in the case of the EU's forceful position on documentation) shape ideational outcomes even when alternative options appear more reasonable to many of the participants in the conversation. A comment by one of the activists sums up the roles of both ideational and material power in the biosafety talks. In a discussion of why the Miami Group achieved some of its demands on LMO-FFPs and documentation, while losing most of its other demands, this activist remarked:

> Canada and the United States lost all of the arguments. The only reason they got anything was because of pure economic power ... They had the sledgehammer ... and therefore there were limits to how far the rest of us could go in offending them. (Interview #5)

Conclusion

The only way to understand many of the outcomes in the Cartagena Protocol is to look at the global relation of force of the GEO issue in 1999 and early 2000, and the relations of force within the meetings. These relations enabled a distinct shift from a biosafety discussion grounded in the discourse of risk to one rooted in precaution. This new context recognizes the inherent risks of GEOs as different from those of non-GEOs; it reverses the implicit cost-benefit analysis for GEOs; it emphasizes uncertainties and complexities, as well as possible long-term, cumulative, and synergistic effects of GEOs; it recognizes that the stringent standards of trade agreements, when it comes to the domestic regulatory measures that they deem acceptable, may not be sufficient for protecting the environment and human health from the risks of GEOs; and it prioritizes consumer over producer risks.

By encompassing each of these elements, precautionary discourse ostensibly creates the space for those who wish to argue that the full environmental implications of GEOs may be simply unknowable, while the negative socio-economic implications of many GEOs are all too clear, and who wish to shift the debate to whether or not the world *needs* GEOs. However, the discourse could also be mobilized to reinforce the notion that all that matters in the context of making decisions on whether to employ GEOs in farming is a (thoroughly researched) quantification of risk. By adopting positions within the emergent discourse of precaution, the Miami Group and the EU were able to narrow the rules on how and when the precautionary principle would apply to LMO imports so that precaution would no longer be tied to the idea of a comprehensive assessment. Most important for the proponents of this latter understanding of precaution – the under-standing that was eventually embedded in the protocol – even precaution-ary import decisions under the protocol will require scientific evidence of risk to biodiversity (although it is unclear how much), and these decisions need to be made on a case-by-case basis. In other words, unless strong sci-entific evidence emerges that suggests that GE might be an inherently dan-gerous process, blanket bans on GEOs taken on a precautionary basis would not easily find support in the Cartagena Protocol.

This narrower interpretation of precaution found in the final protocol was not the result of a strategic focus adopted by the activist CSOs in the Biosafety Protocol talks. Throughout the talks, many of these groups main-tained their call for assessments that took into account the socio-economic, ethical, environmental, and health implications of genetic engineering; they did not leave socio-economic considerations behind. However, as we saw in the case of China, their state allies did not always focus on the issue of socio-economic impacts. In the end, a range of different actors and idea-tional currents shaped what I have chosen to call the Cartagena discourse of precaution. Especially important was the relationship between precau-tion and the earlier discourses of hazards and risks, each of which had pre-viously functioned to sideline the relevance of "non-technical" issues. That precaution could reinforce the technical–non-technical dichotomy, and the importance of this dichotomy to global trade rules meant that in this con-text the normalization of precaution helped to discipline other elements of the comprehensive assessment position.

7

The Politics of Precaution in the Wake of the Cartagena Protocol

The story of the Cartagena Protocol is the story of a relatively simple, but important, change: this protocol gives international legitimacy to a precautionary response towards GEOs. An emergent discourse of precaution towards GEOs was normalized in the very creation of a binding international treaty to govern these organisms before their widespread adoption, and before any specific examples of major and irreversible harms caused by them. This discourse of precaution was also normalized through several articles in the treaty itself, including those on definitions, the AIA procedure, decision-making procedures, and the preambular text on the relationship between this multilateral environmental agreement and other international agreements. That a precautionary response to genetic engineering can be found in each of these places was a considerable achievement for the critics of GEOs, given that the EU, which became the strongest advocate of precaution at the Montreal ExCOP, had removed clauses related to precautionary decision making from the final text that it proposed in Cartagena less than twelve months earlier. The overall result is a major departure from the trend seen in the late 1980s and early 1990s, when a perspective that framed GEOs as minimally risky because of their equivalency to non-GEOs was shaping the global regime of agbiotech safety through organizations such as the OECD and the FAO/WHO.

Although a precautionary protocol does represent a significant development in the emerging global regime for GEOs, one that has already tempered the GE revolution in certain ways, as I describe below, one needs to put this development in perspective. The achievement of a precautionary protocol is only one element in the larger struggle over GEOs in global agriculture. As noted in the Introduction, the 2006 WTO Dispute Settlement Body panel ruling against the EU over its GEO moratorium – a ruling mainly still rooted in the discourse of risk – is one important example that shows that the precautionary approach to GEOs is not yet the norm in all key international institutions (WTO 2006). Furthermore, from the point of view of many in the transnational historical bloc, the Cartagena Protocol simply clarifies that precautionary policies must be based on scientific evidence, as embodied in the principles of WTO agreements. As one Canadian diplomat puts it,

> We created a process through the protocol which would require a [risk] assessment. Like trade rules on precaution, [the protocol] requires good science in making a judgment, and continued pursuit of further scientific evidence [in cases of scientific uncertainty]. (Interview #8)

As a result, although the import decision-making clauses in the protocol may give greater leeway for political decisions on risk than related articles in the SPS Agreement, for example, the protocol also implicitly reproduces many of the basic norms of neo-liberal environmentalism. These norms also include the premise that economic growth and increased trade in GEOs can further the goal of sustainable development, and that trade restrictions are valid only if put in place to deal with scientifically verifiable threats to biodiversity. These norms ignore the possibility that social, cultural, and economic impacts of GEOs could also be contrary to the pursuit of sustainability, and actually reject the idea that there should be room for the consideration of these factors in domestic regulations if such consideration might result in significant restrictions on international trade. In addition to seeing the final protocol as a victory for those who wanted to challenge the biotech bloc's revolutionary ambitions in the early 1990s, then, we should also see it as testimony to the difficulty of exacting concessions from the larger transnational historical bloc. (For an analysis of how the protocol can be seen to reproduce norms of the neo-liberal response to environmental threats, see Bernstein and Cashore 2002). As Gramsci suggests, once an historical bloc is entrenched in both civil society and the state, defining regulatory practices as well as ways of thinking, those who wish to challenge the hegemonic status quo do indeed face an uphill battle.

In this chapter, I examine the outcomes of the Cartagena Protocol negotiations and some of the productive effects of the framing of precaution found in this treaty. The precise implications of the agreement are still difficult to gauge. The protocol entered into legal force on 11 September 2003, ninety days after the fiftieth nation deposited its instrument of ratification (as per Article 37 of the protocol) (CBD 2000a). And, while over 130 countries had ratified or acceded to the protocol by mid-2006, not one of the original members of the GEO-exporting Miami Group of countries has yet done so (CBD 2006a). It is, nonetheless, possible to identify a number of ways that the Cartagena Protocol has already had a significant impact on the global politics of genetic engineering, and on the politics of trade and the environment more broadly.

The Cartagena Discourse of Precaution and Trade Disciplines

In order to have a complete picture of the Cartagena Protocol and its productive effects, we first need to consider how this treaty's framing of precaution relates to the approach to risk in related WTO agreements. This relationship is important because almost all parties to the protocol are also members of the WTO, and various aspects of the Cartagena Protocol potentially overlap with at least three WTO agreements: the 1947 General Agreement on Tariffs and Trade (GATT), the 1994 Agreement on Sanitary and Phytosanitary Measures (SPS Agreement), and the 1994 Agreement on Technical Barriers to Trade. In general, these agreements are designed to ensure that domestic measures that impact on the trade in goods do not discriminate based on the products' country of origin, and that any measures put in place to achieve acceptable policy goals are no more trade restrictive than necessary (GATT 1986, Articles I, III, XX; WTO 1994b, Articles 2.1, 2.2; WTO 1994a, Articles 2.2, 2.3). My focus here is on the SPS Agreement, since it is the one that provides the most detail on the obligations of WTO members when taking measures designed to protect human, animal, or plant health that result in trade restrictions. (For a detailed examination of overlaps between all three agreements and the Cartagena Protocol, see the appendix to Mackenzie et al. 2003). One of the main objectives of the SPS Agreement is to ensure that such measures have a scientific basis (although certain non-scientific standards, such as those for halal and kosher foods, are also recognized) (Hutchinson 2001, 8). My argument regarding the relationship between the protocol and the SPS Agreement is that the framing of precaution normalized through the Cartagena Protocol expands upon and adds new detail to the approach to risk assessment and management found in the SPS Agreement. At the same time, some of the trade disciplines of the WTO agreements must also now be seen, in discursive terms, as disciplines of the Cartagena discourse of precaution.

The need to look at the Cartagena discourse of precaution as closely linked with the SPS approach to risk in the field of LMO trade stems from the compromises reached on the savings clause in the final three paragraphs of the protocol's preamble. To recap, the first of these paragraphs recognizes that trade and environment agreements should be "mutually supportive," the second emphasizes that the protocol should not be interpreted as "implying a change in the rights and obligations" of parties under other international agreements, and the third notes that the previous two paragraphs are "not intended to subordinate this Protocol" to other agreements (CBD 2000a, Preamble). The key issue that the negotiators of these paragraphs were attentive to was the possibility of a trade dispute and of the potential for conflict between the protocol and WTO agreements.

It remains unclear how a dispute over LMO biosafety restrictions involving states that are WTO members and parties to the protocol would be settled. At this point, there has been only one WTO dispute regarding GEOs. This involved the complaints submitted, in early 2003, by the United States, Canada, and Argentina about the EU's moratorium of 1999-2004. The details of that case, and how the Cartagena Protocol relates to it, are discussed later in this chapter. The most important point to note here is that the case did not involve states that were both members of the WTO and parties to the protocol. The EU was beholden to both treaties but the complainants were not. As a result, even though the EU referred to the protocol in its defence, the dispute resolution panel held that states could be held to account only for their obligations to treaties they had formally ratified (WTO 2006, 299).

What would happen in a dispute among states that are both members of the WTO and parties to the protocol? Would the case be heard by a WTO dispute settlement panel, or would it go through the dispute settlement procedure of the protocol? If the latter, this would mean the case would go to the International Court of Justice, the dispute settlement procedure of the protocol's parent treaty, the CBD. This is a method of dispute settlement that has not yet been used (Cors 2000). Even if the case were to go to the WTO, how would the protocol be treated by a panel and the Appellate Body? In the WTO GE case, the panel explicitly avoided taking a position on how it would deal with a situation in which another international agreement, such as the Cartagena Protocol, were relevant in a case in which the parties in the dispute were parties to that agreement, but that agreement was not binding on all WTO members (WTO 2006, 301).

The answers to the above questions are said to lie in the three preambular paragraphs on the relationship between the protocol and other international agreements. But these paragraphs appear contradictory. In fact, their meaning remains a subject of ongoing discussion among academics and trade lawyers. The view of Safrin (2002, 445), legal counsel to the US Biosafety Protocol delegation, is that the middle of the three paragraphs in the preamble is effectively a savings clause giving primacy to WTO agreements. That this text is in the preamble and not in the operational part of the treaty is deemed irrelevant by her. Furthermore, Safrin sees the final paragraph as simply a statement that the presence of a savings clause does not make the protocol any less important than related trade agreements, even though it is beholden to them (446). By contrast, the perspective of Afonso (2002), a member of the Legal Service of the European Commission, is that in a case of conflict between the Biosafety Protocol and the WTO, the non-subordination clause in the preamble means that even a WTO Dispute Settlement Body panel will accept the protocol as the authoritative text. If a panel deems it necessary, it would revisit the records of the negotiations in order to understand the full meaning of what was included in a preamble. These records clearly show, according to Afonso, that a savings clause giving pre-eminence to trade agreements was never the intention of the biosafety negotiators.

A biosafety negotiator from the Philippines believes that "the loose and open-ended language of these paragraphs could create dangerous precedents for all other sustainable development instruments" (Muller 2002, 145). There may be some truth to this statement. Given that these paragraphs are now found in several international treaties and that they will likely find their way into more (the 2001 International Treaty on Plant Genetic Resources for Food and Agriculture also now includes the original three preambular paragraphs taken from the Rotterdam Convention [Barnes 2001, 4-5]), a great deal hangs on how these paragraphs will be interpreted in a trade dispute. Nonetheless, to date no multilateral environmental agreement has ever been directly challenged at the WTO, despite that many have trade implications. Some analysts believe that multilateral environmental agreements (MEAs) are not challenged through the WTO because of the popular support that environmental treaties receive (Isaac, Phillipson, and Kerr 2001). In Gramscian terms, we might say that once an international environmental regime is accepted by international civil society and the community of states, entrenched through supportive ideational, organizational, and material relations of force, legal judgments that go against accepted norms (even if passed by a WTO panel or the Appellate Body) may be less than helpful for those critical of the MEA. A direct challenge to the MEA may only weaken the integrity of the WTO system itself, should it be successful.

For political reasons, then, it is fair to expect that WTO panels and the WTO Appellate Body will look to the Cartagena Protocol for guidance in cases involving LMO trade restrictions when the states involved are beholden to both regimes. At the same time, the first two of the preambular paragraphs mean the protocol's provisions must be read in the context of relevant WTO agreements. Where the protocol's articles provide more specificity than does the SPS Agreement, new rules for international trade are indeed now in place for LMOs. On the other hand, provisions of the SPS Agreement and the disciplines of the other WTO agreements will continue to hold sway over WTO members in areas where the protocol offers no changes or added detail. In discursive terms, this relationship (and the fact that it was front of mind for protocol negotiators) means that the Cartagena discourse of precaution must be understood as having internalized certain WTO disciplines in any examination of the normalizing effects of the new precautionary discourse.

What does the Cartagena discourse of precaution's internalization of WTO disciplines mean in concrete terms? The protocol has established new trade norms, such as the required AIA notification processes for the first transboundary movements of certain classes of LMOs yet to be demonstrated as harmful. At the same time, there are strings attached to the kinds of decisions states may take in the AIA process, based on their obligations as WTO members. Consider the case of precautionary decision making in cases where states are beholden to both the protocol and the SPS Agreement.

As discussed in Chapter 6, the Cartagena Protocol sets slightly different standards that must be met for precautionary decision making on LMO imports from the SPS Agreement. Article 5.7 of the SPS Agreement allows members to "provisionally adopt sanitary or phytosanitary measures on the basis of available pertinent information ... in cases where relevant scientific evidence is insufficient." Furthermore, it requires members to "seek to obtain the additional information necessary for a more objective assessment of risk and review the sanitary or phytosanitary measure accordingly within a reasonable period of time" (WTO 1994a, Article 5.7). In contrast, Articles 10.6 and 11.8 of the protocol state that a "lack of scientific certainty due to insufficient relevant scientific information and knowledge regarding the extent of potential adverse effects of an LMO ... shall not prevent the Party of Import from taking a decision." And this decision is not necessarily provisional, though Article 12.2 states that a party of export may "request that a Party of import ... review a decision [if] a change in circumstances has occurred that may influence the outcome of the risk assessment ... [and] additional relevant scientific or technical information has become available" (CBD 2000a, Articles 10, 12).

Hutchinson (2001) argues that the protocol's phrase "insufficient relevant scientific information and knowledge" is an explicit reference to the potential problem of data gaps and ignorance that can severely restrict the level of scientific certainty about the identification and evaluation of possible adverse effects of GEOs. It appears as if the protocol would permit states to take precautionary measures in the light of these kinds of data gaps, provided that they can provide, at a minimum, a scientific rationale (whether quantitative or qualitative) for the measures. Article 5.7 of the SPS Agreement has never been interpreted so liberally. In addition, in one case when a panel and appellate board were asked to adjudicate on a provisional measure, they concluded that the adoption of such a measure (by Japan) for over three years exceeded a "reasonable period of time" (WTO 1999, para. 93). On this issue, the protocol also appears to have a lower threshold for precautionary action because it creates no obligations for the amount of time a measure can be in place. This comparison shows that the Cartagena Protocol establishes somewhat lower obligations for states to meet when taking precautionary measures to protect against LMOs than does the SPS Agreement. The protocol appears to recognize that precautionary decisions are ultimately political judgment calls made in the face of incomplete science; these decisions cannot be tied to assumptions that more and better science will soon sort the matter out.

Hutchinson (2001, 30) points out, however, that the SPS Agreement can be read to be consistent with the new standards for precautionary decision making contained in the protocol: "Article 5.7 of the Agreement ... may be interpreted to accommodate the low threshold of action mandated under the Protocol." Cottier (2002, 478) concurs that the protocol's overall approach to risk analysis can be seen as adding specificity to the SPS Agreement: "It is conceivable to construe the provisions on risk assessment [found in WTO agreements] in light of the more advanced and better rules on risk assessment and risk management of the protocol." Furthermore, the same appellate body that ruled that Japan's provisional measures were in place for too long stated that "a 'reasonable period of time' has to be established on a case-by-case basis and depends on the specific circumstances of each case, including the difficulty of obtaining the additional information necessary for the review and the characteristics of the provisional SPS measure" (WTO 1999, para. 93). This case-by-case approach suggests that the Appellate Body is open to the idea that some gaps in risk assessment data may take many years to fill, if they are ever filled at all. Such a possibility is consistent with the EU's interpretation of Article 5.7 released in its communication on the precautionary principle:

> According to the SPS Agreement, measures adopted in application of a precautionary principle when the scientific data are inadequate, are provisional and imply that efforts be undertaken to elicit or generate the necessary scientific data. It is important to stress that the provisional nature is not bound up with a time limit but with the development of scientific knowledge. (European Commission 2000, 12)

The SPS Agreement thus can be interpreted in such a way as to accommodate the new rules for precautionary risk assessment and management established in the protocol.

In addition to establishing what appear to be lower thresholds for precautionary action and less onerous expectations of the length of time a provisional measure may be in place, the protocol adds further definition to precautionary action than does the SPS Agreement in a number of other ways. Cosbey and Burgiel show how the protocol better defines the precautionary principle by (1) spelling out the details of a risk assessment for LMOs (found in Annex 2 of the protocol); (2) allowing parties to take into account socio-economic considerations in making decisions (although in a limited way; Article 26); and (3) placing the onus on the exporter to establish the harmless nature of an LMO in question by allowing importers to require exporters to pay for risk assessments (as found in Article 15, paras. 2 and 3) (Cosbey and Burgiel 2000; CBD 2000a). None of these provisions is provided for in the SPS Agreement.

While the protocol expands the international understanding of what a precautionary approach to LMOs looks like in practice, these commentators also point out – and this is a key point in the context of the argument put forward here – that importers wishing to take precautionary action must also fulfill certain requirements of the SPS Agreement that are not covered by the protocol (assuming, once again, that they are parties to both agreements and that the LMO-restricting measure is also an SPS measure). Most important, the SPS Agreement requires members, when setting the level of protection desired, to "take into account the objective of minimizing negative trade effects" (Article 5.4), including ensuring "that such measures are not more trade-restrictive than required to achieve their appropriate level of sanitary or phytosanitary protection, taking into account technical and economic feasibility" (Article 5.6). Members must also "avoid arbitrary or unjustifiable distinctions in the levels it considers to be appropriate in different situations, if such distinctions result in discrimination or a disguised restriction on international trade" (Article 5.5) (WTO 1994a, Article 5).

Taken together, these trade disciplines require a government to proceed carefully in making any decisions on how it will regulate products that are deemed, through the decision framework accepted under the protocol, to have potential impacts on human, animal, or plant health. These disciplines are, as Cosbey and Burgiel (2000, 11) point out, "fairly onerous obligations and clearly balanced toward protecting commercial interests from unfair discrimination." Because of the way that the Cartagena Protocol was negotiated in the shadow of the WTO agreements, in discursive terms this new internationally accepted framing of precaution, or what I call the Cartagena discourse of precaution, must be seen as effectively reproducing these WTO disciplines for measures pertaining to LMOs.

This multi-layered understanding of what a precautionary response to living GEOs entails that resulted from the Cartagena negotiations has clearly evolved from what the authors of Article 19.3 of the Convention on Biological Diversity had conceptualized back in 1992. The Cartagena discourse of precaution singles out GEOs from other organisms for distinct regulatory treatment but requires that regulatory actions treat domestic GEOs and imported ones in the same way. It accepts as legitimate regulatory action taken in the absence of full scientific certainty, thereby erring on the side of protecting consumers over producers. However, it requires such action to be based on scientific assessments of individual GEOs, in this way foreclosing the possibility of blanket bans. This discourse accepts the appropriateness of setting up new hurdles to the first intentional transboundary movement of many classes of living GEOs (through the requirements of AIA) but does not go so far as to put up new hurdles to the international trade in bulk commodities. This discourse also accepts that socio-economic considerations may enter into risk assessment and risk management decisions pertaining to LMOs, but only if countries taking such decisions are still living up to their obligations under the WTO – obligations designed to limit protectionism for economic reasons. What are the implications of this discourse, and the treaty that embeds it in international environmental politics, in the realms of ideas, organizations, and material capabilities?

Ideational Politics

This study of the Cartagena Protocol has demonstrated how ideas function in at least two ways in global politics. The ideas of neo-liberalism, equivalency, and risk, for example, each operated as ideologies, providing ways of thinking about genetic engineering and regulation around which diverse actors coalesced. These particular ideas also functioned discursively once widely adopted, impacting on (but not necessarily determining) regulatory conversations about GEOs by setting boundaries around what was, and was not, relevant. When considering the implications of the Cartagena Protocol in the realm of ideas, then, it is important to examine both the ideological and discursive effects of the ideas embedded in this new international agreement.

The Cartagena Protocol's most significant impact in the realm of ideas relates to its operationalization of the precautionary principle. Ideologically, the protocol as a whole has enabled a precautionary view of the GEO issue to gain further credibility, and thus new adherents, on the international stage. Discursively, the protocol negotiations also contributed to shaping what precaution means on the ground, in the field of GEO safety. Evidence of the productive effects of the Cartagena framing of precaution can be found in the European Commission's 2000 position paper on precaution and the European Parliament's 2001 and 2003 regulations on GEOs. Signs of the disciplinary effects of this discourse can also be found in ongoing struggles in the EU over the implementation of these policies.

Post-Cartagena Interpretations of Precaution in the EU

Within five days of the conclusion of the Montreal ExCOP, the European Commission released a paper outlining its interpretation of the precautionary principle (European Commission 2000). This paper was closely connected to the Cartagena Protocol talks. It had been developed in 1999 in order to achieve the intra-EU consensus on the principle that was needed to bring a common position forward in the biosafety negotiations. Once the talks were completed, and once the EU saw that its interpretation of precaution was consistent with the protocol's text, the Commission's paper was made public (interview #9).

In the paper, the precautionary principle is framed in the context of risk management. In their application, measures taken on behalf of the precautionary principle should be (1) proportional to the level of protection desired; (2) non-discriminatory; (3) consistent with similar measures already taken; (4) based on an examination of the potential benefits and costs of action or lack of action (including, but not limited to, economic costs and benefits); (5) subject to review in the light of new scientific data; and (6) capable of assigning responsibility for producing the scientific evidence necessary for a more comprehensive risk assessment (European Commission 2000, 3-4).

The biosafety negotiations were clearly influential on this elaboration of precaution. Previously, in the European Commission's 1998 appeal in the WTO beef hormone dispute, the Commission's lawyers had argued that the precautionary principle enabled the EU to override its SPS Agreement obligation to base a ban on hormone-tainted beef on scientific risk assessment (Charnovitz 2000). In other words, this earlier position saw the precautionary principle as existing outside the domain of risk assessment and management, as a separate level of political decision making. In its 2000 communication, the precautionary principle is framed as an integral component of risk analysis. This new understanding reflected the EU-Miami Group compromise on precaution that also appears in the decision-making clauses of the Cartagena Protocol, a compromise that must be contextualized in light of both protocol- and WTO-related debates.

Despite its acknowledgment that the precautionary principle's role is within the risk analysis process, the communication paper maintains that "judging what is an 'acceptable' level of risk for society is an eminently *political* responsibility" (Commission 2000, emphasis in original). This statement reflects one of the few EU victories in the beef hormone dispute. In that case, the WTO Appellate Body affirmed that a state can set a regulatory bar as high as it likes, as long as the standard is scientifically justifiable and non-discriminatory (i.e., applies equally to domestic production and imports) (WTO 1998a). The Commission's comment also speaks to the EU's recent experience with the GEO moratorium, which strongly informed its position in the biosafety talks and in the communication paper. The moratorium was a result of European publics' demands for precautionary action vis-à-vis risks that, to many in GEO-exporting countries, appeared fairly insignificant and remote. States more closely allied with the biotech industry were uncomfortable with the idea that risk decisions be based on anything but clear evidence of potential harm. Still, the EU was determined to infuse the overall risk debate with a recognition that political judgment figures centrally in any and all decision about risks, and that this judgment should be proportionate to the level of acceptability of a risk (rather than to the level of potential harm). This perspective was encapsulated in a statement by UK environment minister (at the conclusion of protocol negotiations) Michael Meacher (2002, 232):

Risk, and its management, reflects prevailing scientific views and their perception and acceptance by civil society; these can change over time, and differ from one country to another and from one culture to another. The Protocol recognizes these differences, not as a licence for arbitrary and unscientific decision-making, but as an integral part of a comprehensive and accountable framework for assessing and managing risks to global biodiversity.

From the EU perspective, this reading of risk is now entrenched in the protocol's articles on precaution, and it will continue to figure centrally in the EU's interpretation of the precautionary principle in all other fora, as evidenced by the communication paper.

What the new European reading of risk and precaution meant in practice became clear in March 2001, when the European Parliament passed its new regulatory directive 2001/18/EC on the deliberate release of GEOs (termed "genetically modified organisms") to the environment and to the market (European Parliament 2001). This directive stands in marked contrast to the previous directive, 90/220, which did not even mention the term "precaution." Directive 2001/18 notes that "the precautionary principle has been taken into account in the drafting of this Directive and must be taken into account when implementing it," and draws specific attention to the Cartagena Protocol in justifying its approach (European Parliament 2001, 1). The new regulation moves beyond 90/220 by giving only time-limited consent for releases of GMOs. It also requires the labelling as well as traceability of GEOs introduced into the environment and the market. The regulation calls for extensive scientific evaluation of GEOs that must take into account "potential cumulative long-term effects associated with the interaction with other [GEOs] and the environment" before their release. Assessments must consider any direct or indirect, immediate, delayed, or unforeseen effects from GEOs. It also notes that "it is important not to discount any potential adverse effect on the basis that it is unlikely to occur." This regulatory directive calls for verification of the genetic and phenotypic stability of the insert and the GEO, and it calls for a phase-out of antibiotic resistance markers in new GEOs. There are also provisions for taking into account ethical considerations, and for public consultations, which "shall be taken into consideration." The regulatory directive also states that the details of environmental risk assessments must be accessible to the public. Finally, 2001/18 requires monitoring of all GEOs once they are released to the environment.

In 2003, the European Parliament and Council passed two more regulations that spell out their precautionary approach to GEOs: Regulation EC 1829/2003 contains regulations on genetically modified food and feed, while

EC 1830/2003 concerns the traceability and labelling of genetically modified organisms, and the traceability of foods derived from genetically modified organisms (European Parliament and the Council of the European Union 2003a and 2003b). These two pieces of regulation were designed to supplement or amend 2001/18, the 1997 regulations on novel foods, and the 1998 rules on labelling (European Parliament 1997a and 1998). One important feature of 1829/2003 is that it rejects the simplified notification procedure based on "substantial equivalence" for foods that are products of GEOs but no longer contain DNA that was present in the 1997 Novel Food Regulations. The new food and feed regulations state:

> Whilst substantial equivalence is a key step in the procedure for assessment of the safety of genetically modified foods, it is not a safety assessment in itself. In order to ensure clarity, transparency and a harmonised framework for authorisation of genetically modified food, this notification procedure should be abandoned in respect of genetically modified foods. (European Parliament 2003a, Preamble, para. 6)

Another difference between the new regulations and previous iterations is that all products derived directly from GEOs will now have to be labelled as such, while oils from GE crops, among other products, escaped previous labelling laws if they did not contain the proteins produced by recombinant DNA. The 2003 regulations also include the EU's interpretation of acceptable thresholds of GE ingredients in labelled non-GE products. Such GE ingredients, which must still be approved in the EU, may make up to 0.9 percent of a non-GE product as long as they are "adventitious" (technically unavoidable).[1] An important procedural change found in the EU regulations of 2003 is that applications to have GEOs approved in Europe will be directed to the European Food Safety Authority (EFSA; established in 2002). Under earlier regulations, applications first went to a Competent Authority in one of the EU member states for an assessment, and the information was then shared with other National Competent Authorities.

The EU regulations of 2001 and afterwards appear to be more precautionary in the way that many critics of GEOs had wanted to see. These regulations are designed to meet the high expectations of European consumers at the expense of putting new costs on industry (for labelling and traceability, for example). Through detailed risk assessments and measures to include public participation in regulatory decision making, these regulations are also designed to rebuild consumer confidence in food after the BSE debacle of the 1990s and other food scares. Nonetheless, the institutionalization of new regulations in Europe has not brought the debate over GEOs to a close on that continent. Even as the first GEOs were approved for cultivation and food use in 2004 under the new regulations, these authorizations came only through the controversial comitology process (Greenpeace 2005c). Comitology is controversial because of its undemocratic structure. In the case of GEO approvals, this process gives unelected civil servants in the EFSA the right to make final decisions on behalf of the European Commission when the Environment Council, the body made up of the elected environment ministers of member states that is ostensibly responsible for decisions on GEO authorizations, does not reach a qualified majority. In the spring of 2006, the controversy around the use of comitology to have GEOs approved for food and agricultural uses led EU Environment Commissioner Stravos Dimas and Health and Consumer Protection Commissioner Markos Kyprianou to introduce new measures for ensuring that the EFSA will work more closely with member states on GEO risk assessments (European Commission 2006). It remains unclear whether such procedural amendments will appease those members that want to continue to take a more cautious approach to the adoption of GE crops.

Furthermore, while environmental CSOs have largely supported the efforts to strengthen the EU's regulatory system, they continue to hound European regulators on several issues. These issues include the access to environmental assessment information promised in Directive 2001/18 and the question of labelling thresholds. CSOs also remain concerned about the adequacy of risk assessments that, as in North America, continue to rely primarily on data provided by the biotech companies themselves. European CSOs are concerned too that animal products (such as meat and eggs) raised on GE feed will not have to be labelled and traced through the system as are foods made directly from GEOs (Greenpeace 2005b; FOEE 2003).

The negotiation of the Cartagena Protocol clearly played a supportive role in the creation of more precautionary – though still contested – GEO regulations in Europe. The protocol negotiations did this by establishing greater international legitimacy for precautionary regulatory assessments of GEOs, the concept on which the new EU regulations are built. The disciplines of the Cartagena discourse of precaution have also affected subsequent political developments concerning GEOs in Europe, as illustrated by the debates over country-specific and regional GEO bans.

By early 2006, five EU member states (France, Austria, Germany, Greece, and Luxembourg) continued to have in place bans on specific GEOs for cultivation, food, feed, or processing based on invocations of the safeguard clauses in Directive 1990/220 or Directive 2001/18. In each of these cases, the GEOs in question had received European-level approval, but individual member states raised environmental or human health-related safety objections. At the time of this writing, these bans remain the subject of intense debate. On the one side are the European Commission and the EFSA, which argue that the bans are not scientifically justified (Pew Initiative on Food and Biotechnology 2005). On the other side are the European environment ministers and many CSOs that seek to continue the GEO bans "if there are continued questions over their safety" (Anonymous 2005a, 6).

A second, related debate has emerged about the ability of states or regions to ban GEO production entirely, in order to protect traditional or organic crops from being contaminated (via pollen flow and seed mixing) with transgenes. In 2003, for example, the Austrian province of Upper Austria passed the Genetechnology Prohibition Law that prohibits the cultivation of GE crops in the region, "based on a precautionary approach," because GE crops could potentially damage ecosystems (Anonymous 2005b). This law was struck down in the EU Court of First Instance in late 2005 as being scientifically unjustified (Anonymous 2005b). To proponents of GEOs, these types of regional bans appear to be directed at protecting markets for GE-free products. In other words, they represent (socio-)economic restrictions on the production of GEOs that cannot be justified by assessments of health and environmental risks.

The disciplines of the Cartagena discourse of precaution are evident in these debates over GEO bans, even though the protocol itself is only one of a set of international and domestic law towards which EU states are beholden. The Cartagena discourse accepts as legitimate case-by-case examinations of the risks of GEOs. A blanket ban on all GEOs, by contrast, is not an acceptable precautionary action. It is this position that the European Commission, the very body that negotiated the precautionary clauses in the protocol for the EU, has been trying to get EU member states to accept since the new EU regulations were put in place, though with only limited success:

> Based on legal and scientific considerations, and the case-by-case approach of the environmental legislation, the adoption of blanket policy aimed to make Member States or any particular part of it "GM-free" would not be acceptable when seeking to impose conditions that would not be justified in terms of protection of human health and the environment. (European Commission 2004)

A desire for consistency with the Cartagena Protocol, and in particular the compromises this negotiation involved on precaution decision making, is one factor that has influenced the European Commission's stance against blanket bans in intra-EU debates. This desire was also related to another, more immediate, political factor at play in these debates: the WTO trade dispute over GEOs between the EU and the United States, Canada, and Argentina that began in 2003. In that trade dispute, the complainants argued that country-wide bans on specific GEOs already approved at the community level were not justifiable on scientific grounds, and the dispute resolution panel agreed with the complainants on this issue in its 2006 interim ruling (WTO 2006). Even before the ruling was publicized, however, the European Commission was making every effort to show its trading partners that the EU was moving towards a clear, transparent, and non-discriminatory (albeit stringent) regulatory regime that was consistent with the Cartagena Protocol on Biosafety on risk assessments pertaining to environmental safety (WTO 2006; European Communities 2004).

The example of the European Commission's communication on the precautionary principle, along with the EU's new regulatory framework for GEOs and disputes about its implementation in relation to national and regional bans, all point to the continued influence of the idea of precaution in the EU since the protocol's inception. This influence has been buoyed by public concern and official scientific skepticism. The specific shape that precaution has taken regarding GEOs has also clearly been informed by the compromises reached between the EU and other OECD countries in the Cartagena Protocol. These compromises reject blanket GEO bans and GEO restrictions based primarily on potential socio-economic impacts – two examples relevant here. In other international fora related to GEO regulation, we can also identify productive impacts of the Cartagena Protocol's framing of precaution. Consider the case of the Codex Alimentarius.

Codex Alimentarius Principles and Guidelines
In July 2003, lengthy debates in a Codex Alimentarius task force on GEO issues resulted in the formal adoption of three documents. The first is a statement of principles that provides a global framework for evaluating the safety and nutritional aspects of GEOs. The second and third are guidelines on the safety assessments and labelling of foods produced from GE plants and food produced with the aid of GE microorganisms (Codex Alimentarius Commission 2003a, 2003b, 2003c). Taking place as they did in the wake of the Cartagena Protocol, the negotiations that led to these principles and guidelines might be expected to also have entrenched the discourse of precaution. This type of outcome is precisely what many of the CSO participants, such as Consumers International, sought by citing the example of the Cartagena Protocol (Consumers International 2002). In the end, the Codex principles and guidelines were influenced by the Cartagena Protocol and its framing of precaution in a number of ways, though not to the extent that Consumers International had hoped for.

The best example of the direct influence of the protocol in the Codex Principles is in its scope, which is focused on "foods derived from Modern Biotechnology." Like the protocol, the Codex principles single out the products of genetically engineered organisms for distinct regulatory treatment rather than grouping them with wider classes of novel organisms (as seen in North American legislation). In making this move, the Codex principles adopt, word for word, the definition of modern biotechnology found in the Cartagena Protocol (Codex Alimentarius Commission 2003a, para. 8).

The Codex principles and guidelines break with the equivalency assumptions underpinning regulatory frameworks in the United States and Canada in a number of other ways. The guidelines go farther than the voluntary consultation process in place at the FDA in the United States, for example, by requiring "a premarket safety assessment following a structured and integrated approach ... on a case-by-case basis" for all GEOs. Such assessments are expected to be based on data from various sources (e.g., the applicant's data, published scientific literature, and data from independent scientists) and include consideration of both intended and unintended effects of genetic engineering, such as gene silencing (Codex Alimentarius Commission 2003b, paras. 14-17).

The mention of unintended effects and gene silencing exemplifies that complexity talk has been gaining ground in international regulatory circles. Another example of the move away from equivalency assumptions is the principles' requirement that authorities take into account the uncertainties identified in the risk assessment and implement measures to manage these uncertainties including, "as appropriate, food labelling conditions for marketing approvals and post-market monitoring" (Codex Alimentarius Commission 2003a, paras. 16-18). The principles also call for safety assessments to be "fully documented and open to public scrutiny," which is more open than current practices in the United States and Canada (Codex Alimentarius Commission 2003a, para. 23). Furthermore, they specifically state that, for the foreseeable future, no GEO can be considered a "conventional counterpart" for the purposes of establishing the substantial equivalence of subsequent organisms in the safety assessment process, despite efforts by US delegates to have already-approved GEOs accepted as "conventional" (Codex Alimentarius Commission 2003a, para. 8 and n. 5; Bereano 2002).

These examples and others show how far the Codex principles and guidelines have moved away from what the biotech bloc would have liked to see implemented by a WTO standard-setting body in the mid-1990s. Still, the guidelines do not go quite as far as the Cartagena Protocol in several areas. For example, there are no specific references to precautionary decision making in these documents.

The principles do note that they should be read in conjunction with the *Codex Working Principles for Risk Analysis*, and that document, also approved in 2003, states that "precaution is an inherent element of risk analysis" and that "the degree of uncertainty and variability in the available scientific information should be explicitly considered in the risk analysis" (Codex Alimentarius Commission 2004a, paras. 11, 23, 24). Furthermore,

> the report of the risk assessment should indicate any constraints, uncertainties, assumptions and their impact on the risk assessment. Minority opinions should also be recorded. The responsibility for resolving the impact of uncertainty on the risk management decision lies with the risk manager, not the risk assessors. (Codex Alimentarius Commission 2004a, para. 25)

Despite these articles, the Codex iteration of precaution is not nearly as strong as that found in the decision-making articles of the Cartagena Protocol. The guidelines do appear to give room for risk managers to take precautionary decisions in the face of limited or uncertain risk assessment data, and the Codex principles do state that "other legitimate factors relevant for the health protection of consumers" can be taken into account in deciding on food standards (Codex Alimentarius Commission 2004b, para. 2). But whether states can adopt "precaution" in decision making as one such factor remains the topic of contentious debate in Codex meetings. These debates occur along familiar fault lines, with the United States, Canada, and Argentina on one side of the debate and the EU on the other (Anonymous 2002a).

The centrality of "substantial equivalence" in the guidelines is also cause for concern. Although it is acknowledged, as in the new EU regulations, that "substantial equivalence is not a safety assessment in itself" (Codex Alimentarius Commission 2004b, para. 16), the use of a comparative approach based on this concept can still lead to a focus on the potential hazards associated with the intentionally introduced trait, at the expense of a wider study of the hazards associated with the GE food as a whole. This is one of the main concerns with the use of substantial equivalence in North American GEO regulation, discussed in Chapter 3. As the Royal Society of Canada Expert Panel on the Future of Food Biotechnology pointed out in 2001, a rigorous assessment of "substantial equivalence" can be used to support conclusions about food safety only if it is based on a detailed examination of the novel organism and its conventional comparator at four levels: DNA structure (including a search for unanticipated insertions); gene expression; a proteomic analysis; and secondary metabolite profiling (which provides more relevant detail than a standard proximate analysis) (S. Barrett et al. 2001, 187-89). The Codex documents' emphasis on unintended effects (to which five paragraphs are devoted), along with detailed technical requirements of premarket safety assessments, suggest a wide-ranging approach that will evaluate the new GEO as a whole on each of these levels.

Knowing whether this has actually taken place in regulatory practice, however, depends on the ability of independent scientists to scrutinize decisions, and the risk assessment data on which they are based, under the principles' transparency requirements.

Another example demonstrating that GEO critics may have had only limited success in the Codex negotiations is the ambiguous outcome on the traceability debate. France sought Codex reinforcement for the approach being established in EU regulations, which sees the traceability of all GEOs, from seed to plate, as an inherent part of a precautionary approach that allows regulators to determine their origins if unexpected health problems develop. The United States and its allies, on the other hand, argued that traceback – a tracing system for tracking a product back to its source once a hazard is established – is more than adequate as a risk management tool (Bereano 2002). The final outcome in this debate states that specific tools may be needed to facilitate risk management, including the "tracing of products for the purpose of facilitating withdrawal from the market when a risk to human health has been identified or to support post-market monitoring" (Codex Alimentarius Commission 2003a, para. 21). "Tracing" has not appeared as a term in regulatory circles before this document and is clearly designed to be an ambiguous compromise between traceability and traceback.

As Bereano, a CSO representative at the negotiations, notes, "the compromise wording satisfies both sides – each can pretend they got their way – but actually just avoids the real controversy" (Bereano 2002, 6). In an effort to ensure consistency between the Codex outcomes and the CSO victories in the Cartagena Protocol, this same CSO representative made sure that the draft report of the sessions recorded – against the US delegation's objections – that "identification requirements of Article 18 of the Cartagena Protocol [have] relevance to the use of traceability" and that "applications on product tracing would also need to be consistent with the provisions of the Cartagena Protocol after its entry into force" (Bereano 2002, 8).

The Codex principles and guidelines are clearly more precautionary than the biotech bloc would have accepted only a few years earlier, and the Cartagena Protocol's negotiations, along with concomitant developments in Europe, were important in achieving this outcome. Again, however, these documents do not go as far as GEO critics such as Consumers International were looking for. The differences between the Codex and Biosafety Protocol outcomes demonstrate that the dynamics of the biosafety negotiations were indeed specific to those circumstances. The actors, discourses, timing, and the particularly institutional setting of the CBD all contributed to the precautionary outcomes realized there; relations of force were different in the Codex. I cannot undertake a detailed analysis here, but will note that one major difference is the extent to which transnational food and agri-chemical companies directly participate in Codex committees in general, in contrast to the enormous impact of public interest CSOs in the biosafety realm (Avery, Drake, and Lang 1993; Pardo-Quintillan 1999).

Together, these examples from EU regulation and the Codex Alimentarius negotiations show that the Cartagena Protocol has had an influence in the field of international GE regulation, but that results are mixed. These outcomes, especially when juxtaposed with North American regulatory models, demonstrate the ongoing lack of unity in the global regime for GEOs. In the wider multilateral environmental regime, we can also find evidence that the Cartagena Protocol's framing of precaution has had distinct impacts. The 2001 negotiations on an international treaty regulating persistent organic pollutants (POPs) offer a case in point.

Stockholm Convention on Persistent Organic Pollutants

Persistent Organic Pollutants, or POPs, are chemicals that remain intact in the environment for long periods, become widely distributed geographically, accumulate in the fatty tissue of living organisms, and are toxic to humans and animals. Examples include DDT, dieldrin, and PCBs. During negotiations on an international POPs treaty in South Africa in December 2000, it was widely accepted that at least twelve chemicals (the "dirty dozen," which include the above-mentioned three) should be banned or severely restricted because of their human health and environmental impacts as endocrine disruptors. The EU, along with many environmental CSOs active on the issue, wanted the POPs treaty to also set out procedures for identifying and controlling new chemicals before their widespread use, if they were likely to exhibit some of the same characteristics as the dirty dozen. These arguments were based on the precautionary principle as a rallying ideology and often drew on, as an example, the operational precautionary clauses in the Cartagena Protocol (e.g., Blojkovac and Muldoon 2000, 12). In response, the United States and Canada, along with industry CSOs, opposed the principle in anything but the preamble to the POPs treaty.

In the end, the precedence of the Cartagena Protocol allowed the EU and environmental CSOs to win this part of the POPs debate. Precaution was again accepted as the common sense. Article 8.7(a) of the final text of the Stockholm Convention on Persistent Organic Pollutants, concluded in May 2001, notes that "lack of full scientific certainty shall not prevent" a committee set up to review additional chemicals from moving forward with its recommendations. Article 8.9 then states that the "Conference of the Parties, taking into account the recommendations of the Committee, including any scientific uncertainty, shall decide, in a precautionary manner," how to control these chemicals (Stockholm Convention 2001). To achieve the precautionary language in the final text, the EU had to accept a US proposal that this committee not begin to function until after the entry into force of the Stockholm Convention (this event took place in May 2004; Stockholm Convention 2006). Now that the convention is operational, this committee's precautionary mandate should have an impact on the way new chemicals are dealt with.

Institutional Politics

The Convention on Biological Diversity and the SPS Agreement of the WTO had profound influences on the formation of the Cartagena Protocol. In the same way, this new international agreement can be expected to impact on a wide variety of future international and domestic institutions. One of the ways the protocol can be expected to have long-term productive effects is through its transformation of the concept of prior informed consent (PIC) into an advanced informed agreement (AIA) procedure. AIA, as elaborated in the protocol, moves beyond PIC by applying to substances that *may* be risky, rather than those already widely accepted as dangerous. This shift can be expected to set a precedent for the development of similar procedures for pharmaceuticals and other industrial products.

That AIA is based on explicit consent also represents a major development in international law. Historically, governments from industrialized countries have had little regard for the environmental and human health effects of products created in their jurisdictions but consumed elsewhere in the world. More recently, international treaties on the trade in hazardous waste and dangerous chemicals have broadened the responsibilities of exporters. The protocol takes another step in that direction.

There are, of course, important exceptions to the AIA procedure in the Cartagena Protocol that will limit its effectiveness. Most important, bulk commodities such as corn and soy intended for food, feed, or processing are not covered by AIA. For these shipments, importing countries have the responsibility to follow up on risk assessment information provided by exporters to a web-based clearing-house in order to decide whether to accept shipments. This exception is substantial because grain sold as food or feed can still introduce transgenes into the environment. Nonetheless, the shift to exporter responsibility for LMOs intended for direct introduction into the environment is an important development in the international politics of risk. This move represents one more step towards internalizing the onus for assessing the potentially negative impacts of a technological change in the states and companies promoting that change.

Another way that the protocol will have long-term institutional effects is through its articles on capacity building.

Building Biosafety Capacity

That the AIA procedure is based on explicit consent means that LMOs will be approved only once domestic laws and regulations are in place and functioning. This, coupled with Article 22 of the Biosafety Protocol, which calls for cooperation in the building of regulatory capacity in developing countries and in countries with "economies in transition," has meant that these nation-states have been able to leverage funds and technical support from wealthier states to establish LMO regulatory systems.

What does technical support and funding for biosafety capacity development mean in practice? Canadian civil servants state that they recognize the responsibility AIA places on the shoulders of LMO exporters, ethically and legally, and they say they are willing and prepared to take this responsibility seriously (interviews #1, #7, #13). While laudable, these admissions should tip us off to the fact that capacity building is far from a value-neutral exercise. Some in the Canadian biotech industry believe that Canada's participation in regulatory capacity building would be "paying for our own guillotine," since, under the Cartagena Protocol, "Canada's customer countries ... can use socioeconomic reasons to reject Canadian exports" (Jacqueline Duckering of Ag-West Biotech, quoted in Clapp 2002, 31). Given the deep involvement of agricultural scientists and government risk assessors from the United States, Canada, and elsewhere in the biotech bloc, it is just as possible that capacity building will result in the establishment of minimalist domestic regulatory requirements in the global South, regardless of the precautionary language of the protocol. The evidence from capacity-building efforts to date suggest that this area has indeed become a site of struggle between regulatory minimalists and those who argue for a more comprehensive and precautionary approach.

The United Nations Environment Programme, together with the Global Environment Facility (GEF), took the lead in promoting capacity building under the auspices of the Cartagena Protocol in November 2000 (UNEP 2002, 6). UNEP's experience in capacity building in developing countries goes back three years earlier, to similar work under the 1995 UNEP technical guidelines (UNEP 2005). The UNEP-GEF Pilot Biosafety Enabling Activity project assisted eighteen countries to prepare national biosafety frameworks and also conducted regional training workshops. In 2000, this project, now known as the Project on Development of National Biosafety Frameworks, was extended to include up to one hundred countries for the entry into force of the Cartagena Protocol. It has since broadened again to include up to thirty more countries.

The main focus of the development project is on strengthening capacity to develop national biosafety frameworks (UNEP 2005, 1-2). These frameworks incorporate regulatory systems, administrative systems, decision-making systems, and public information mechanisms (UNEP 2002, 6).

Another project was initiated in 2002 alongside the development project to implement national biosafety frameworks in twelve countries (eight of which are funded through UNEP-GEF, and two each by the United Nations

Development Programme and the World Bank). The goal of this project is to ensure that countries have in place a policy on biosafety; an operational regulatory regime for biosafety in line with the Cartagena Protocol or other relevant international obligations; a workable and transparent system for handling notifications (related to AIA) or requests for approvals; workable and transparent systems for public information; and a national biosafety website (UNEP 2005, 3). A third UNEP-GEF project was initiated in 2004 to assist eligible countries in using the Biosafety Clearing-House, the web-based information hub of the protocol. This project is intended to strengthen capacity by providing training to key stakeholders and by providing computer hardware and software so that countries can participate effectively in the clearing-house (4).

Altogether, US$50 million had been channelled into these three projects by May 2005 (UNEP 2005, 1). This money has gone a long way towards building domestic regulatory regimes for GEOs. Still, a number of issues are raised by these projects that need to be addressed. A survey of the work of the UNEP-GEF shows that the vast majority of the resources in this project are being devoted to creating legal, policy, and administrative frameworks in developing countries for regulating GEOs. There is no doubt that these are important goals. Parties to the protocol are already receiving notifications from parties of export requesting the "intentional transboundary movement" of LMOs as per Article 8 of the protocol. They need to have the administrative mechanisms in place to respond to such notifications, and they need to make decisions about import within a specified time.

The problem is that delegates from the global South demanded, from early on, that capacity building focus on developing scientific capacity for GEO safety assessment (interview #14). While UNEP-GEF does note that the biosafety framework project aims to "strengthen risk assessment and management capabilities," African participants in UNEP-GEF workshops have noted that capacity needs in this area have hardly been addressed in UNEP-GEF's work to date (interview #23). UNEP-GEF reports suggest that even workshops focused on risk assessment and risk management deal with these topics only at a superficial level, though participants recognize there is "world-wide divergence of opinions on, and approaches to, the potential risks of LMOs" (UNEP 2003, 3). That UNEP-GEF is intent on getting domestic notification and decision-making processes in motion, without an emphasis on building safety assessment capacity, may help explain the eagerness of GEO exporters to be involved in their projects: it was the Canadian, Swiss, and US governments that contributed software to developing countries in the Biosafety Clearing-House project (UNEP 2005, 4).

The UNEP-GEF project has been supplemented by private sector and bilateral initiatives aimed at capacity building. Since the conclusion of the biosafety negotiations, members of the Global Industry Coalition in particular, along with US and Canadian regulators, have had a strong role in organizing workshops on legal and scientific capacity building in developing countries. Several Canadian initiatives have been funded through the Canadian International Development Agency and led through the industry CSO BIOTECanada. Also involved are AgrEvo, Monsanto, DuPont, Pioneer Hi-Bred, and other transnational agri-chemical and biotech companies. These companies have undertaken capacity-building workshops in the Philippines, Thailand, Vietnam, Kenya, Mexico, and Malaysia, among other countries (CBD 2000b).

What does the focus on regulatory frameworks instead of the scientific methodologies necessary for undertaking precautionary risk assessments, and the participation of biotechnology industry partners in capacity-building exercises, mean for the import decisions that developing countries will take? These decisions will likely be based on data coming from the North and provided by the corporations applying for regulatory approvals themselves, rather than on locally produced data. And when it comes to assessments of that data, even where there is scientific capacity, it is quite likely that the pattern seen in Canada and the United States will be repeated in the South. This pattern is summarized in the following comment made by a Canadian official:

> If you look at who the regulators are, if you look at who does the risk assessments ... primarily ... They're not of the ecological community. They're molecular biologists ... They frame the discussion. And still to this day I don't think there is much of an ecological dimension brought to [that discussion]. (Interview #6)

It is also unlikely that regulatory discussions will place much emphasis on the social and economic risks of GEOs. Although participants in UNEP-GEF workshops on risk assessment put forward the idea that "socio-economic, cultural, religious and other factors should be assessed and taken into account in the decision making process" for LMO imports, these training workshops did not provide very much guidance on tools for incorporating these factors (other than mechanisms for public participation) in regulatory frameworks. In this way, the UNEP-GEF projects reproduce a discourse that considers socio-economic considerations to be irrelevant to GEO import decision making.

The idea that socio-economic considerations are irrelevant to risk assessment appears to have already been internalized by some of the people who undertake risk assessments in the South. Consider the case of India. In Chapter 3, it was noted that biosafety evaluation in the Indian context must include not just ecological and human health considerations but also responsiveness to the needs, constraints, and priorities of Indian agriculture. It was this broad mandate that led the Indian government to announce it would not allow the terminator technology to be used in that country. In 1997, this mandate also brought Indian diplomats to the biosafety negotiations, arguing that "users [should] carry out assessments prior to the use and release of LMOs as regards to the socio-economic welfare of societies" (Lewis 1997a, 42).

In contrast to this earlier position, the factors considered in risk assessments undertaken in 2001 and 2002 on transgenic cotton and soy in India related only to environmental safety and human health. When asked about his country's earlier position, one Indian scientist involved in those risk assessments stated that socio-economic considerations were "rubbish." He noted that India did prepare a cost-benefit analysis for Bt cotton, but that those calculations "cannot be part of decision making under the Cartagena Protocol on Biosafety" (interview #16). This example demonstrates how the marginalization of socio-economic considerations in the Cartagena Protocol is helping normalize a narrow understanding of risk assessment in the global South. This understanding differs in important ways from the "comprehensive" framing of GEO risks that delegates from these same countries brought to the biosafety talks only a few years earlier.

The Indian experience, together with the heavy involvement of regulatory and industry arms of the biotech bloc, suggests that capacity building under the protocol may still become a critical process in the normalization of the biotech bloc's narrow reading of GE regulation as defined by the risk discourse, even though the protocol also *allows* for a more precautionary approach. Capacity building is an important example of the day-to-day actions that consolidate power that Gramsci and Foucault emphasize. It is such actions that will define over the long term the impact of the Cartagena Protocol on global and domestic regimes of governance for agricultural biotechnology. This evidence confirms the view of one US CSO activist, who, when asked how he felt about the final outcomes of the Montreal ExCOP, stated:

> I was euphoric, because we were up against some very powerful forces, but that is not the end of the game. It's just a redefinition of the game ... They lost a lot in terms of the formal language of the protocol. The actual implementation is where power has power ... Power is exercised by the day-to-day dripping of the water on the rock. (Interview #2)

Capacity building in the South is crucial to the long-term impact of the Cartagena Protocol. Although both industry capacity-building efforts and the UNEP-GEF may leave some things to be desired, there are also signs that a more precautionary and holistic framing of GEO risks is being shared with the growing cadre of biosafety regulators in the global South. An important example is the work of the International Project on GMO Environmental Risk Assessment Methodologies (GMO ERA; originally known as the GMO Guidelines Project).

GMO ERA is the initiative of a network of 260 public sector scientists, 70 percent of whom are located in developing countries, associated with the Global Work Group on Transgenic Organisms in Integrated Pest Management and Biocontrol of the International Organisation for Biological Control. This group is certainly not as influential as UNEP-GEF, but it does have considerable credibility in the international biosafety community and draws on the support of Hamdallah Zedan, the executive secretary of the CBD secretariat, among others.[2] The goal of GMO ERA is to develop detailed scientific methodologies for supporting environmental risk assessment and management of genetically engineered organisms (GMO ERA 2005). In terms of the dichotomy presented in Chapter 3, GMO ERA's framing of risk assessment is solidly rooted in complexity talk, likely because this group is composed of ecologists, entomologists, and other scientists, and not only molecular biologists (interview #24). The group is coordinated by two scientists who have been on the cutting edge of developing new tests for ecological risk assessment (see Andow and Hilbeck 2004).

The risk assessment and management methodologies being developed by GMO ERA consider a wide variety of possible ecosystem effects of transgenic organisms, including impacts of gene flow (on target and non-target biodiversity); the implications of the evolution of resistance (to Bt proteins in pest populations, for example); and the impacts of altered crop management practices. GMO ERA also stresses the importance of recognizing that results of the imprecision of genetic engineering processes (such as multiple transformations, duplications, inversions, and mixtures of native and transgene DNA) can result in expression of unintended transgene products, altered expression of native plant genes, and plants with new and unexpected traits (GMO ERA 2005). The group believes that risk assessment should never focus on simply the intentionally introduced trait; it must consider the new organism as a whole.

An interesting dimension of GMO ERA's approach is the way a precautionary approach is infused in all the steps of its risk assessment and management framework. Precaution is evident in the elements described above. It is also found in the emphasis GMO ERA places on identifying and prioritizing knowledge gaps as a critical step in the assessment process (GMO ERA 2005, 8). Most important, however, is the way GMO ERA conceptualizes the initial steps in the process. Rather than see risk assessment simply as a test of the environmental safety of the GEO itself, the GMO ERA approach begins with problem formulation and options assessment (PFOA).

GMO ERA sees PFOA as a process that should be undertaken by a multistakeholder group. The first step, problem formulation, requires the group to clearly define the problem, or unmet basic need, needing change. The second step, options assessment, involves considering the various options that exist other than, or complementary to, the GEO in question that might enable the need to be met. Any subsequent assessment of the GEO takes place in the context of this broader discussion of needs and alternatives. As GMO ERA (2005, 3) notes, the goal of systematically comparing alternative technological futures is to encourage "the option or options that meet the needs and values of a given society" (for detailed examples of the application of the GMO ERA approach, see Hilbeck and Andow 2004, and Hilbeck, Andow, and Fontes 2006). Needless to say, this process may lead to the conclusion that there are better approaches to meeting the identified need than the GEO that prompted the analysis.

Beginning a risk assessment with an examination of goals and alternative means of reaching those goals is precisely what advocates for more comprehensive and precautionary approaches to regulation have sought for years. This idea guided Canadian anti-biotech activists in the mid-1990s, when they asked (somewhat rhetorically, given that Canada's dairy production is supply managed in order to avoid excess milk production): "What is the problem that recombinant bovine growth hormone is designed to solve?"[3] This idea also led the (officially silenced) UNEP Expert Panel IV to conclude in 1993 that "when the use of a GMO is not clearly seen to offer an advantage, it would make sense that the traditional technologies and systems continue" (UNEP 1993, para. 86). More recently, K. Barrett and Raffensperger (2002) have argued that the first steps in implementing a precautionary approach towards GEOs would be setting long-term goals for social, economic, and environmentally sustainable agriculture, followed by a full assessment of alternative means of achieving those goals against the specific technologies in question. In making this argument, they draw on the "alternatives assessment" methodologies spelled out in detail by O'Brien (2000). Significantly, approaches that explicitly compare a GEO with all other options available for attaining specific societal goals are even more cautious than the EU's latest regulations. The EU's approach mandates science-based assessments of GEOs, a process that (implicitly) compares only the new technology with the status quo.

These examples show that although the Cartagena discourse of precaution may have succeeded in driving a wedge between the (now widely accepted) logic of precaution and the idea of a comprehensive assessment that fully examines potential socio-economic consequences of GEO introductions, CSOs, international scientific bodies, and academics are all continuing to develop methodologies framed by the assumption that these approaches are inseparable. To what extent these individuals and groups will shape the future governance of GEOs around the world remains to be seen.

Unfinished Business: Socio-Economic Considerations in Decision Making

One of the reasons the issue of socio-economic considerations can and does remain on the table in discussions on GEO import decision making is because the final outcome on this issue in the Cartagena Protocol is somewhat ambiguous. In full, Article 26.1 states:

> Parties, in reaching a decision on import under this Protocol or under its domestic measures implementing the Protocol, may take into account, consistent with their international obligations, socio-economic considerations arising from the impact of living modified organisms on the conservation and sustainable use of biological diversity, especially with regard to the value of biological diversity to indigenous and local communities. (CBD 2000a)

It is clear that Article 26.1 has a mini-savings clause that states WTO obligations cannot be overriden in an effort to consider socio-economic factors in import decisions. But what, exactly, does this article allow? In an academic paper published shortly after the protocol's negotiation in 2000, Stabinsky argues that Article 26.1 means that potential socio-economic impacts from the introduction of an LMO that will have secondary impacts on the environment can be taken into account in import decision making. As mentioned earlier, in the case of coffee, a secondary environmental impact could be the ecological impact associated with the shift from small- to large-scale production that is expected to accompany the introduction of a (hypothetical) GE coffee bean. Stabinsky argues that a measure such as a ban on the use of the coffee bean is not only consistent with Article 26.1, it could also be considered WTO-legal as long as certain conditions are met. The measure must be non-arbitrary, non-discriminatory, and not a disguised restriction on international trade. Stabinsky (2000) argues for the incorporation of rigorous, peer-reviewed social-impact assessment methodologies into a country's risk assessment framework in order for it to meet the expectation that a risk assessment leading to a ban on the GEO in question is not arbitrary. Stabinsky points out that the ban must also satisfy Article XX(g) of GATT 1994, which accepts "measures ... relating to the conservation of exhaustible natural resources if such measures are made effective in conjunction with restrictions on domestic production or consumption" (WTO 1994c, quoted in Stabinsky 2000, 278).

Stabinsky does note, drawing on Nissen's work, that WTO panels have previously interpreted Article XX(g) as requiring that a measure is "primarily aimed at" conservation of the resource and that the GEO ban must be seen as the "least trade restrictive" measure a state could take (Nissen 1997 in Stabinsky 2000, 280). These expectations are tall orders, especially if the ban is based on a potential secondary impact arising from the adoption of the GEO. As a result, the Cartagena Protocol is not perfect from the point of view of advocates of a more comprehensive assessment process. Still, Stabinsky concludes that environmental impacts that result from socio-economic outcomes related to the potential introduction of a GEO *may* be consistent with both the Cartagena Protocol and a state's obligations under the GATT.

Mackenzie and her colleagues (2003, 163-64) at the International Union for the Conservation of Nature and Natural Resources (IUCN), in their influential *Explanatory Guide to the Cartagena Protocol*, agree that parties to the protocol may take socio-economic considerations into account in making import decisions, as qualified by the mini-savings clause. What is interesting about their interpretation of Article 26.1 is that they read the same sentence very differently from Stabinsky. Whereas Stabinsky understands the article to allow consideration of downstream environmental impacts of socio-economic changes caused by a GEO's introduction, Mackenzie and colleagues (163-64) state:

> Where the introduction of LMOs under the Protocol affects biological diversity in such a way that social or economic conditions are or may be affected, a Party can use Article 26 to justify taking such impacts on its social and economic conditions into account for purposes of making decisions on imports of LMOs or in implementing domestic measures under the Protocol. Such social or economic impacts are generally referred to as secondary or higher order effects in technology assessment literature.

To put it succinctly, for Mackenzie and her IUCN colleagues, secondary socio-economic impacts that might result from an LMO's introduction can be taken into account in their own right. Stabinsky, on the other hand, remains focused on the ecological effects that would follow from the socio-economic changes caused by an LMO's introduction.

An example of the kind of impact the IUCN group is thinking about is the potential loss of an indigenous community's ability to practise traditional livelihoods as a result of ecological changes brought about by an introduced LMO. Mackenzie's group states that information about this kind of impact may be taken into account in procedures for assessing and addressing socio-economic impacts in risk assessment and management. It could also be considered in procedures for public consultation with communities that will be directly affected by an import decision (Mackenzie et al. 2003, 165).

This discussion of just two interpretations of Article 26.1 shows that it is possible to understand the article in various ways. The uneasy relationship between it and the WTO agreements also results in interpretive challenges.

Mackenzie and her colleagues (2003, 238-39) point out that "further guidance on the implementation of this provision might be expected from the COP/MOP in due course." Indeed, some discussion on the issue of socioeconomic considerations did take place at COP-MOP-2 in Montreal in 2005. This discussion was not focused on Article 26.1, but on Article 26.2. Article 26.2 encourages parties to "cooperate on research and information exchange on socio-economic impacts of LMOs, especially on indigenous and local communities" (CBD 2000a, Article 26.2).

In the Montreal meetings, negotiators debated a decision to invite parties to the protocol to share information through the Biosafety Clearing-House on research results, both positive and negative, involving socio-economic impacts of LMOs. One point of contention in this discussion was whether parties should also be asked to share information pertaining to "possible modalities for taking socio-economic considerations into account in decision-making." This phrase was sought by many representatives of the same states that originally called for Article 26.1 (without its mini-savings clause), including Malaysia, Norway, and Ethiopia. The 2005 proposal was rejected by states speaking on behalf of the biotech bloc, however, on the basis of the argument that "there is no international agreement on what constitutes socio-economic considerations" (an argument presented by an Australian delegate in Working Group II of COP-MOP-2 on Thursday, 2 June 2005). The larger issue, of course, is that countries in favour of minimizing restrictions on international trade in LMOs remain dead set against further dialogue on the modalities mentioned above, since this dialogue could eventually lead to the development of an international consensus on methodologies for incorporating socio-economic impacts into risk assessments. The Malaysians and others "agreed to delete the request for information on modalities of incorporating socioeconomic considerations into import decisions, with the understanding that the wording does not prejudge nor limit information to be submitted" (Chasek 2005, 12).

Despite the outcome of the Cartagena Protocol negotiations in 2000, the debate about the role of socio-economic considerations in decision making on LMO imports is continuing at the Meetings of the Parties to the Protocol and in processes related to the protocol's implementation. This ongoing discussion is due in part to the ambiguity of Article 26.1. Various options and methodologies continue to be brought forward, especially by academics and CSOs, to show that socio-economic considerations can and must remain central to the debate over LMOs. One more example that deserves brief discussion is a 2005 World Resources Institute White Paper entitled *Integrating Socio-Economic Considerations into Biosafety Decisions: The Role of Public Participation.*

The WRI paper discusses in broad strokes socio-economc benefits and costs associated with the introduction of GEOs (Fransen et al. 2005). Included in this discussion are implications of intellectual property rights for traditional seed savers, impacts of GEOs on potential markets for organic crops, and effects on labour demands. The WRI paper acknowledges that parties to the Cartagena Protocol are restricted in the extent to which they can consider socio-economic issues in import decisions under the protocol, stating that "it could be prudent if Parties are to avoid disputes with their trading partners, such as complaints under the World Trade Organization" (Fransen et al. 2005, 9). It is somewhat surprising, therefore, that the paper provides a detailed discussion of various methodologies for incorporating socio-economic considerations into decision making, including economic modelling, cost-benefit analysis, social impact assessment, and a sustainable livelihoods framework.

The WRI's rationale for its extensive discussion of these methodologies is that "socio-economic considerations are relevant to domestic biosafety decisions and not just to transboundary movement of LMOs" (Fransen et al. 2005, 9). It is only this second group of decisions that is covered by both the protocol and the WTO. Furthermore, the WRI points out that socio-economic considerations should not only be relevant to regulators but should inform every stage of the research, development, and regulation of GEOs. To this end, while the paper does put forth recommendations for governments, it also includes recommendations for the scientific research community, the biotechnology industry, the public agricultural sector, and civil society.

The WRI's efforts in this regard are important because they point to what Gramsci and Foucault would both suggest is really the key site in the struggle over the relationship between socio-economic considerations and biosafety: the internalization and normalization of this emergent discourse in the day-to-day practices of civil society. It is true that the WRI paper is just one paper, but it is also important to recall the discursive impact of an earlier WRI document, "Global Biodiversity Strategy" (1992), as a key site in the framing of biodiversity issues in terms of equity, conservation, and sustainable use. This framing came to underpin the CBD and, because of its institutional heritage, the Cartagena Protocol as well.

Material Politics

The goal of the genetic engineering revolution has been to normalize the use of GE seeds around the world. In the early 1990s, it appeared as if conversations about biotechnology under the auspices of the Convention on Biological Diversity would further this goal, given their focus on technology transfer. As it turned out, the shift to a conversation about biosafety, organized first through the discourse of risk and later one of precaution, meant that the ensuing talks became more hostile towards the biotech industry than at first anticipated. It is difficult to determine the precise implications of the Cartagena Protocol itself on the global production of GEOs. What is clear is that the rise of a precautionary response in the late 1990s was accompanied by a slowdown in the rate of adoption of GEOs in global agriculture. Since then, the protocol's establishment as a new international institution has played an important role in legitimizing the time and effort it is taking for states to develop competent regulatory systems. Once these regulatory systems are in place, GEOs can expect to receive greater scrutiny, especially for potential environmental hazards, than would have been the case if the biotech bloc's assumptions about GEO equivalency had been normalized in the global regime for GEO governance.

A Precautionary Slowdown

I have already documented in detail many aspects of the global rise of a precautionary response towards GEOs in the late 1990s. This response was manifested in the EU moratorium of 1999-2004. This response was also seen in the institutionalization of GEO labelling policies in various countries (including European countries, but also South Korea, Australia, Japan, and China), in the renewal of CSO resistance to agricultural biotechnology in North America in late 1999, and in the negotiation of a precautionary protocol on biosafety. Evidence that this precautionary response resulted in a slowdown in the growth of uptake of GEOs can be found in the statistics compiled by the International Service for the Acquisition of Agri-Biotech Applications (ISAAA).

According to the ISAAA, between 1995 and 1999, the adoption of commercial GE crops was growing at an enormous rate, from only a few test plots in 1995 to almost 40 million hectares in 1999. The year 1997 even saw a growth in area planted of over 500 percent in comparison with 1996 (James 2002). It is striking, therefore, to see that there was almost no growth in the adoption of commercial GE crops in developed countries between 1999 and 2000. And, even after all countries of the world are included in the calculation, the growth in the adoption of GE crops was less than 17 percent per year between 2000 and 2003. More recently, in 2004 and 2005, growth rates have risen again to over 20 percent per year. Still, 94 percent of the total global area devoted to commercial GE crops in 2005 was found in only five countries: the United States, Argentina, Brazil, Canada, and China (James 2005).

Aside from these numbers showing a slowdown in the uptake of GE crops around 2000, other signs in the early part of this decade demonstrated active global resistance to GEOs. In 2001, several countries in different regions of the world, including Croatia, Bolivia, and Sri Lanka, announced plans to ban genetically engineered crops entirely (Villar 2002). Then, in the summer of 2002, when food shortages racked a number of African states, four countries even rejected US food aid because it contained genetically engineered corn (Vint 2002). (Three of these countries, Mozambique, Malawi, and Zimbabwe, eventually accepted the GE shipments on the condition that the grain was milled to prevent planting [Glover 2003].)

These events all point to the impacts of growing skepticism towards GE crops and foods in the late 1990s and early 2000s. That commercial GE seed plantings never actually declined, however, also points to the success of the biotech crop developers in creating seeds with traits of interest to farmers, as discussed in Chapter 2. How has the institutionalization of the Cartagena Protocol impacted on global production of GEOs since this initial slowdown?

As an institution, the Cartagena Protocol has had only limited direct influence on global production and trade in GEOs. With regard to LMO-FFPs, the category that constitutes the vast majority of globally traded transgenic organisms, the protocol's main function is to provide the Biosafety Clearing-House (BCH). The BCH is the website where countries that approve GEOs for domestic production or food list those decisions so that other states are aware of the products that may find their way into the global commodity trade. To date, there are just over four hundred records on the BCH pertaining to LMO-FFPs. A large number of these records are approvals for GEOs for domestic use in three of the main countries producing these organisms: the United States, Argentina, and Canada. Another ten states, including Australia, New Zealand, Japan, the EU, South Korea, and Mexico, have provided the BCH with information on whether specific LMO-FFPs are permitted in their countries, and for what purposes those organisms are permitted (e.g., only for food use or for feed) (BCH 2006a).

Another major reason for the existence of the BCH is to record decisions made according to the advance informed agreement procedure of the protocol. This section of the website reveals how little activity has actually taken place under the protocol's most important mechanism. To date, there has been only one AIA decision posted to the BCH. This is a Norwegian decision made in 1997 (but posted to the BCH only in 2006) that rejects a variety of Bt corn on precautionary grounds. The corn is rejected because it was engineered with the assistance of an antibiotic resistance marker gene. The decision by Norway cites inadequate scientific data on the potential of horizontal gene transfer from the corn to pathogenic bacteria (BCH 2006b). The Norwegian decision also notes that risk assessors found no net benefit to the use of this Bt crop, and thus no contribution to sustainable development, a key criterion of its Gene Technology Act.

That there are no other AIA decisions listed on the BCH should not be taken to indicate a lack of protocol influence on international trade in GEOs. In fact, the exact opposite is the case. The negotiation and subsequent implementation of the Cartagena Protocol, and especially the capacity-building measures being undertaken because of the protocol, have given several countries the institutional and discursive space necessary to take their time on the domestic adoption of GEOs. More AIA decisions are likely to appear on the BCH in the future, but only once countries have frameworks in place to manage notifications, do risk assessments, and make decisions. The process of negotiating the Cartagena Protocol raised the awareness of many delegates from developing countries about the potential hazards of GE crops, and these countries are now taking their time to ensure that regulatory frameworks are operational before giving consent to GEOs intended for direct introduction into the environment.

Before the conclusion of the Cartagena Protocol negotiations, a few developing countries already had domestic regulatory systems for GEOs or were in advanced stages of developing them. Argentina, South Africa, and India are examples of the former, while Taiwan, which introduced its regulatory system for GE foods in early 2001, is an example of the latter (Jaffe 2004). The vast majority of the world's countries, however, had little regulatory and scientific capacity for dealing with GEOs according to the expectations of an instrument such as the Cartagena Protocol. While developing this capacity (especially the scientific capacity) is a long process, this process is at least underway around the globe. What these emerging regulatory systems will look like, and whether they will be as precautionary as the protocol allows them to be, are important questions. But such questions lie beyond the scope of this study. Given that so few regulatory systems existed before the conclusion of the protocol's negotiation, it is nonetheless likely that GEOs will receive greater scrutiny under the guidance of a precautionary protocol than they would have had the biotech bloc's equivalency framing of GEOs been normalized in its place.

Coercion and the WTO

Even if they have not all ratified the protocol, GEO-exporting countries have cooperated with the regulatory capacity-building efforts of the Cartagena Protocol. Even the United States, which is not expected to become a party to the protocol anytime soon, since it is not a party to the CBD, has made use of the BCH to report domestic decisions about the intentional release of GEOs for commercial purposes. These states realize that the protocol and the BCH are part of the new global reality for GEO trade and are anxious to see the mechanisms working smoothly so that trade will be hindered as little as possible.

Acting in good faith towards the BCH is one way that GEO exporters are trying to show that they really have nothing to hide when it comes to international trade. In Gramscian terms, cooperation with the BCH by exporters can be seen as a key step for the biotech bloc in the process of building consent for GEO trade in the post-Cartagena world. Gramsci's work also reminds us, however, that historical blocs with hegemonic ambitions will use both consent and coercion to get their way. Coercive actions on the part of the biotech bloc were evident in earlier efforts by the United States to protect intellectual property rights through the TRIPS agreement. In the field of biosafety, the biotech bloc also has resorted to coercion to have its interests met since the conclusion of the protocol's negotiation, in the form of threats of trade sanctions through the WTO. Such threats were faced by each of the countries that had considered imposing bans on GEOs in 2001: Croatia, Sri Lanka, and Bolivia. Similar threats resulted in the biotech trade dispute at the WTO, initiated in 2003, between the United States, Canada, and Argentina, on the one hand, and the EU on the other.

In Croatia's case, a November 2001 letter from the Embassy of the United States in Croatia to the Croatian Ministry of Environmental Protection and Physical Planning states: "If a ban is implemented, the U.S. Government must consider its rights under the WTO" (Villar 2002, 2). Similar US government threats led the Sri Lankan government to indefinitely defer its 2001 Food Act, which would have banned GEOs. In South America, the Bolivian government's resolution to ban GEOs was revoked in October 2001 because of pressure from Argentina, which had also threatened to take Bolivia to the WTO for unjustifiable restrictions on trade (2).

The EU (technically the "European Community" in the context of this trade dispute) never formally placed a ban or even a moratorium on GEOs. Still, between 1998 and 2004, no new GEOs were approved for use in the EU while the regulatory system was being updated. The only new products to enter Europe during this period were products of GEOs no longer containing DNA and which could therefore be determined to be "substantially equivalent" under the novel food regulations of 1997 (Pew Initiative on Food and Biotechnology 2005). In April 2004, the EU's new regulatory framework, rooted in directives 2001/18, 1829/2003, and 1830/2003, became fully operational. Shortly thereafter, the new regulatory system approved its first GE seeds and foods, and a "number of applications for marketing" of GEOs were said to be at "an advanced stage of examination" by EU officials (European Commission 2003, 6).

Despite that moves were underway to reopen the EU market to GEOs, on 13 May 2003, the United States, backed by Canada and Argentina, launched a WTO complaint against the EU under the SPS Agreement and the Agreement on Technical Barriers to Trade, as well as the GATT (although the focus of the complaint was the SPS Agreement), arguing that the EU's delay in approving new GEOs since 1998 was not based on sound science, and that it is harmful to those developing countries that have followed the EU's lead in rejecting GEOs (USDA 2003a). Shortly after the WTO complaint was launched, the EU met with the United States, Canada, and Argentina to try to resolve differences. In those meetings, the EU argued that its regulatory system was fully compliant with the SPS Agreement, being clear, transparent, and non-discriminatory. The complainants, however, decided to move the process forward. In August 2003, they formally asked for a WTO dispute settlement panel to hear their case, including appeals (USDA 2003b).

The interim ruling of the dispute settlement panel on the EU "biotech products" case, released to the parties in the dispute in early February 2006, was leaked to the press and the public by the Friends of the Earth International later that month (FOEI 2006a). The panel's final ruling was released to the parties in May 2006. That ruling has not yet been made public at the time of this writing; however, government insiders have indicated that it differed little from the leaked interim ruling (Reuters 2006). The analysis presented here is based on the publicly available positions presented by the parties in the dispute, and the interim ruling of February 2006.

The WTO dispute between the United States, Canada, and Argentina, on the one hand, and the EU, on the other, is important to our discussion here for three reasons. First, this case tested the willingness of the WTO to accept delays in GEO approvals between 1998 and 2004 that the EU argues were closely tied to its precautionary stance towards GEOs. As a result, the outcomes of this case provide observers with a clearer sense of how a dispute settlement panel understands the notion of a precautionary response towards GEOs under the SPS Agreement alone. Second, because of arguments raised by the EU, this case led the dispute settlement panel to make comments about the relevance of multilateral environmental agreements such as the Cartagena Protocol to WTO disputes. These comments reveal the ongoing tension between multilateral environmental agreements (MEAs) and global trade rules. Third, this case is significant because of what it demonstrates about the degree of uptake of the precautionary discourse in the WTO and in the heartland of the biotech bloc: the United States, Canada, and Argentina.

The central argument put forward by the United States and its allies in this dispute was that the EU contravened provisions (Article 8 and Annex C) of the SPS Agreement that are intended to ensure that any SPS measures taken by member states are carried out "without undue delay" (United States 2004). In broad strokes, the complainants also claimed that the EU (1) failed to notify its moratorium as an SPS measure (as required under Article 7 and Annex B); (2) failed to provide risk assessments on the hazards of biotech products as required by Article 5.1 of the SPS Agreement; (3) maintained a moratorium without "sufficient scientific evidence" as required by Article 2.2 of the SPS Agreement; and (4) arbitrarily discriminated against imported

GEOs by defining different "levels of protection" for domestically produced foods made with GE processing aids (such as the GE enzymes use in cheese production) in comparison with imported GE foods.

The EU response to the allegations of "undue delay" was to say that while it may have appeared to the world that a moratorium on GEOs was in place, there were actually no delays in approvals that could not be accounted for. In each and every case noted by the United States, the EU submission to the panel argued that applications were still moving forward through the regulatory system. Where delays had taken place, the EU showed that these were because of either the lack of response or incomplete responses from applicants to requests for more information put to them by EU regulators. These requests sought more detail on issues such as molecular characterization, compositional analysis, toxicity, allergenicity, herbicide use, and residue behaviour (European Communities 2004).

In response to the rest of the US claims, the EU's main argument was that the other provisions of the SPS Agreement cited by the complainants relate to the development of SPS measures (including their new regulatory framework), and not to their application. As a result, claims about the insufficiency of scientific evidence and the lack of provision of risk assessments, as two examples, were irrelevant to arguments about delays in the application of the SPS measure. In the words of the EU, "The alleged delay in completing the approval procedures for certain applications does not constitute, itself, a sanitary or phytosanitary measure" (European Communities 2004, para. 469). These delays, as well as member-state bans of specific products (using the safeguard clauses of directives 90/220 or 2001/18), should be seen as interim or provisional measures under Article 5.7 of the SPS Agreement and not be held to the standards of a final SPS measure (para. 586; Suppan 2005). Article 5.7 allows provisional measures in cases where the "relevant scientific information is insufficient" while members seek to obtain "a more objective assessment of the risk ... within a reasonable period of time" (WTO 1994a, Article 5.7). The EU's argument was basically testing whether a precautionary product ban – even if only temporary pending a full risk assessment – would be deemed acceptable under Article 5.7 of the SPS Agreement.

The dispute settlement panel ruled against the EU on each of these points. In its view, there was indeed a general de facto moratorium on product approvals in the EU between June 1999 and August 2003, when the panel was established, and the moratorium violated several articles of the SPS Agreement (WTO 2006, 580, para. 7.1277). This moratorium resulted in "undue delays" in the completion of twenty-four product regulatory applications in Europe over that period (of the twenty-seven mentioned by the complainants) (689-805). Furthermore, while the panel accepted that EU member states can put in place provisional measures, including product bans, in the face of insufficient scientific evidence to undertake risk assessments, it held that neither the moratorium nor the member-state bans could be justified on this basis. In regards to the safeguard bans, the panel took note that the EU's own scientific committees had already reviewed and approved each of the nine products facing member-state bans, and in several cases these approvals were even reviewed in light of specific member-state objections. This was proof, the panel felt, that there was "sufficient evidence" to undertake risk assessments for each of those products, and that those risk assessments did not justify product bans as provisional measures under Article 5.7 (970-1017).

In its ruling, the dispute settlement panel made several references to precaution that deserve to be examined. First, the panel rejected the EU's argument that the precautionary principle should be considered a principle of international law relevant to this case. This ruling was not surprising considering the outcome in the beef hormone dispute described in Chapters 5 and 6. In fact, the panel quoted the Appellate Body's comment made during the beef hormone dispute that "the status of the precautionary principle in international law continues to be the subject of debate among academics, law practitioners, regulators and judges" (WTO 2006, 304). The biotech products panel stated: "It appears to us from the Parties' arguments and other available materials that the legal debate over whether the precautionary principle constitutes a recognized principle of general or customary international law is still ongoing." Furthermore, there "has been no authoritative decision by an international court or tribunal which recognizes the precautionary principle as a principle of general or customary international law" (305).

Despite their rejection of the precautionary principle as a customary principle of international law relevant to the case, the panel repeatedly mentions that certain EU member states were taking efforts to implement a precautionary approach in GEO approvals, and it does not fault them for doing so. Instead, the panel appears to accept the implementation of a precautionary approach when deciding on how to act on the basis of risk assessments, as long as that approach does not cause "undue delays." For example, in the case of Spain's risk assessment for a particular variety of transgenic maize, the panel notes, "We are not convinced that the Spanish CA [the national authority responsible for the file] could not have identified its needs for additional information and forwarded appropriate requests for information to the applicant sooner than it did, even while following a precautionary approach" (WTO 2006, 754). Elsewhere, the panel notes, "We perceive no inherent tension between the obligation set out in Annex C(1)(a), first clause [of the SPS Agreement], to complete approval procedures without undue delay and the application of a prudent and precautionary approach to assessing and approving GMOs or GMO-derived products" (635).

Put most simply, the dispute resolution panel believes that a precautionary approach to GEO risk management is acceptable under the SPS Agreement, as long as the risks upon which a decision is taken have been properly assessed and documented. This risk management strategy could even legitimately take the form of moratoria, of the type put in place by the EU. According to the panel,

> Our conclusion ... should not be construed to mean that it would under no circumstances be justifiable ... to delay the completion of approval procedures by imposing a general moratorium on final approvals of biotech products. We consider that there may conceivably be circumstances where this could be justifiable. For instance, if new scientific evidence comes to light which conflicts with available scientific evidence and which is directly relevant to all biotech products subject to a pre-marketing approval requirement, we think that it might, depending on the circumstances, be justifiable to suspend all final approvals pending an appropriate assessment of the new evidence. The resulting delay in the completion of approval procedures might then be considered not "undue." (WTO 2006, 637)

Furthermore, the ruling makes clear that even a provisional measure, such as any taken under Article 5.7, must eventually also meet the requirement of being justifiable on the basis of a quantitative risk assessment, whether that assessment is taken before or after the measure is put in place.

Given the panel's apparent acceptance of precaution as a risk management tool, how is its interpretation of acceptable precautionary action different from what might be justified under the Cartagena Protocol on Biosafety? In my view, even a hypothetical dispute under the Cartagena Protocol over the EU moratorium of 1999-2004 would have reached some of the same conclusions that this panel reached. Parties to the protocol would also expect the EU to have followed due process, and for decisions to have been made to approve or reject individual GMOs at appropriate bureaucratic levels within the time frames set out in the protocol. The WTO panel's concerns about the lack of movement on many of the applications before the EU would thus have also elicited concerns under the protocol. However, this ruling does exemplify the issue, raised earlier in this chapter, of potentially differing interpretations of acceptable thresholds for precautionary action under the SPS Agreement and the protocol.

Recall that Articles 10.6 and 11.8 of the Cartagena Protocol hold that a "lack of scientific certainty due to insufficient relevant scientific information and knowledge regarding the extent of potential adverse effects of an

LMO ... shall not prevent the Party of Import from taking a decision." This language appears to allow precautionary measures even in cases where the exact magnitude and potential severity of a hazard remain unclear. By contrast, the WTO panel has confirmed the SPS Agreement's more restricted reading of the type of hazard that could merit risk management (whether precautionary or otherwise). In its analysis of Austria's safeguard ban on T25 maize, for example, the panel notes that Austria had put forward a list of reasons for its provisional ban, including concerns about the spread of pollen to cultivated surrounding fields, long-term ecological effects, and the possible development of antibiotic resistance, as well as allergenicity and toxicity. The panel accepted each of these issues as potentially legitimate reasons for taking measures under the SPS Agreement to restrict the use of specific products. However, the panel observed that a risk assessment must include, at a minimum, the identification of a hazard and an evaluation of the likelihood of its occurrence. The panel found that the studies cited in the Austrian documents supporting that country's ban did "not indicate relative probability of the potential risks it identifies, but rather makes reference to possibilities of risks." As a result, "given the lack of evaluation of likelihood in the ... study, we consider that the study does not meet the definition of a risk assessment as provided in Annex A(4)" (WTO 2006, 929). The stronger emphasis on precaution in the Cartagena Protocol, along with the wording of Articles 10.6 and 11.8 of that treaty, means that an evaluation of Austria's rationale for its safeguard ban on T25 maize may very well have been more sympathetic under the protocol than when that rationale is scrutinized under the SPS Agreement alone.

The second reason that this WTO ruling is relevant to this analysis of the Cartagena Protocol is because of the comments the panel made on the relevance of the protocol, along with other MEAs, to international trade law. Although this dispute took place in the context of the WTO, the Cartagena Protocol on Biosafety was invoked in the arguments put forward by various parties, and it was mentioned in the panel's ruling. These references to the protocol reveal the current status of the relationship between MEAs in general and the WTO in global politics.

In each of its submissions to the panel, the EU raised the relevance of the Cartagena Protocol to the proceedings. In fact, one of the EU's main legal arguments was that the environmental concerns that lay at the heart of the "delays" in GEO approvals in Europe fell outside the scope of the SPS Agreement entirely. The EU argued that the SPS Agreement is concerned with measures relating to a narrow set of risks to animal, human, and plant health. The "environment," to which regulatory directive 90/220 refers at least twenty times and 1829/2003 refers at least twenty-nine times, is not mentioned once in the SPS Agreement. Furthermore, environmental risks are not necessarily easily comparable with the kind of risks covered by the SPS Agreement:

The word ... environment ... does not focus on a short term risk to the life or health of a particular animal or plant. It is based on a long term perspective acquired by stepping back from the problem and considering the big picture. It takes into consideration the situation of a whole range of living organisms and other matters. It is concerned not just with each one of these different things, but also, at the same time, with the plethora of relationships between these different things. It is concerned with the overall balance and equilibrium of natural systems over time. (European Communities 2004, paras. 416-17)

This definition matters, the EU contended, because a regulatory instrument designed to protect the environment might take a different form in comparison with a typical SPS measure. Many of the environmental risk issues dealt with in EU regulations, such as the allergenic potential of GE pollen, are not typical sanitary or phytosanitary measures. The EU argued that such measures should be judged according to international standards of environmental law. In this case, that international standard is clearly set by the Cartagena Protocol, an agreement that permits decisions on GEO imports to be based on both risk assessments and the precautionary principle (European Communities 2004, para. 454).

The EU acknowledged that none of the complainants in this case is a party to the protocol, though Argentina and Canada had signed the agreement. Indeed, as the panel pointed out in its ruling, the protocol came into force two weeks after the panel's inception. Nonetheless, the EU argued that WTO agreements, quoting an Appellate Body decision in the US shrimp case, "cannot be read in clinical isolation of public international law" (European Communities 2004, para. 453). In that case, brought against the United States by India, Malaysia, Pakistan, and Thailand, a measure intended to protect sea turtles was found to be incompatible with the GATT because the United States discriminated against WTO members in how it implemented the measure. The United States lost the case on this measure, which had been enacted to protect the natural environment, but in making its ruling the Appellate Body specifically pointed out that "we have *not* decided that sovereign states should not act together bilaterally, plurilaterally or multilaterally, either within the WTO or in other international fora, to protect endangered species or to otherwise protect the environment. Clearly, they should and do" (WTO 1998b, para. 185; emphasis in original). In the case of a dispute over environmental regulation of GEOs, the EU understood this 1998 WTO decision to mean that the dispute resolution panel should take into account the norms of the protocol, even though all parties to the dispute were not parties to the protocol.

The first thing to note about the WTO panel's responses to these arguments is its discussion on the nature and scope of SPS measures. Contrary to EU expectations, the WTO panel ruled that all the reasons raised by the EU for its member state environmental measures concerning GEOs, including the allergenic risks of pollen and even the economic risks of GE contamination of non-GE fields, were potentially acceptable SPS measures falling within the scope of Annex A of the SPS Agreement, the annex that lays out the definition for SPS measures (WTO 2006, 842-43; 1994a, 77). These observations bring to an end the idea, often raised in the context of the Cartagena Protocol negotiations, that the SPS Agreement does not really take into account the full range of environmental measures a state might put in place to protect against risks to biodiversity.

This decision to consider a wide range of environmental concerns as potentially legitimate SPS measures appears progressive. The problem, as the EU pointed out, is that the challenge of predicting and assessing some environmental risks means they are not easily subjected to the disciplines of more traditional SPS measures. These risks may not, for example, be as easily quantifiable as the risks of bacterial contamination of food. This is precisely why the SPS Agreement was considered insufficient, from the point of view of many environmental activists, for the emerging threats to biodiversity from GEOs. It is why these activists lobbied for the Cartagena Protocol as a more precautionary treaty that, at the very least, would eventually be read in tandem with the SPS Agreement. Unfortunately, the panel's responses to the EU's arguments over the relevance of the Cartagena Protocol to this particular case do not give any indication that the protocol will eventually assume its place, in the eyes of WTO panels, alongside the SPS Agreement for matters pertaining to living GEOs.

As noted earlier in this chapter, the dispute settlement panel rejected the idea that an MEA was relevant to a dispute among parties when only some of those parties are parties to that MEA. In response to the EU's argument that a past Appellate Body decision had paid attention to MEAs outside the scope of WTO agreements, the panel notes that this practice had been followed only to help shed light on the accepted meaning of specific terms: "We think that, in addition to dictionaries, other relevant rules of international law may in some cases aid a treaty interpreter in establishing, or confirming, the ordinary meaning of treaty terms in the specific context in which they are used" (WTO 2006, 307). That this practice can extend to examining the meaning of terms in treaties to which not all the parties in a particular dispute are beholden does not mean, according to the panel, that those parties should be understood to be ruled by the remaining conventions of that treaty.

This interpretation was fairly predictable. It would be extremely difficult to hold states to the terms of treaties that they had not ratified. Still, at least one civil society activist argues that the panel could have taken the EU's arguments about the relevance of the protocol into account and "declined

to rule given the lack of consensus about risk assessment and risk management options in multilateral agreements" (Suppan 2006, 3). What was particularly lamentable, from the point of view of advocates of harmonization
between trade law and MEAs, was the panel's decision to completely avoid
the question of when an MEA becomes relevant under the WTO regime:

> It is important to note that the present case is not one in which relevant
> rules of international law are applicable in the relations between all parties
> to the dispute, but not between all WTO Members, and in which all parties
> to the dispute argue that a multilateral WTO agreement should be inter-
> preted in the light of these other rules of international law. Therefore, we
> need not, and do not, take a position on whether in such a situation we
> would be entitled to take the relevant other rules of international law into
> account. (WTO 2006, 301)

By continuing to treat WTO law as separate from the growing body of
global environmental law, the dispute resolution panel has, in the words of
Suppan (2006, 3), "reinforced the schism between the WTO and the United
Nations system." It has done so at a time when, arguably, an integration of
these two bodies of international law is what the international community
needs to move forward. There are no major surprises here, given the disciplinary role WTO dispute resolution panels and its Appellate Body have
typically played on the part of the transnational historical bloc. Nonetheless, this case could have contributed to opening the door towards greater
accommodation of MEAs in WTO dispute resolution.

The third reason that this WTO dispute is relevant to a discussion of the
Cartagena Protocol is what it demonstrates about the status of precautionary discourse in the heartland of the biotech bloc (Canada, the United States,
and Argentina) and in the WTO. On the surface, the arguments brought
forward in this case by Argentina, Canada, and the United States, and by
the ruling of the panel itself, all appear to be embedded in the discourse of
risk rather than that of precaution. In terms of the complainant's case, for
example, the United States argued that there are no unique risks associated
with GEOs deserving a higher degree of regulatory scrutiny when compared
with other organisms:

> If scientific uncertainty concerning the risks of biotech plants had been as
> great as claimed by the European Communities, it is unlikely that any of
> these products would have successfully completed the regulatory process in
> any country. The assertion that the complexities – and uncertainties – of
> assessing the risks of the biotech plants currently in the EC system are far
> greater than non-biotech products is not born out by experience. (WTO
> 2006, 100)

All three complainants also rejected the relevance of the Cartagena Protocol on Biosafety to the case at hand, and Canada argued that even if the protocol were relevant, it does not entitle the EU to take measures that disregard the conclusions of its scientific risk assessments. In Canada's reading of the protocol, "decisions regarding the importation of LMOs should be made on a case-by-case basis, and be based on a transparent, scientifically sound risk assessment." This statement appears to be firmly grounded in the discourse of risk, ignoring the compromises on precaution in the protocol's final text. And when it comes to the panel's ruling itself, some representatives of the biotech bloc have actually framed it as a victory of "sound science" over a more precautionary approach. Consider this statement by a representative of the Grocery Manufacturers of America: "The WTO's decision makes it clear that biotech regulations must be based on sound science and that the EU's approach to biotech crop approvals is unwarranted" (Sarah Thorn, quoted in Grocery Manufacturers Association 2006).

These examples from the WTO dispute appear to suggest that precautionary discourse has yet to make inroads into the biotech bloc's heartland and that it has had little impact on this particular ruling. However, such a reading misses a big part of the story. Even in this trade dispute, there is evidence to illustrate the growing strength of a discourse of precaution associated with GEOs. The panel's recognition of a precautionary approach to GEOs as legitimate points in this direction, as does the panel's comment that a risk assessment may even be completed after an SPS measure has been put in place, in order to justify that measure in the context of a trade dispute. The panel's statement that "it is of no particular importance whether a specific risk assessment which is claimed to serve as a basis for a safeguard measure was performed before or after the adoption of that safeguard measure" (WTO 2006, 927) recognizes that protective measures may sometimes be put in place before a quantitative risk assessment fully establishes the severity of a risk. It is also significant that the panel did not actually deem the EU's regulatory approach "unwarranted," as the industry representative quoted above claims. The EU's current regulatory approach, which is firmly embedded in precautionary norms and which sets a fairly high regulatory bar for GEO safety, was not even at issue in this particular dispute, nor have the outcomes of this dispute required any legislative changes.

The panel's comment about the potential legitimacy of a general moratorium on "all biotech products" is also an indicator of the impact of precautionary discourse towards GEOs on this ruling. Even though the panel was only speculating on the possibility of such a moratorium, that it could envision a situation in which a risk was identified that implicated "all biotech products" reveals an assumption that these products are indeed a distinct class of organisms embodying a shared set of risks. This assumption has become common sense through the growing acceptance of a precautionary framing of GEOs, yet is a far cry from gene talk and the discourse of equivalency promoted by the biotech bloc as recently as the late 1990s.

This point is also relevant in terms of what the WTO dispute tells us about the uptake of precautionary discourse in Canada, the United States, and Argentina. One of the key signs of the impact of the Cartagena discourse of precaution towards GEOs is often missed by analysts looking at the details of this case, because it has to do with what was *not* at issue in this trade dispute. The SPS Agreement states that governments "shall avoid arbitrary or unjustifiable distinctions" in the level of protection they consider appropriate in different situations "if such distinctions result in discrimination or a disguised restriction on international trade" (WTO 1994a, Article 5.5). In Canada, the United States, and Argentina, GEOs are treated as equivalent to non-GEOs. As a result, these states could have based their case, at least in part, on unfair discrimination against GEOs, in relation to non-GEOs, in European regulation. If this case were brought forward in the mid-1990s, when the equivalency framing of GEOs was still gaining ground internationally, an argument along these lines would likely have been made, and the EU may have been hard pressed to argue for the distinctive treatment that GEOs receive under its regulations. However, it was through the negotiation of the Cartagena Protocol, and then the Codex principles, that international precedents were set for treating "genetically modified organisms" as a distinct regulatory category. It is the existence of these precedents, and the larger discursive framing of GEOs that they came to embody, which has meant that the question of whether GEOs deserve different regulatory scrutiny is no longer at the heart of the trans Atlantic debate over GEOs. That the equivalency discourse did not even come to the fore in the EU biotech products case at the WTO is final proof that the biotech bloc no longer sees this position as tenable in an international setting. In the context of the WTO's SPS Agreement, the debate now appears to be about just how precautionary measures towards GEOs can indeed be, and about the nature of the scientific evidence that is needed to support such measures. Together, these examples show that even the WTO and the nations at the heart of the biotech bloc are not immune to the gradual shift towards a precautionary framing of GEOs in global environmental politics.

Documentation Diplomacy

Two of the most important compromises that advocates of a strong protocol made towards GEO exporters in the Cartagena Protocol negotiations were the exemption of LMO-FFPs from AIA and the lack of detailed documentation requirements for this class of LMOs (the "may contain LMOs" clause of Article 18.2(a)). Were LMO-FFPs to have been made subject to AIA, a major slowdown in the global grain trade could have occurred the day the protocol came into force, if it were adhered to. And even if LMO-FFPs were made subject only to stringent documentation requirements (rather than fully beholden to AIA), the requirement to segregate and account for LMOs that this would entail could have resulted in the shunning of GE varieties by the international grain traders. Neither of these scenarios – each of which could have split the biotech bloc between the GE seed companies and the grain traders – came to pass because of exceptions made for LMO-FFPs in the protocol text. It is important to recall, however, that the outcome on documentation for LMO-FFPs was not a final one. As a result, a central debate during meetings of the Intergovernmental Committee for the Cartagena Protocol (ICCP; the interim body set up to oversee the protocol's implementation before its entry into force) between 2000 and 2002, as well as the first three Meetings of the Parties to the Protocol, focused on issues of documentation for LMO-FFPs.

In full, Article 18.2(a) states that each party shall take measures to require that documentation accompanying

> living modified organisms that are intended for direct use as food or feed, or for processing, clearly identifies that they "may contain" living modified organisms and are not intended for intentional introduction to the environment, as well as a contact point for further information. The Conference of the Parties serving as the meeting of the Parties to this Protocol shall take a decision on the detailed requirements for this purpose, including specification of their identity and any unique identification, no later than two years after the date of entry into force of this Protocol. (CBD 2000a, Article 18.2(a))

The issue of documentation requirements for LMO-FFPs surfaced at all three ICCP meetings. During the ICCP's second meeting, in Nairobi in 2001, a group of technical experts was established to consider "all relevant aspects and disciplines for the implementation of paragraph 2(a) of Article 18." The report of this group, tabled at ICCP's third meeting in The Hague in 2002, reveals that the basic positions brought forward during the protocol's negotiation had not shifted at all (CBD 2002).

GEO-commodity–exporting countries such as Canada and the United States still refused to accept the need for anything more than the "may contain" language found in the article. They believed that documentation requiring detailed lists of the specific LMOs in a grain shipment would require costly segregation systems, and that this outcome would be out of proportion to the risks presented by LMO-FFPs, given that this class of LMOs is not intended for direct introduction into the environment. The strongest advocates of detailed documentation requirements were the African countries, with Egziabher of Ethiopia once again speaking on their behalf. These countries sought detailed documentation in order to facilitate risk assessment and management practices for LMO-FFPs. If a shipment were to arrive with documentation that listed unapproved LMOs, for example, it could easily be stopped at the border and quarantined pending a risk assessment decision. These countries also sought to ensure that documents would cover all LMOs that might be present in a particular shipment, whether or not they were there intentionally. This inclusion of "adventitious" materials under the scope of documentation requirements would ensure that the costs of testing for accidental presence of LMOs would be borne by exporters rather than importers. The wider rationale for these demands, as Bereano (2006, 4) notes, is that these states

> often lack dedicated biosafety legislation, fiscal resources and the complex biosecurity infrastructure that developed countries possess. For these nations, the Protocol is a collective way to achieve regulations, which each alone might be too weak to enact or enforce.

Initially, the EU also wanted detailed documentation requirements alongside the African states, in part because this practice would fit with labelling and traceability expectations being established under the new EU regulatory framework. However, the EU position shifted between 2000 and 2006.

Discussions on documentation at the third ICCP meeting made little progress, partly because of the intransigency of GEO exporters on this issue, but also because the ICCP had no mandate to take binding decisions on the future of the protocol. In The Hague, one European diplomat noted that he was looking forward to the first Meeting of the Parties to the Protocol, because none of the former Miami Group members was expected to ratify the treaty before then. In formal negotiations between parties, the major GEO exporters would no longer be in a position to block consensus on the documentation issue (interview #25).

Given such expectations, it came as a surprise to many delegates to COP-MOP-1, in Kuala Lampur in 2004, when progress on LMO-FFP documentation remained a major challenge. Among other items, COP-MOP-1 dealt with two issues related to the implementation of Article 18.2(a). First, the meeting discussed interim measures for the implementation of the "may contain" language. Second, it discussed a framework for deliberating upon the "detailed requirements" in the lead-up to a decision on this matter at COP-MOP-2 in 2005 (Chasek 2004). The Kuala Lampur meeting did result in decisions on each of these items. For example, COP-MOP-2 "urges" and "encourages" parties and other governments to take measures ensuring that details on LMO identities (including unique identifier codes) are provided on documentation accompanying shipments that "may contain" LMO-FFPs. One of the reasons that the COP-MOP's language was no stronger (many of the parties present were asking for this information to be "required") was because GEO-exporter interests were present among the parties to the protocol in Kuala Lampur, despite the lack of Miami Group ratifications.

In a clever strategic move that would have major implications for the direction of COP-MOP discussions on documentation issues, the United States and Canada signed a trilateral agreement with their NAFTA partner, Mexico, in early 2004 that sets out documentation requirements for LMO-FFPs traded among these three countries (AAFC 2004; Mackenzie 2004). The trilateral arrangement requires that shipments only be accompanied by documentation that states they "may contain" LMOs. The agreement also deals with two other important issues that remained undecided in protocol deliberations. It states that a normal commercial invoice can serve as the accompanying document, rather than a "stand-alone document," the format many developing country delegates, along with Norway, had argued for in the ICCP meetings. In the protocol discussions, the stand-alone document was opposed by commodity traders because of worries it could stigmatize those shipments (Falkner and Gupta 2004). The agreement also states that there is a 5 percent threshold for the unintentional presence of LMOs before a shipment must be tagged with the phrase "may contain LMOs." This is a threshold level for adventitious materials that many CSOs believed would lead to further contamination of Mexican maize varieties by imports from the United States, even if those imports were not labelled as transgenic (FOEI 2006b; GE contamination of Mexican maize varieties was confirmed to have already occurred by Quist and Chapela 2001).

This trilateral agreement became important in Kuala Lampur and subsequent COP-MOPs because the United States and Canada were not parties to the protocol at that meeting, but Mexico was. Having recently signed this agreement with the two countries at the heart of the biotech bloc, Mexico came to Kuala Lampur unwilling to agree to any decision that was not in line with its trilateral arrangement. In addition to seeking language that would do no more than "encourage" states to adopt more detailed documentation requirements, Mexico opposed the introduction of any new stand-alone documents for LMO-FFPs.

Mexico's support for weak documentation requirements for LMO-FFPs at Kuala Lampur and afterwards shows that the positions of specific states on issues of GEO trade and biosafety are continually evolving in response to ideational, material, and institutional forces. In Chapter 5, it was noted that Mexico refused to join the Miami Group because it wanted a strong international instrument to back the precautionary moratorium it had put in place in the mid-1990s against GE corn imports for planting (Galvez 2002). At the biosafety negotiations in Montreal in 2000, Mexican delegates sought operational language in the Cartagena Protocol that would allow import bans even in the absence of clear evidence of harm (interview #14).

By Kuala Lampur in 2004, however, trade relations with the United States (its largest trading partner) had evidently become a higher priority than standing firm with most other protocol parties on detailed documentation requirements.

Another party to the Cartagena Protocol that adopted a position in line with North American GEO exporters in Kuala Lampur was Brazil (Falkner and Gupta 2004). Brazil, for example, stood alone with Mexico as party to the protocol proposing the use of commercial invoices instead of stand-alone documentation (Chasek 2004). The main reason Brazil adopted this position, in 2004, was because it *had* actually become an exporter of LMO-FFPs since the biosafety negotiations were concluded in 2000, despite efforts to remain GE-free, and this situation had led the country's government to frame its biosafety interests in a new light.

As recently as 2002, Brazil was recognized as an exporter of choice for European food companies seeking GE-free ingredients. The Brazilian government had not approved any GE soy or corn plantings, so the country's supplies were officially GE-free (Glover 2003). In October 2003, however, that government passed an emergency measure authorizing the planting of GE soy after it formally recognized what Brazilian farmers already knew: there was widespread illegal planting of Roundup Ready soybeans that had been traded on the black market, smuggled across the border from Argentina (Zarrilli 2005). Initially, this emergency measure was to be in place for one year only, but it was renewed again in 2004. By the summer of 2005, Brazil had 10 percent of global area devoted to GE crops (James 2005). Brazil's move from being proudly GE-free to speaking on behalf of exporter interests shows how rapidly shifting material factors can affect the international politics of biosafety. It also demonstrates that the illegal trafficking of GEOs has become an important mechanism for the spread of the biotech revolution.

COP-MOP-1 made two sets of decisions on documentation for LMO-FFPs. The first concerned the implementation of the "may contain" language, as discussed above. The second concerned the process for developing the "detailed requirements" that Article 18.2(a) states must be in place within two years of the protocol's entry into force. The COP-MOP decided to establish another open-ended technical expert group on identification requirements of LMO-FFPs, asking this group to prepare a draft decision for COP-MOP-2 on numerous issues, including identification requirements; the form of that documentation (e.g., the commercial invoice or a stand-alone document); and the threshold for "adventitious" or unintentional presence of LMOs that would trigger identification requirements (Chasek 2004). This technical experts group met in Montreal in March 2005. After three days of negotiation, still no compromise had been reached on the key LMO-FFP documentation issues. As a result, a chair's text, which many delegates felt overrepresented the minority position seeking to retain "may contain" language in any final decision, was transmitted to COP-MOP-2 for further deliberation. This text was forwarded with a caveat "recognizing that this text does not represent consensus" (Lin and Ching 2005).

The issue of documentation for LMO-FFPs was again at the centre of debate at COP-MOP-2 in Montreal in May 2005. Despite criticism from some of its former Like-Minded Group colleagues, Brazil brought forward the GEO-exporter position that "may contain" language was sufficient and that commercial invoices were suitable documentation (Chasek 2005). In a surprise move, New Zealand held a position that correlated with the interests of biotechnology companies and large GEO-exporting countries when it rejected any clauses on adventitious materials that might result in importers requiring documentation for shipments containing only traces of LMOs.

New Zealand's position on this issue was a surprise because that country has had a GEO-labelling regime in place since 2001, requiring a product to be labelled as genetically modified unless individual ingredients are less than 1 percent genetically engineered and that that fraction is "unintentional" (FSANZ 2003). New Zealand's concerns about adventitious presence thresholds – expressed in a statement by a New Zealand spokesperson in Working Group I at COP-MOP-2 in Montreal on 2 June 2005 – were focused on the idea that the protocol must ensure "meaningful and practical requirements" for the international grain trade. In New Zealand's view, any outcome that might allow a country to establish a zero-risk threshold on adventitious presence of LMOs in shipments not labelled as containing LMO-FFPs, as some developing countries were saying should be their right under the protocol (Chasek 2005, 8), would be highly impractical for the global grain trade.

The lack of movement by Brazil and New Zealand in negotiations on documentation at COP-MOP-2 was eventually followed by their formal objection to a draft decision put forward in the meeting's final plenary (CBD 2005, para. 163). As a result, the proposal prepared by the chair of the Working Group was sent forward to COP-MOP-3, to be held in Curitiba, Brazil, in March 2006, as a draft requiring further negotiation. The draft decision held that the type of documentation (whether a stand-alone document or a commercial invoice) would be up to the choice of the importer. The importer would also be able to decide on the threshold for adventitious or technically unavoidable materials. The draft decision held that documentation should clearly state that a shipment may contain LMOs approved by the party of import and specify which ones these could be, or it should clearly state that the shipment contains LMO-FFPs and specify which ones they are (CBD 2005, Annex 3). The first option noted in the draft decision is a far cry from the detailed documentation sought by developing countries and the EU in Montreal in 2000, yet it was still too detailed for Brazil, speaking on behalf of LMO exporters, to accept. For New Zealand, the flexibility this proposal offered countries on setting thresholds for adventitious material was the major stumbling block.

That there were still no detailed documentation requirements in place for LMO-FFPs under the Cartagena Protocol by the end of 2005, after the deadline established in Article 18.2(a) and more than five years since the protocol's initial negotiation, was a sign of the ongoing lack of coherence in the global regime for biosafety. This situation can be traced back to differences in both material interests and risk perception among GEO exporters and importers. At the ideational level, the debates about documentation and thresholds, and related debates about traceability and labelling, are closely tied to ongoing idcational struggles over the precautionary response to GEOs. One of the central tenets of the precautionary principle is that a precautionary action should be proportional to the risks at hand. The question, in the cases of both documentation requirements and labelling, is whethei an impoiting state's right to know, in detail, what is coming across its borders in a particular shipment justifies changes to the international grain trade that may introduce new costs for exporters, and thus ultimately consumers too.

The issue of thresholds for technically unavoidable presence of GE grain in shipments not labelled as containing them is particularly important to this debate because thresholds that are too low could require changes in the way grain is transported and stored, introducing new costs. Many grain traders argued that anything under a 5 percent threshold would be too low (Falkner and Gupta 2004). On the other hand, thresholds that are not low enough could lead to contamination in cases where LMO-FFPs end up being used as seed despite their intended use. This situation is known to commonly occur in Mexico and other developing countries and has already led to contamination of traditional landraces in the centre of origin for maize (Quist and Chapela 2001). It is also true that many of the same grain traders that argue for a 5 percent threshold have been meeting EU tolerances of 0.9 percent for GE ingredients in "non-GE" labelled products, as well as the 1 percent threshold in New Zealand.

Is the imposition of new documentation, segregation, and testing costs on exporters proportional to the additional risks imposed on importer environments by simpler "may contain" documentation requirements with high thresholds for adventitious materials? The answer to this question is what biosafety negotiators had not yet agreed on by late 2005, and despite a decision finally being made on the LMO-FFP documentation issue at the ensuing meeting in Brazil in early 2006, this basic question still does not have a shared international response. I return to a discussion on the concept of proportionality as a new fault line within the discursive politics of precaution below.

The Curitiba Rules

Another temporary truce was finally reached in the ongoing struggle over documentation requirements to accompany LMO-FFPs in early 2006 at COP-MOP-3 in Curitiba, Brazil. The biggest surprise of that meeting was the turn-around in the host country's position on the documentation issue. Rather than maintain its outright support for continued "may contain" documentation requirements, on the second day of negotiations, Brazil's president Luiz Inácio Lula da Silva announced a new proposal that contained these three elements:

1. In cases where LMO-FFPs are subject to identity preservation in production systems (a "procedure for management of production, transport, processing and distribution, with a view to ensuring a product's integrity and purity with respect to specific characteristics"), the document should state that the shipment "contains" LMOs and specify which ones are in the shipment;
2. In cases where the LMO-FFPs are not subject to identity preservation, that the documents accompanying the shipment state that it "may contain" LMO-FFPs, and specify which ones these could be;
3. By 2010, documentation for all shipments that contain LMO-FFPs specify that the shipment "contains" LMO-FFPs. (Chasek 2006, 6)

That the Brazilian proposal would require all LMO-FFPs to be segregated and labelled as such by 2010, thereby ruling out the "may contain" option after that time, was a major shift in position for that country.

What happened to lead to Brazil's new position, its second major shift within three years? There are several factors to consider. At the level of ideas, it is important to recognize that Brazil's president Lula da Silva had been at the head of a group of twenty-two developing nations causing tensions at WTO meetings by demanding a reduction in US and European subsidies for agricultural producers and increased market access for products from the South (Rapoza 2003). Even though this stance appears to align closely with neo-liberalism, Lula has developed his own hybrid ideological perspective, one that does not fit in easily with the neo-liberalism touted by the nations at the heart of the transnational historical bloc. For example, Brazil linked the North's interests in discussions about investment rules and intellectual property rights (and implementation of the TRIPS Agreement, in particular) with his group's demands for subsidy reduction. In 2003, this stance led to a collapse of the WTO talks in Cancun, Venezuela (Padgett and Downie 2003). Clearly, Lula is not a strong ally of the transnational historical bloc, and it is unlikely that he wanted to be perceived as a spokesperson for US-based Monsanto, which is precisely how Brazil was presented by many activists following COP-MOP-2 in Montreal in 2005. It is true that the Brazilian stance towards the WTO also existed a year earlier, when Brazil took a position against detailed documentation in Montreal. However, the intervening ten months, combined with the fact that COP-MOP-3 was taking place on Brazilian soil (thereby shining the spotlight onto Lula even more strongly), as well as the organizational and material circumstances laid out below, culminated in Brazil having a change of heart.

In terms of organizational relations of force, Brazilian activists had spent the ten months between the second and third COP-MOPs orchestrating an international online protest that resulted in a deluge of emails to President Lula asking Brazil to reconsider its stance (Bereano 2006). Many of these same CSOs also put their point of view forward through the extensive internal consultations on the Brazilian position in late 2005 and early 2006 – consultations that were said to have brought the Brazilian trade and environmental ministries into conflict (ICTSD 2006; Bereano 2006; Garton, Falkner, and Tarasofsky 2006).

In terms of material relations of force, that Brazil occupied a unique position in the global biotechnology revolution helps explain its shifting stances on the LMO-FFP documentation issue. As noted above, Brazil had become one of the world's largest GEO exporters by 2005, having already established itself as one of the world's largest exporters of GE-free grains not long before that. This situation meant that many Brazilian producers were already working through segregated supply channels. Some companies were selling conventional soy shipments with GE content, while others remained specialized in GE-free soy and soy products, in addition to organic GE-free products, primarily for European markets that paid a premium (USDA 2005a). Brazil's experience with the coexistence of these two supply streams demonstrated that it was not as difficult as the grain-trading associations were suggesting to know exactly what particular shipments contain.

This stance was further confirmed for Brazil by Monsanto's actions in that country. To ensure that farmers were paying royalties for using Roundup Ready soybeans bought on the black market, Monsanto had instituted variety testing for all shipments of soy arriving at 95 percent of the grain elevators in two Brazilian states (FOEI 2005). Monsanto's actions demonstrated that the testing of grains to determine their genetics was not as difficult as some biotech proponents made it out to be. If only implicitly, Monsanto had also shown Brazil, and the world, that it could be in the interests of the biotech purveyors to pay for this testing themselves.

Finally, it is important to recognize that Brazil was potentially going to be at a competitive advantage were its proposal to phase in "contains" documentation by 2010 accepted at COP-MOP-3, since the infrastructure for segregating and labelling GE soy (its major GE export product) was already well established in that country. That it was the only major GEO-exporting country to have ratified the protocol would also help Brazil become a preferred supplier to the soy importers among the 130 other (mostly importer) states that had done the same. This analysis is supported by T. Young (2006, 7), who writes: "Brazil might be counting on gaining a competitive advantage, in particular vis-à-vis other Latin American countries, by being able to put in place systems that will allow Brazilian exporters to segregate biotech from conventional products."

With this major shift in Brazil's position at COP-MOP-3 in early 2006, Mexico came to the fore once again as the main proponent of a position in line with the grain-trading companies. During the debate over Brazil's proposal, which was accepted as the basis for negotiation, Mexico argued against any "contains" language that did not have a "may contain" option, and against a timeline for implementing such language. As at COP-MOP-1, the Mexican position was closely tied to its trilateral agreement with the other NAFTA countries. The head of the Mexican delegation at COP-MOP-3 stated that his country's position in these talks was based on "the possibility of maintaining a series of trade agreements with other countries, and our commitments to the United States and Canada" (Marco Antonio Meraz, quoted in Osava 2006, 1). Garton and his colleagues (2006) point out that in recent years, Mexico had also become a small producer of GEOs, so it is conceivable that Mexico was beginning to see its GEO interests as an exporter. However, in my view, it is questionable whether this had a significant impact on that country's negotiating position, given that Mexico had actually been a producer of GE crops as early as 1996, and by 2005 its total GE cropland was still less than 100,000 hectares (James 2001, 2005).

At COP-MOP-3, another country voicing a position aligned with the corporate grain traders was Peru, though it caved to the emerging international consensus earlier than Mexico (Belmonte 2006). Similar behaviour on the part of the Peruvian delegation was witnessed at COP-MOP-2 in 2005. Garton and his colleagues, drawing on a United States Department of Agriculture report of 2005, note that Peru does not yet produce commercial GE crops, but that it has been drafting new regulations in line with the regulatory framework in place in the United States. The USDA reports, for example, that Peru recently "changed its position on labeling from a restrictive perspective which established the use of GMO in a product to a more flexible view using wording such as 'may contain GMO'" (USDA 2005b, quoted in Garton, Falkner, and Tarasofsky 2006, 5).

Another surprise in the COP-MOP-3 debate over documentation requirements was Paraguay's position. Hitherto rarely heard from in the protocol negotiations, Paraguay presented a position aligned with Mexico at the Curitiba meetings and held this position throughout the week. This move included backing Mexico during a tense moment in the final stages of the talks. On the fifth day of negotiations, Mexico, backed by Paraguay, almost derailed the emerging consensus by insisting on reopening the debate over documentation requirements to insert a paragraph stating that the specific documentation requirements being negotiated would not apply to trade with non-parties to the protocol (Chasek 2006; Belmonte 2006). Most observers feel that this phrase simply reiterates the understanding on trade with non-parties included in Article 24 of the protocol.

Garton and his colleagues (2006) believe that the position held by Paraguay at COP-MOP-3 also can be explained by that country's recent entrance to the international club of GEO exporters, growing 1.8 million hectares of GE soy in 2005. Paraguay first officially reported its production of GE soy in 2004 after approving several varieties of Roundup Ready soybeans in October of that year. However, Paraguay had been in a similar position to that of Brazil for several years, with GE soybean production taking place by farmers using illegally imported seeds. As a result, in the first year that GE soy was officially grown in that country, it already represented 60 percent of the soybean crop, and Paraguay found itself in the position of being the world's fourth largest exporter of GE soy (James 2004).

The final outcome on the question of "may contain" versus "contains" at Curitiba, largely because of an unwillingness by Mexico and its allies to consider any language that would eventually require "contains" documents for all LMO-FFP shipments, was the following: Where the identity of the LMO is known through means such as identity preservation systems, the shipment should be labelled as containing LMO-FFPs. This clause remained relatively unchanged from the Brazilian proposal. Where the identity of the LMO-FFPs is not known, the "may contain" label can still be used. In either of these cases, exporters will be expected to provide the common, scientific, and, where available, commercial names, as well as a transformation event or unique identifier code, for all the LMO-FFPs that are, or may be, contained in the shipment (CBD 2006b, 60-62). Instead of a phasing out of the "may contain" language, as proposed by Brazil, the decision states that the parties will review experience with these practices at COP-MOP-5 (which will take place in 2010), "with a view to considering" a decision at COP-MOP-6 (in 2012) to "ensure" that all relevant shipments clearly state that they "contain" LMO-FFPs (CBD 2006b, 61).

The final decision on the implementation of Article 18.2(a) "acknowledges" that the phrase "may contain" does not require listing of LMOs of species "other than those that constitute the shipment." This statement is generally understood to mean that in cases where GE soy, for example, may be adventitiously present in shipments that "may contain" GE corn, the soy does not have to be listed in the accompanying documents. However, the potential adventitious presence of GE soy in a conventional soy shipment would trigger the labelling requirement (Garton, Falkner, and Tarasofsky

2006; Lin and Ching 2006). This statement was included in the decision in the place of any specific thresholds for adventitious materials in shipments that "contain" or "may contain" LMOs, as sought for by African countries in particular. The lack of specific thresholds for adventitious materials, unlike the draft decision at COP-MOP-2, appeared to be enough to appease concerns raised by New Zealand at that earlier meeting. This statement in the final text in Curitiba meant that New Zealand did not oppose consensus from being reached this time around (Chasek 2006, 7). It also meant that thresholds for adventitious presence of GEOs in shipments will continue to be set at the domestic level by importing states (Lin and Ching 2006).

The other remaining issue of contention from earlier rounds of debate over documentation was that of whether existing commercial invoices would suffice, or whether a stand-alone document would be necessary. The final decision in Curitiba also avoided answering this question, leaving it up to states to establish their own requirements.

Overall, the Curitiba rules show that the forces that have been resisting detailed documentation requirements for LMO-FFPs maintained a strong hand in negotiations, even without the direct participation of countries such as the United States, Canada, and Argentina. Mexico and Paraguay, together with Peru's support, prevented the phasing in of full "contains" documentation by 2010 proposed at the outset of COP-MOP-3. Mexico went even farther than blocking detailed documentation requirements by demanding the clause that acknowledges that the documentation decisions will not apply to trade with non-parties to the protocol. In sum, Mexico and its allies have ensured that "may contain" will continue to be used in the global grain-trading system until at least 2012, and possibly longer. As a press release from a biotech industry organization put it, the only significant adjustment the industry has to make following the Curitiba decision is to "include a list of the biotech events that may be contained in the shipment" after the "may contain" LMO-FFPs label that most traders are already using when shipping to states party to the protocol (CropLife International 2006).

Despite these victories on the part of Mexico and its allies, the final decision on Article 18.2(a) at COP-MOP-3, as with every previous decision on these questions, is still inching in the direction of the possibility of full documentation for all LMOs, including those destined for food, feed, or processing. One of the outcomes we can expect to see from this decision is the further development of testing, segregation, and tracing techniques in both exporting and importing countries, since these countries will want to be prepared for the possibility of more detailed requirements in 2012 or afterwards. The decision also "requests Parties and urges other governments and relevant international and regional organizations to take urgent measures to strengthen capacity-building efforts in developing countries" specifically in order to assist them "to meet and benefit" from the new documentation requirements for LMO-FFPs (CBD 2006b, 62). Notably, the calls for capacity building came from countries of the global South that are importers as well as from those that are exporters, such as Brazil and Paraguay (ICTSD 2006, 8). The decision states that these efforts are to be examined as one of the considerations leading to the decision, expected at COP-MOP-6, on whether to bring in full "contains" documentation for all LMO-FFPs. Together, these clauses are going to encourage increased surveillance of GEOs in global trade, making the adoption of full documentation for all LMO-FFPs more likely in the future (though still not an eventuality).

What should we make of the fact that the Curitiba rules on documenta-tion for LMO-FFPs were realized, with their continuation of the "may con-tain" language, without the direct participation of the original countries at the heart of the biotech bloc, especially the United States, Canada, and Argentina, to block consensus? If these countries were at the heart of a biotech bloc through the 1980s and 1990s, what does the dynamic of the years 2000 through 2005 tell us about the biotech bloc in 2006?

The original constellation of forces at the heart of the biotech bloc have remained active and have continued to have a strong impact in the COP-MOPs. The trilateral agreement with Mexico and the centrality of this agree-ment to the outcomes in the LMO-FFP documentation debate demonstrate that the point of view of the United States, Canada, Argentina, and their industry allies was never absent from these deliberations. At the same time, the playing field has changed in important ways from 2000 to 2006. In 2000, the central line of division at the Montreal ExCOP was North-North, between the United States and Canada, on the one hand, and the EU on the other. In the final moments of those negotiations, the views of coun-tries in the global South had little immediate impact. By 2006 in Curitiba, the line of division was South-South. On most of the issues of contention Mexico, Paraguay, and Peru stood in opposition to the African countries, with Brazil occupying a position between the two poles. Rather than un-dermining the usefulness of the "biotech bloc" as a descriptor, what this dynamic has shown observers are the new faces of the biotech bloc. For various material, institutional, and ideational reasons, states such as Mexico, Paraguay, and Peru have aligned themselves with the United States, Argen-tina, and Canada, and the biotechnology industries in those countries. These states are building regulatory systems that are harmonized with their hemi-spheric partners, and they are also becoming exporters of GEOs to the world. In doing so, they are adopting the same exporter mindset that occupied the Miami Group of countries in the Cartagena Protocol negotiations of 1998-2000.

Brazil's position in the Curitiba negotiations also shows us a new face of the biotech bloc, one that is particularly interesting given earlier perceptions of the exporter interest in the protocol context. Brazil's position goes farther than any other country with a strong connection to the biotech industry (with the exception of some countries in the EU) to accommodate the demands of the critics of genetic engineering. The Brazilian position at Curitiba demonstrates that exporter interests, in a post-Cartagena world cautious of GEOs, can be aligned to a large extent with the demands for detailed LMO-FFP documentation put forward by the African Group and other importer states. This position, which recognizes the eventuality of segregated supply chains for GEOs and non-GEOs, and seeks to find a competitive advantage in that system, shows how far the biotech bloc as a whole will likely have to go, whether sooner or later, to accommodate the precautionary response to GEOs embedded in the Cartagena Protocol on Biosafety. In the context of the biotech bloc, at this point Brazil is still a marginal figure, but its position at Curitiba is another sign of how the global politics of biotechnology are shifting in relation to the precautionary discourse.

In earlier chapters, the biotech bloc was characterized as an historical bloc with hegemonic ambitions of global proportions. In Gramscian terms, it is still appropriate to see the biotech bloc as an expanding hegemonic formation, but only if we understand the emerging environmental regime for GEOs, of which the Cartagena Protocol is one key element, as exacting significant concessions from the bloc's leading actors as its norms are internalized. Over the short term, the protocol, together with a growing allegiance to the ideology of precaution among consumers and governments, has slowed the biotech industry's growth in real material terms. And while the biotech bloc has not yet fully compromised on detailed documentation requirements for all LMO-FFPs, it still looks like it is headed in that direction. Labelling was a major hit against the biotech bloc in Europe and many other countries. As a result, some LMO exporters can be expected to keep fighting against the possibility of any international agreement that could be seen to legitimize detailed documentation requirements for LMO commodities (in the eyes of the world's citizens as well as a future WTO dispute resolution panel). However, Brazil's shift on this issue at COP-MOP-3, combined with its new status as a major LMO-FFP exporter, demonstrates that exporting countries no longer have a unified perception of what the "exporter interest" entails in a post-Cartagena world.

When it comes to the material politics of GEOs more generally, consumer resistance in Europe and elsewhere remains at the heart of the challenges faced by the biotech bloc. This resistance has resulted in a very different biotech revolution than was expected only a decade ago. At the same time, the forces at the heart of the biotech bloc continue to create some of their own biggest challenges through the ways they choose to respond to resistance.

Consider the case of the WTO dispute over the EU moratorium. Gramsci would take note that the pro-GE complainants resorted to the coercive tool of a WTO dispute without, once again, paying sufficient attention to the need to build consent among Europeans (let alone those who live in the rest of the world) as consumers and as citizens. That the United States won its dispute at the WTO does not necessarily mean that European citizens will have any more confidence in GEOs in 2006 and afterwards than they did in early 2000. If anything, WTO rulings that ignore other emerging norms of international law will simply serve to further discredit that institution in the eyes of many.

Canadian Implications of a Precautionary Protocol
In the early 1990s, Canada adopted a regulatory system organized around the concepts of novelty, familiarity, and substantial equivalence. As GE products approached commercialization, the Canadian government faced ongoing battles with environmental, farm, and consumer organizations over whether the regulatory system for plants with novel traits and novel foods was adequate. What is the Canadian position on a precautionary Cartagena Protocol, and what impact has this international agreement and the concomitant rise of precaution had on the politics of GE in Canada?

Canada signed the Cartagena Protocol in April 2001 and, even though industry organizations have vehemently urged the Canadian government not to ratify, government officials say that they will likely ratify the protocol (Clapp 2002; interview #1). The official government position is that the protocol should be welcomed as good for biodiversity and not perceived as a threat to the vibrant Canadian biotech industry (Canada 2001). Notably, this position does not mention the precautionary clauses of the treaty. Instead, it reinforces the view held by the Miami Group at the end of the negotiations that the protocol "should be implemented in a manner that is consistent" with the international trade regime. Canada's signing of the Cartagena Protocol means that the protocol can be used as a tool for critics of Canada's own regulatory system for PNTs and novel foods, especially given that this regulatory system has always been touted as being based on concepts developed by international expert bodies. A 2001 report from the Canadian Institute for Environmental Law and Policy, entitled *Mixed Messages,* took exactly this angle in its critique of Canada's regulatory system. The report argues that Canada's domestic regulatory system for GEOs contradicts basic principles underlying the protocol, especially the precautionary principle. Contradictions exist, according to MacRae (2001), because the Canadian system does not have the precautionary approach as a specified objective; relies on familiarity and substantial equivalence to deliberately limit the scope of risk assessment; makes no effort to evaluate the societal benefits of GEOs (assuming, instead, that they are worth the risks);

undertakes no independent, peer-reviewed scientific assessment of GEOs; and interprets the absence of evidence of risks as evidence for the absence and/or manageability of risks.

This report demonstrates the way that the emerging international discourse of precaution can provide a footing for activists demanding more scrutiny of novel GEOs in North America. Similar critiques were made when Canadian environmental CSOs participated in consultations on whether or not Canada should ratify the Cartagena Protocol in September 2002 (CEN Biotechnology Caucus 2002). Significantly, however, these critiques carry no legal weight. While the protocol does reference the precautionary principle in its objective and decision-making clauses, it implies that countries *may* take precautionary action on LMOs, not that they are *required* to. This discretionary language is also common in Canadian environmental laws and is indicative of the lack of importance given to them (Boyd 2003). The discretionary nature of this language points to the need, as Foucault would note, for a precautionary approach to GEOs to become accepted as truth among the wider networks of experts engaged in the governmentality of GEOs in Canada. Were a precautionary approach to be accepted as the norm, Canadian regulators would be more restricted in how they implement the protocol.

In the absence of such precautionary norms, the official Canadian position is that the protocol "leaves the regulation of the import of bulk LMO commodities to parties' existing domestic regulatory regimes" (Ballhorn 2002, 112). This position was reflected in the draft Living Modified Organism Regulations circulated by Canadian government departments for consultation in August 2002. These regulations focus on the new export responsibilities that Canada would hold should it ratify the protocol and do not mention any changes the protocol might imply for granting domestic approvals to PNTs, novel foods, or feed (Environment Canada 2002).

Impacts on Pesticide Regulation
While the Cartagena Protocol has had little direct impact on Canadian biotech policy, the protocol's precautionary framing has had effects outside the field of biotech in Canada. These implications are the wider productive effects of discourse that Foucault would highlight. The June 2001 Canadian Supreme Court decision on *Spraytech v. Hudson* provides an important case in point. In this decision, the Supreme Court affirmed the right of a local municipality to ban cosmetic use of pesticides on private property. Even though the lawn-care pesticides being sprayed are approved by federal agencies in Canada, and even though their precise health and environmental impacts remain controversial, the court agreed that municipalities are empowered to set their own desired level of protection. Significantly, the court noted that

in order to achieve sustainable development, policies must be based on the precautionary principle. Environmental measures must anticipate, prevent and attack the causes of environmental degradation. (Supreme Court of Canada 2001, paras. 30-32)

In making this claim, the court drew specific attention to the Cartagena Protocol, using it as an example when stating that "there may be currently sufficient state practice to allow a good argument that the precautionary principle is a principle of customary international law." This is an interpretation that the Canadian government fought throughout the protocol's negotiation.

The StarLink Incident

An example of the wider productive effects of precaution in the field of agbiotech – an example that signals the entrenchment of this discourse in the governmentality of biotech despite the Canadian government's formal resistance – can be found in the responses of the government and of transnational food companies to the StarLink incident of September 2000. StarLink is a variety of Bt corn developed by Aventis CropScience. The protein produced by this variety (taken from the Bt subspecies tolworthi Cry9C) is heat stable and resistant to degradation in gastric juices, two important indicators of possible allergenicity (Dawkins 2000b). In the United States and Canada, StarLink was on the market because it had been approved for animal feed. It had not yet been approved for human consumption in either jurisdiction when traces of StarLink were discovered in taco shells by anti-GEO activists in the United States and then elsewhere in the global food system. News of the incident was rapidly picked up by the media, and consumers demanded answers. The response of the Canadian government and the food industry was to remove all corn and corn products from the human supply chain if there was a possibility that they contained StarLink, and to put in place a new regulatory protocol that would prevent a novel organism from being approved in Canada without meeting the expectations of Health Canada for food safety (even if not intended for food) and those of the CFIA for feed safety and environmental introduction (interviews #7, #17).

These responses are important on two levels. First, through the Cartagena Protocol process, the Canadian government had fought the notion that LMO-FFPs needed to be evaluated for their possible environmental impacts, because these organisms were *only* intended for food, feed, or processing. The StarLink incident led the government to adopt, in practice, the more precautionary strategy of evaluating all novel organisms through all three regulatory channels, regardless of their intended uses. The rapid move by the food-processing industry to remove all traces of StarLink from the human food supply is also telling. Throughout the biosafety debates, both the food industry and the Miami Group had been warning overseas importers (and thus consumers) that they would have to be able to accept from 1 to 5 percent of "adventitious [GEO] materials" in shipments of non-GEO commodities (interview #10). This allowance, they argued, was simply the price of a global food system that includes GEOs. When hit with the StarLink controversy, however, the food industry sought to eradicate any possibility that a consumer might be affected by a GEO-induced allergic reaction, and it dumped millions of dollars worth of Bt corn and corn products as a result. This move was a precautionary response. The Cry9C protein had *not* been established to be dangerous, yet the risk of a public relations disaster led the industry to go to great lengths to protect the reputation of its products. This action reveals, once again, the different relationship that food processors have to GE when compared with that of the seed developers. When a controversy erupted, the food processor's promises to seed companies to stand by them vanished momentarily. The processor's fickle allegiance to the biotech revolution may yet create a major schism in the industry arms of the biotech bloc.

The Royal Society of Canada Expert Panel on the Future of Food Biotechnology

A further example of the productive effects of the discourse of precaution on the politics of agbiotech in Canada can be found in a report written by an expert panel of the Royal Society of Canada, released in 2001, entitled *Elements of Precaution: Recommendations for the Regulation of Food Biotechnology in Canada* (S. Barrett et al. 2001). This report was the result of a risky initiative taken by the Canadian government to try to rebuild public confidence in GEOs when anti-GE sentiment was at its peak in the fall of 2000. This venture was risky because, in Gramscian terms, the government was not merely drawing on the organic intellectuals of the biotech bloc itself in this move, as had been its practice in the past. Instead, the government was reaching out to the cadre of what Gramsci would call "traditional" intellectuals. These are intellectuals positioned outside an emergent historical bloc, remaining from an earlier social formation (Gramsci 1971, 7). In Foucauldian terms, this was a move to re-establish the credibility of the government of biotech within the wider governmentality of GEOs in Canada, so outreach was made to Canadian experts in molecular biology and related disciplines. In this case, the CFIA, Health Canada, and Environment Canada turned to the Royal Society of Canada to elicit formal consent from these "independent" scientists and policy experts for the Canadian GEO regulatory system.

The result of this initiative was a critical report, released in January 2001. What the Canadian government did not anticipate was that growing

European backlash against GEOs, combined with the reinvigorated public skepticism in North America in 2000, had truly shifted the discursive terrain of GE governmentality. In its report, the Royal Society of Canada's Expert Panel on the Future of Food Biotechnology in Canada espoused a "precautionary" framing of GEO risks. This led the panel to raise specific concerns about the ambiguity, in Canadian regulatory practice, of the concept of substantial equivalence (S. Barrett et al. 2001, 36). The panel provided what it thought would be a scientifically justifiable (re-)interpretation of substantial equivalence, which

> should require *rigorous scientific analysis* which establishes that, despite *all* changes introduced into an organism as a result of the introduction of novel genes, the organism poses *no more risk* to health or the environment than does its conventional counterpart. (183, emphasis mine)

The expert panel argued that government regulators do not appear to be consistently applying this high standard of equivalence. Instead, assumptions of equivalence (the assumptions espoused in gene talk and reproduced through the discourse of equivalency) appear to have shaped regulatory practice in Canada. The result, the panel wrote, are cases where "substantial equivalence is determined based on the assumption that no changes have been introduced into the organism other than those directly attributable to the novel gene" (182).

In a clarification of its position, the chairs of the Royal Society panel state: "It did appear that examination of molecular biological data during the Health Canada assessments did not routinely extend to possible pleiotropic impacts of the transgene" (Brunk and Ellis 2001, 1). This situation is particularly problematic, from the panel's point of view, because risk assessments are

> based solely on data and information provided by the petitioner, and decision documents describing and validating outcomes are ... not readily available to either the scientific community or general public. (1)

In a private response to the Royal Society panel report, regulators at Health Canada argued that "there are never assumptions of Substantial Equivalence" in the regulatory process (Green 2001, 2). This letter listed the variables that are looked at to evaluate a new food for safety, including compositional and nutritional information. It also noted that Health Canada and the CFIA "consider the potential for secondary or unexpected impacts," though there is no mention of any experimental protocols designed to reveal such effects.

The focus of the Royal Society panel report's criticism, and the urgency with which the regulators jumped to the defence of regulatory practices in Canada, demonstrates that the question of a GEO's equivalence to a non-GEO remains at the heart of the controversy over genetic engineering today, as it was when these issues first erupted fifteen years ago. In light of the emergent discourse of precaution, however, greater attention is now being paid to establishing that equivalence as a matter of fact, rather than accepting it as an assumption. While the CFIA and Health Canada officials might curse the Royal Society panel for increasing their workload, the new ideational context calls upon regulators to redouble their efforts to understand all the possible implications of GEOs before giving approvals for commercialization. For critics of GEOs, this is a progressive step. However, Foucault would also remind us that this dynamic is an important example of the way that biopolitical conflict often leads to the development of new practices of regulation and surveillance that ultimately allow industrial capitalism to make nature even more docile and productive.

In a demonstration of the way that wider processes of governmentality do indeed place disciplines on government, Health Canada and the CFIA affirmed in 2001 that they will implement all fifty-three of the expert panel's recommendations (Health Canada 2001). In 2004, the Canadian government was still far from achieving its target (despite its own claims to the contrary), though some progress had been made. Canadian CSOs critical of biotech continue to press companies to fully implement the expert panel's recommendations, though the regulatory culture in Canada appears slow to accept the implications of a more precautionary approach (Andrée and Sharratt 2004; Andrée 2006). These CSOs are helped in the process by the ability of European CSOs to access risk assessment data, under the new EU regulations, that has always remained hidden as confidential business information in Canada (Greenpeace 2005b). These data help shine a light on the limitations of Canadian regulatory reviews for novel foods and plants with novel traits.

The Cartagena Protocol, the Supreme Court pesticide decision, and the Royal Society panel report, among other developments, are forcing the Canadian government to pay more attention to precaution. As an example of the prevalence of the new discourse and the necessity for all political actors to mobilize within it, the Canadian government released a position paper on precaution in 2003 (Canada 2003). In "A Framework for the Application of Precaution in Science-based Decision Making about Risk," the government attempts to reframe precaution in the light of its perceived interests (in biotech, among other substantive areas). When contrasted with the EU's communication on precaution from 2000, this paper illustrates the way that the international debate over precaution has moved on from earlier debates on risk.

The New Canadian Position on Precaution
Like the EU, the Canadian government recognizes the precautionary principle as a tool within risk management; that precautionary measures should
be based on the identification of a potentially negative effect from a phenomenon, product, or procedure; and that these measures should be non-
discriminatory, transparent, and accompanied by an allocation of
responsibility for producing additional scientific information (Canada 2003).
However, whereas the EU emphasizes that precautionary decisions should
be proportional to a society's chosen level of protection, Canada stresses
that decisions should be proportional to the chosen level of protection *and*
the potential severity of the risk. Canada further states that precautionary
measures must represent the least trade-restrictive option, while the EU document makes no mention of this requirement.

The similarities and differences between the Canadian paper and the earlier EU communication, as with the positions brought forward by these
actors in the WTO trade dispute, show how the official trans-Atlantic conversation on the precautionary principle has indeed moved in the direction
suggested by my analysis of the final year of the Cartagena Protocol talks.
On the one hand, the application of precaution *has* become central to any
discussion about risk, and the EU and Canada, at least, have moved closer
in their understanding of what a precautionary approach implies. On the
other hand, the differences between the EU and Canadian interpretations
of precaution point to two major ideological divides that still exist in this
field.

The first division concerns the question of whether the "least trade-
restrictive" principle is necessary to meet the (still widely accepted) goal of
liberalized trade, or whether the principle of non-discrimination is sufficient to address these considerations. The second division is over the issue
of proportionality. Should the "severity of risk" continue to be a legitimate
test for judging the proportionality of a precautionary risk management
decision? The Canadian paper assumes that the "severity of risk," based on
the "magnitude and nature of a potential harm," can be objectively determined on a "sound scientific basis" (Canada 2003, 8, 11). The European
Commission's report, on the other hand, goes into considerable detail in its
description of the nature of scientific uncertainty in order to back its asser-
tion that, in the end, risk management decisions should be proportional to
the "acceptability" (or lack thereof) of a risk, given that potential severity is
not always clear (European Commission 2000, 16).

The Biopolitics of Precaution
This study has allowed us to see the contours of a biopolitical struggle in
the negotiations on the Cartagena Protocol, and in the debates over GEOs
more generally. In this struggle, discursive practices were adopted as truth

claims, and arbitrated as such, though they were clearly tied to the values and interests of the actors involved. Beginning in the biosafety debates of the early 1990s, we saw a network of actors with a range of concerns about GEOs adopt precaution as a discourse of resistance in their struggle against the biotech bloc's GE revolution. Then, because of material, organizational, and ideational relations of force, this discourse grew in its strength and influence through the biosafety talks. These factors included new scientific facts but were certainly not limited to them. Eventually, by late 1999 and early 2000, precaution became what Gramscians would define as the hegemonic ideology of biosafety, establishing the "common sense" that shaped Cartagena Protocol outcomes on import decision making and related articles. However, while accurate, this Gramscian framing does not fully articulate the power of precaution in the protocol negotiations. It is for this reason that I framed Chapter 6 in specifically Foucauldian terms, focusing on the productive effects of the discourse of precaution in the biosafety talks.

Precaution moved from being a discourse of resistance, empowered within the risk framing but wielded by only vociferous critics of GEOs in their calls for a comprehensive assessment of these technologies, to becoming the norm. As a discourse of normalization in this context, precaution established a new playing field for the biosafety debate, with new truths around which all the actors, and not just GEO critics, came to frame their interests and strategies. As a normalizing discourse, precaution exacted disciplines on the biosafety conversation that would affect all the actors.

This shift in the role of precaution in the biosafety context occurred first in formal policy positions of the EU, where the precautionary principle was brought forward by politicians who wished to remain true to the principle of liberalized trade while also addressing the demands of a cautious electorate for safe food. Then, in the final round of talks, the Miami Group's representatives began to internalize a precautionary framing in relation to their own interests and concerns. In this move, the Miami Group worked to narrow the understanding of what precaution would mean in practice so as to minimize the possibility that LMO-import decisions would hamper trade in GEOs or interfere with already existing rules of trade. The final text of the Cartagena Protocol reflects the compromise reached between the EU and the Miami Group on how precaution could be integrated with a neo-liberal reading of sustainable development and ecological modernization. Still, beyond the protocol, this compromise is not yet stable, as we see in the different interpretations of the protocol (and of the precautionary principle) held by EU, Canada, and US representatives, and in the greater traction that risk discourse still appears to have over precaution in the states at the heart of the biotech bloc and in the WTO.

Through its uptake in EU regulatory practice, the Cartagena Protocol on Biosafety, and related instruments (such as the Codex principles), the precautionary response to GEOs is gradually being normalized in international politics. This process is necessarily both productive and disciplinary. By being internalized in the GEO risk analysis practices of the EU, precaution will likely produce more detailed risk assessments, and more in-depth study of the long-term implications of GEOs in agriculture and food. Precaution's growing role in ecological governmentality in Canada, as recognized in the Canadian government's formal adoption of the precautionary "approach" and the discourse's adoption by scientists in the Royal Society of Canada, means that over time Canada may also see more careful scrutiny of GEOs.

Even in the United States, a precautionary discourse appears to have had an impact. Consider the case of the Biotechnology Industry Organization's voluntary moratorium, initiated in 2002, on planting crops such as corn that have been genetically engineered to produce proteins for pharmaceutical uses in areas where those same crops are of major economic importance. Environmentalists may have concerns that this moratorium does not go far enough, but it is still an example of the biotech industry's adoption of a more precautionary approach to the cultivation of potentially hazardous GEOs. The incident that led to this moratorium was the StarLink debacle of 2000. Efforts to clean up StarLink contamination reportedly cost the food industry hundreds of millions of dollars (Anonymous 2002b). The BIO voluntary moratorium, like Brazil's new position on documentation at COP-MOP-3, demonstrates that a precautionary framing of GEOs is gradually being internalized by states exporting GEOs and by the companies that develop and market these products, even if only for economic reasons. This kind of internalization by a range of actors is a central dimension of the normalization process that accompanies the rise of a new discourse.

In terms of its disciplinary effects, precaution will limit opportunities for raising certain arguments by rendering them irrelevant in relation to the new regime of truth. This discipline was felt by the Miami Group and industry organizations putting forth the equivalency position in the biosafety talks. It was also felt by those actors who brought forward arguments for a protocol that would include socio-economic assessment as part of a comprehensive approach to risk management but who were not able to have their views incorporated in the protocol's final text. Even though these were the same actors who trumpeted precaution in the first place, their positions on socio-economic considerations were deemed irrelevant because of the strength of a perceived split between technical issues and non-technical concerns. This division was inherited from the earlier hazard and risk discourses, and it reflects the central dynamic of biopolitical struggle described here: concerns about genetic engineering were taken seriously only if framed as scientific truths. Ironically, this result means that more precautionary biosafety regulations, like those being developed in the EU, will focus on uncertainty, complexity, and long-term – even if only hypothetical – risks, while not addressing immediate social and economic impacts from the introduction of GEOs to agriculture and food.

On the international stage, precaution is realigning the GEO debate. As it grew in its impact, many critics of GEOs adopted the new discourse as their primary strategy of resistance. Through the normalization of precaution in the biosafety discussion, it would appear as if these critics became consensual participants in a biotech revolution that will now work *through* precautionary regulations to get products to market. However, this reading misses the subtleties of biopolitics. Foucault would stress the way that discourse always produces new opportunities for resistance, and resistance to GEOs does indeed continue through new debates within the discourse of precaution. This resistance occurs both inside the technical parameters of the conversation on what true precautionary risk analysis might mean and through readings of precaution that attempt to challenge the underlying technicist assumptions of a governmentality rooted in risk assessment and management.

In risk analysis, the precautionary discourse offers critics of GEOs a means for demanding that regulators always take the assessment one step farther in order to reveal the full effects of genetic engineering. They ask, for example, what about data from crops grown in a wider variety of environments, or crops subjected to different kinds of selection pressures? In Europe, broad acceptance of the need for precautionary assessment has triggered massive new ecological studies, and efforts, using the emerging sciences of proteomics and metabolomics to characterize many of the minute differences between GE and non-GE foods through detailed profiling techniques (Kuiper 2000, 2). Similarly, in the realm of risk management, the Curitiba rules are spurring on the development of new technologies to make identification of the genotypes of grains in international trade easier and less costly.

Together, these scientific and technological developments arising out of the struggle to develop a precautionary response to GEOs confirm the Foucauldian observation that biopower, in day-to-day practice, involves an intensification of surveillance and intervention at two levels: the social (and ecological) body, and the individual (that is, organismic) body. The result of these interventions will be a vast new reservoir of ecological and molecular knowledge. Whether this knowledge will lead to more accurate prediction and control of the impacts of genetic engineering remains to be seen. Foucault (1978, 143) would note that when it comes to attempts at surveillance and administration, "life constantly escapes them."

The precautionary discourse has also been deployed to further a different kind of resistance, one that continues to challenge the very idea of science-based risk assessment in order to bring the wider social, ethical, and economic issues associated with GE into the conversation, as evidenced by the

GMO ERA project. In its articulation, the precautionary principle is an explicit acknowledgment of the relationship between norms and science in risk decision making. This acknowledgment has led to a growing recognition among academics and policy analysts of the closed nature of the processes through which these norms are usually determined, involving tacit judgments made by small groups of scientific experts themselves (Wynne 1994). The arguments for more precautionary regulation have thus been accompanied by demands for more openly democratic decision-making processes on risk issues, as opposed to simply a reversal of the dominant norms (see the contributions to Raffensperger and Tickner 1999).

This desire to open up risk-based decisions in the context of the precautionary frame also came through in the Royal Society panel's report. The expert panel expressed concern about the very narrow mandate it was given by the federal government, in which it was specifically asked to review the preparedness of the regulatory system for the kinds of risks that will be posed by the products of biotechnology over the coming ten years. The panel noted that even though the public often articulates concerns in terms of risks, there is actually a much wider set of issues that people associate with genetic engineering. The report then introduced the concepts of socio-economic, philosophical, and metaphysical risks associated with GEOs as a way to reframe the discussion (S. Barrett et al. 2001, 1-9). It suggested that each of these kinds of risk is part of the wider public debate over genetic engineering, and that the Canadian government should not lose sight of them in an effort to define a purely technical solution to the genetic engineering controversy.

The Royal Society panel's move to introduce socio-economic, philosophical, and metaphysical concerns into the language of risk and precaution is similar to the move by CSO activists in the biosafety talks to frame socio-economic considerations as a risk issue. In both cases, these actors are working within accepted discursive (and institutional) boundaries in their attempts to broaden the conversation. There are dangers inherent in this move, because all of the assumptions about a narrow, objective, technical solution that is implied by risk (as currently understood) may taint the ensuing conversation. At the same time, however, I believe that the Royal Society's move (and that of the CSOs in the biosafety talk) to reframe risk and precaution is necessary and, over the long term, a promising path to a different conversation. Ultimately, the objectivist framing of risk and precaution is not tenable. This framing denies the value judgments implicit in risk decisions. The normalization of precaution offers tools to begin to pry apart the objectivist box around risk, because even as it is narrowly interpreted in the Cartagena Protocol, precaution assumes a value position rather than the lack thereof – the value of being careful in the face of ignorance or uncertainty. That this assumption is explicitly accepted in operational articles of this treaty is an important step toward a wider discussion of values in GEO regulation.

This idea of opening up GE crop and food regulation to more than a particular version of science-based decisions is still officially resisted by the Canadian government, as it is by other governments at the heart of the biotech bloc. However, Canada's recent experience with Roundup Ready wheat put the government in a position where it was forced to start considering other options.

In the spring of 2004, Monsanto, which had been developing Roundup Ready wheat (with support from Canadian government funding) since 1997, decided to withdraw its application to the CFIA and Health Canada to have the product approved for commercial planting. This came about because Canadian wheat farmers currently export significant quantities of wheat to ·Europe, and most wheat growers, in an effort to protect these markets, did not want to see engineered wheat varieties introduced into Canada. The earlier introduction of GE canola had destroyed markets for non-GE canola (through cross-pollination), and wheat producers were worried that they would find themselves in a similar situation. The opposition of these farmers, especially through the Canadian Wheat Board, was so strong that Monsanto decided to stop its efforts to commercialize the GE wheat. However, Monsanto's withdrawal took place only after Canadian civil servants had started working on contingency measures that would enable them to prevent the product from getting to market (on socio-economic grounds) if it passed environmental and health safety assessments (Jeffs 2003).

Monsanto's decision allowed the company and the Canadian government to avoid the solution proposed by the Canadian Wheat Board, which was to allow market considerations, and the status of segregation capability, to be part of regulatory decisions on whether Canada will register new engineered crops (CWB 2001). Nonetheless, such considerations were clearly part of Monsanto's decision. A growing call to include these kinds of considerations in regulatory decision making from within branches of the agri-food industry itself may yet force the governments at the heart of the biotech bloc to acknowledge that regulatory decisions are inherently political, and not just technical. Such an admission, should it ever come, could provide the space for a more accessible debate about the values implicit in regulation, in place of the current biopolitical struggle over scientific truths, in which only groups with access to expert resources are able to participate.

Reflections on Theory
Rather than rely on a single existing theoretical perspective, this study of the Cartagena Protocol brings together elements of neo-institutionalist regime theory with insights from Gramsci and Foucault. This strategy raises

two questions: Does this integrated approach explain the emergence and significance of the Cartagena Protocol in a way that would not have been possible within any of these traditions on its own? And, what does this integrated approach, informed by the case study, contribute to the theoretical traditions from which it draws?

Neo-institutionalist regime theory on its own would have told us only part of the story. This interpretation would have stressed, for example, the roles of new scientific developments, the institutional context of the CBD, and Mayr's innovative Vienna setting in achieving the final outcome. However, even when they define regimes broadly to include the cognitive and normative dimensions of international environmental governance, regime theorists tend to look at multilateral environmental agreements (MEAs) as the heart of a regime, drawing attention to outside influences only as required to tell the story of the MEA. My approach specifically avoids this narrow gaze by placing the Biosafety Protocol in the context of the material, organizational, and ideational relations of force as they shaped a biotech regime both inside and outside the protocol negotiations. This study of the politics of agbiotech illuminates the importance of the Biosafety Protocol more clearly than would a narrower focus, with an analysis of developments at each of these three sets of relations of force explaining the struggles that led to the protocol we have today.

In the arena of material capabilities, the commercial successes of the biotech revolution in the United States, Argentina, and Canada, thanks to well-funded scientific research; clever product and marketing strategies; horizontal and vertical integration in the food system; and close interrelationships among states, academia, and industry, enabled the rapid establishment of agricultural biotechnology as a transnational industry with global ambitions in the 1990s. The emergence of this industry, that it was strongly established in only three countries in the late 1990s, and that the rest of the world would be LMO importers during the protocol's negotiation created the material context for the political struggles over a biosafety regime discussed here. Other material developments, such as the illicit trafficking of GE seeds into Brazil and Paraguay, have clearly also been important developments for the global politics of agbiotech.

In terms of organization, the establishment of the Convention on Biological Diversity (and the work of the World Resources Institute), followed by the UNEP Expert Panel IV majority report and the Third World Network report, framed the initial rounds of biosafety discussions. However, the precedent of the UNEP international guidelines allowed a schism to emerge between EU countries that were still closely allied with the biotech bloc in 1994 and 1995 (such as Germany and France), and the United States, Canada, and Argentina. Further institutional developments in Europe, including the BSE crisis, the GE labelling law, the EU response to the WTO beef hormone dispute, the EU moratorium, and the context of the SPS Agreement all proved critical in structuring the precautionary protocol we see today. However, although these European developments are necessary to explain the Cartagena Protocol, they do not on their own provide a sufficient explanation for its emergence, as evidenced by the lack of identical developments in the context of the Codex talks over principles and guidelines that ended in 2003.

Shifting ideational relations of force, including the emergence of new ideas about the gene (and how GEOs are, or are not, equivalent to non-GEOs), about the environment and its proper governance, of liberalism, and of risk, each contribute to shaping the terrain on which a particular precautionary approach to biosafety could (eventually) be established as the new common sense. The idea of a precautionary approach to GEOs solidified through the biosafety talks is a compromise between GEO exporters, importers, and GEO critics. This particular version of precaution provides the footing for ongoing debate and struggle over GEOs even as its institutionalization is reshaping environmental law at domestic and international levels.

My strategy of looking at each of these three sets of relations of force in order to conceptualize international politics is adapted from Gramsci's work. I also borrow from Gramsci an attention to the relationships between civil society and the state in extended states and then (when linked with material capabilities and anchored by ideology) in historical blocs. Attention to these relationships enables an explanation of the success of the biotech industry in the United States and Canada (thanks to its grounding in a wider biotech bloc and still wider transnational historical bloc), the EU moratorium (which started in civil society but required government support), and the relationships between activist CSOs and those representatives of the G-77/China who initially called for and then put forward the prototype of a biosafety protocol. These Gramscian concepts allow for a more accurate representation of the close state-civil society-industry relationships that can exist in domestic and international politics than does a typical neo-institutionalist focus on the influence of nongovernmental organizations on governments. Nonetheless, while Gramsci's concepts help develop a more comprehensive account of the biosafety process and its significance than neo-institutionalism would on its own, an exclusively Gramscian reading of this process would have its shortcomings, especially when it comes to analysis of the power of ideas.

In this study, I make an effort to identify those instances when ideas functioned as ideologies, interpreted in a Gramscian sense as ways of understanding the world that cement a particular group's interests as the common interest, thereby providing a rallying point for alliance building. Examples include the ideologies of equivalence, of precaution, and of neo-liberalism. I also identify instances when ideas functioned as discourses, with discourse understood in a Foucauldian sense as normalizing particular understandings as truth and propelling the debate in certain ways rather than others. Examples include gene talk and the discourses of sustainable development and precaution. Based on just these examples, it is clear that some ideas operate as both ideologies and discourses, making my distinction (based on function) at times difficult to follow. Still, I maintain this distinction to enable a clear identification of those instances when discursive power was at work. The effects of the term "living modified organism," the implications of definitions of GEOs (whether as part of a class of novel organisms or as a distinct class of GMOs), and the conclusions on socio-economic considerations are each examples that demonstrate the productive and disciplinary powers of discourse in the politics of agbiotech. Recognizing these effects enables a deeper understanding of the power of ideas than that offered by typical Gramscian scholarship.

That ideas (such as the idea of precaution) are sites of struggle that direct the course of debate but that are never controlled by only one set of actors (therefore being inherently unpredictable) becomes evident in this work through the adoption of a Foucauldian theoretical lens. Foucauldian concepts also enable a recognition, in the global politics of agbiotech, of the dynamics of a wider class of biopolitical struggles over nature and its control that are characteristic of contemporary ecopower. In the field of environmental politics, which appears to be knowledge driven but which is clearly much more, a theoretical framework that helps elucidate the relationships among facts, values, and the day-to-day practices of environmental governance has proven enormously useful. Still, on its own, a Foucauldian analysis would not have given us the picture of the Cartagena Protocol presented here. As interpreted in the work of Litfin (1994), for example, Foucault's concepts would have enabled us to see the emergence of precaution as a discourse, and the implications of this for the field of agbiotech and for global trade and environment issues more generally. But this story also shows the limits of discursive power in the field of biosafety. There was only so far that the Miami Group would move on the basis of precaution at the particular historical juncture documented here, as the debate on documentation to accompany LMO-FFPs demonstrates. These limits may have been missed in a story focused narrowly on discursive change. However, a focus on the three arenas of political activity, or relations of force, allow us to see where discourse was the primary determinant of outcomes in the protocol text (on AIA, definitions, decision-making clauses for LMO-FFPs, for example) and where it was not (on the exclusion of LMO-FFPs from AIA, documentation, "products thereof," and pharmaceuticals). In answer to my first question, then, the integrated theoretical framework developed here does present a more complete story of the Cartagena Protocol and its implications than any of the three theoretical frameworks would on its own.

The main purpose of this book, and the research that went into it, is to tell the story of the Cartagena Protocol on Biosafety as a way of illuminating both global and Canadian biotech politics. However, empirical research is never simply the application of theory; it establishes a dialogue between theoretical constructs and the case being considered. This study offers insights into the politics of material capabilities, institutions, ideas, and especially their interrelationships, from which each of the three theoretical traditions I draw upon can learn.

To begin with, this research confirms that material capabilities really do matter in global politics. The relationships between states and the biotech industry influenced the positions of the United States, Canada, Argentina, the EU, and others in the biosafety talks. However, I also show that it was not the objective state of the industry that mattered as much as subjective perceptions of material interests. In the mid-1990s, several European seed companies had already advanced biotech sectors and even products on the market. Nonetheless, European countries eventually adopted the position that they were, first and foremost, importers and consumers of agbiotech. The biotech industry argued that this was a misperception but, in the end, it was the (mis)perception that mattered. This research shows that material capabilities are necessarily bound up with institutions and ideas: business organizations and product strategies, scientific bodies and research agendas, government agencies and political programs. These institutions and ideas make it possible for material interests to be realized. This analysis suggests that when studying the role of material power in political developments – which is the central preoccupation of most political economists and not just Gramscians – it is important to understand that we are always already looking at webs of material capabilities, perceptions of those capabilities, and the ideas and institutions that bind them together. Material capabilities never exist independently of the other two relations of force. At the same time, I am not suggesting that these relationships are unidirectional, as Marx would have us believe. The metaphor of a web of power that contextualizes the role of material capabilities, a metaphor clearly drawn from Foucault, offers a starting point for further dialogue between Foucauldian and Gramscian scholars.

Institutions and organizations such as the CBD, the EU, the Third World Network, and the Global Industry Coalition (GIC) are clearly important in this narrative. However, the story of the Cartagena Protocol shows that the power of institutions does not lie in the size of their budgets, their personnel, or their connections, though each of these factors can make a difference. Rather, the power of institutions lies in their relationships with powerful ideas. Wright (1994, 10) defines organization as the "mobilization of bias."

When an organization's bias loses its discursive footing in a political struggle, as we saw with the GIC's equivalency position at the Montreal ExCOP, the organization itself loses its influence in consensus-based negotiations, even if it has enormous economic clout. In contrast to the GIC, which held on to the equivalency position despite that it was falling on deaf ears, the Miami Group was able to remain engaged in the discussions in Montreal by mobilizing within the precautionary discourse. This strategic shift allowed it to maintain an influence on the final outcomes of the talks. The recent literature on international regimes recognizes the importance of the cognitive and normative dimensions of international regimes. The analysis presented here suggests that it would be worth revisiting existing regimes with a view to understanding the ways that international institutions are enabled (and disabled) *within* discursive fields.

This study of the Cartagena Protocol emerged from my interest in understanding the ways in which interests, facts, and norms together impact on the global politics of biotech regulation. To understand these dimensions of political life, I developed a theoretical framework focused on hegemony and discourse, concepts that are designed to help account for what might be termed "structural" influences on politics. It was somewhat surprising, then, to see how much the agency of particular actors appears to have influenced the outcome of the protocol talks. Foucault specifically avoids the subject of agency, believing that an identification of individuals with political change misses the wider discursive webs that empower them. Regime theorists don't give much attention to individuals either, seeing them as little more than the figureheads of institutional bias. Gramsci, though, does refer to individuals in his conceptualization of organic intellectuals.

The idea of organic intellectuals offers a useful starting point for understanding the role of actors such as Egziabher, van der Meer, Marquard, Giddings, and Mayr in the politics of biosafety. These actors were organic in that they emerged from particular ideological, material, and organizational movements. They were also intellectuals who consciously redefined ideas, leaving their marks on the fields in which they operated. The actions of these individuals could not be reduced to the function of hegemony or discourse. At the same time, this study showed that agency is limited in very real ways by discursive and material power. An important example is the way that individuals within the Canadian delegation sympathetic to the concerns of the Like-Minded Group were constrained by the pervasiveness of the discourses of neo-liberalism, equivalence, and sound science that said "corn is corn."

The relationship between agency and discourse identified here deserves more attention than Foucault or neo-institutionalists give it. From this research, it is clear that the actors who had the greatest impact on the Cartagena Protocol were not only politicians but also civil servants and representatives of civil society organizations, and that their power was not rooted in their institutional location so much as in their authority in the discursive realm. These observations on organic intellectuals in the biosafety debate fit well with Litfin's conclusions on "knowledge brokers" in the international negotiations on the Montreal protocol designed to protect the ozone layer. Knowledge brokers' most important asset, according to Litfin (1995, 253-54),

> is their flair for translating science, often with a "spin," into a language accessible to decision-makers. Their influence derives from the plausibility of their interpretations, the loudness of their voices, and the political context in which they act.

This study of the Cartagena Protocol differs from the work of Litfin through my emphasis on three sets of relations of force, rather than simply discursive practices. I also delve more systematically into the ability of discourses to produce and discipline political deliberations in ways that cannot be reduced to the interests or intentions of the agents who operate within them. Nonetheless, Litfin's general observations on the role of knowledge brokers complement mine on organic intellectuals. Together, these works on the Montreal and Cartagena protocols offer Foucauldian scholars in the field of environmental politics new tools for conceptualizing the relationships between discourse and agency.

The politics of ideas, functioning as ideologies and discourses, figured centrally in this work. However, as Foucault and Gramsci would emphasize, ideas are never free-floating entities. Both ideologies and discourses can be defined, in general terms, as ideas enabled through networks of actors. This study confirms these definitions by showing how the ideas that influence international politics must already be institutionalized in one way or another in order to have an impact, and how their impact grows through even further institutionalization. (This observation is the corollary of the conclusion drawn above, that institutions are powerful only when tied to authoritative ideas.) The institutionalization of ideas can take different forms. It can take the form of credibility within restricted networks of experts in a field (the most basic configuration of power/knowledge) or it can take the form of formal adoption in international treaties. As we have seen, these different forms of institutionalization are not the same, even though each are forms of power/knowledge as Foucault defines it. Actually, Foucault does not really provide a way of distinguishing between those discourses that have a wide uptake and those that are universalized, being accepted by all participants in a given conversation as the norm. This shift is precisely what happened to the precautionary discourse in the biosafety talks: precaution moved from its status as a vaguely defined academic response to risk issues accepted in specific expert communities to becoming the norm in the biosafety conversation. Through this process and its institutional outcome, precaution became considerably more influential, as is demonstrated by the examples of the Cartagena discourse of precaution's impacts on the POPs treaty and the Canadian Supreme Court's ruling on municipal pesticide bylaws. Gramsci would identify this shift as the moment when the ideology of precaution became hegemonic. Interestingly, Foucault adopts the same language in a passage in his first volume of *The History of Sexuality*. In a discussion of the many relationships in the social body that form a general matrix of power, Foucault (1978, 94) writes:

> These then form a line of force that traverses the local oppositions and links them together; to be sure, they also bring about redistributions, realignments, homogenizations, serial arrangements, and convergences of the force relations. Major dominations are the hegemonic effects that are sustained by all these confrontations.

Foucault chooses to speak of hegemonic effects rather than hegemony, presumably because he refuses to identify a hegemon. Still, his statement offers the space for thinking about different degrees of power/knowledge, and the possibility that discourses can have more or less influence on a given field. He doesn't pursue this train of thought, instead focusing on the way that all discourses are normalizing. However, this study suggests that some discourses are more normalizing than others, and that we need language to describe this possibility. I do not think it is useful to start speaking of hegemonic (and non-hegemonic) discourses, since hegemony brings with it more baggage than is necessary in this context. Instead, I make the distinction between a discourse of resistance and a normalizing discourse. The latter is a discourse that has been effectively universalized within a specific context, while the former has not. Both have productive and disciplinary effects, but the extent of a discourse's effects is often not evident until a discourse of resistance becomes the new norm. An advantage of this choice of terminology is that it speaks to the relationships that necessarily exist between discourses that are universal and those that are not.

The distinction between the function of ideas as ideologies and their function as discourses, often at the same time, suggests a need to think anew about the power of ideas in both the Foucauldian and Gramscian literatures. Foucauldians generally theorize only in terms of discourse, thereby missing the way that some ideas are actively constructed and mobilized by actors to meet their interests. Among Gramscian political economists, there have been efforts to incorporate an understanding of discursive power. Levy and Newell (2002), for example, write specifically of discourses in their work on business in international environmental governance. These authors see discourses as both constituted within and constitutive of relations of force that also have material and institutional dimensions. However, the constitutive effects of discourses are not systematically theorized. Levy and Newell do draw attention to one of the productive effects of discourse through their emphasis on the way discursive struggles can force political actors to make accommodations to one another (see also Levy and Egan 2003). These accommodations help explain the growth of historical blocs and the establishment of hegemony. The notion of accommodation speaks, to a certain extent, to the way discursive power is inherently relational, binding actors with different values and interests together even as they take divergent positions on a given issue. Nonetheless, this Gramscian reading of accommodation does not fully characterize the ability of discourses, as truth claims and turns of phrase embedded in cultural narratives, to lead policy conversations in directions that none of the actors involved initially envisioned.

Were Gramscians to take Foucault's observations on discursive power seriously, they would have to recognize that many acts of accommodation are about more than building alliances. These acts irrevocably change all the actors involved, often setting new chains of events in motion that over the long term reorganize the political field across all three sets of relations of force. The biotech bloc's accommodation of precaution is an important case in point. While we can refer to the biotech bloc as the victor in the global struggle over GEOs, because the market for these products will likely continue to grow, the gradual internalization of precaution means that this historical bloc is becoming a different constellation of actors, institutions, interests, and ideas from what it was when it first emerged. The constantly shifting political landscape in the debate over documentation to accompany LMO-FFPs between 2000 and 2006 reveals how the biotech bloc is evolving as it expands to include new actors and territories.

Gramsci did not explicitly theorize change within hegemonic formations, focusing instead on the establishment and maintenance of hegemony in the face of conflicting interests. This hole in Gramsci's work has led some Gramscians to frame political struggles in terms of hegemonic versus counter-hegemonic forces (e.g., D. Humphreys 1996). Unfortunately, this approach suggests that political conflict takes place between static, dichotomous poles, a situation we don't see in most fields of politics and certainly not in the politics of GEOs. My reading of Gramsci's texts, with their emphasis on transformism (co-optation) and the give-and-takes that enable hegemonic relations, finds a more fluid view of the power dynamics within an historical bloc. Levy and Newell (2002) contribute to this understanding through their discussions of the real material accommodations that leaders in an historical bloc must concede to other social groups in order to build historical blocs. The analysis of discursive politics developed above suggests that Gramscian theory can go one step farther: so-called counter-hegemonic forces can be understood as being internal to a hegemonic formation. This statement is not meant to imply that counter-hegemonic forces, or what Foucauldians would call tactics of resistance, are silenced and have no power. In fact, I am suggesting the opposite conclusion. Throughout their emergence, hegemonic formations are profoundly influenced by ideational, material, and institutional conflicts and accommodations with forces of resistance. The resultant historical blocs may still represent oppressive structures (to some), but they are also very different entities from what their leaders initially envisioned. As one example, while the transnational historical bloc is a vehicle for disciplinary neo-liberalism and the wide range of negative social and economic impacts this implies, it is also a vehicle for the furtherance of human rights and discourses of corporate responsibility. These dynamics must be taken seriously by Gramscian scholars intent on understanding the full effects of hegemony, and the incorporation of a Foucauldian reading of discursive relations can help Gramscians achieve this understanding.

This study emphasizes the unpredictability of discursive and hegemonic politics. In the early 1990s, the equivalency discourse was shaping regulatory policies in North America, and it was expected that the international regulatory regime would reflect the same framing. Instead, by early 2000, a precautionary framing had taken hold in the Biosafety Protocol discussions, even though there had been no major disasters associated with GE, and its influence continues to grow. Along the way, precaution was decoupled from the comprehensive assessment position brought forward by GE critics, taking on a form that would cause minimal upset to the neo-liberal trade regime. Some would read the story of the emergence of precaution delinked from comprehensive assessment as a sign that the neo-liberals were victorious in these negotiations (e.g., Bernstein and Cashore 2002). However, this assessment misses the fact that the precautionary clauses in the Cartagena Protocol are themselves still a major contravention and innovation for neo-liberal ideology. If neo-liberalism can be made to accommodate precaution, I would argue, so eventually could it be made to accommodate the comprehensive assessment position as a whole. This move may have been too much to expect in the Biosafety Protocol negotiations themselves, where the evidence suggests that the comprehensive assessment position was never put forward as a clear and workable option. Still, the broader dialogue on the economic and social implications of global trade, and biotechnology products in particular, continues, and discursive politics are never over. This conversation may yet gain the ideational and institutional power needed to change the norms of global trade. The ongoing debate in Canada about opening up the biotech regulatory process to socio-economic considerations points to this possibility. Stabinsky and others have also made strong arguments that institutional spaces do exist at domestic levels, as well as in the Cartagena Protocol and WTO agreements, for socio-economic assessments to become a part of decision making around GEOs, were the international community to pursue this option (Stabinsky 2000; Mackenzie et al. 2003; Fransen et al. 2005).

Gramsci and Foucault are often read as theorists bent on describing oppressive, debilitating structures. I choose to read them differently. This discussion of the unpredictability and lack of finality in discursive and hegemonic politics should be read as a statement of hope and possibility. The emergence and impact of the discourse of precaution in the Cartagena Protocol negotiations demonstrate that, together, ecologically and socially concerned CSOs, politicians, media, scientists, and others are capable of realizing a major change to the way global environmental issues are dealt with, despite vociferous opposition from well-entrenched powers. The change they helped bring about was dependent on the adoption of a plethora of strategies across many fronts. Major compromises were made, and the process was long and slow. Nonetheless, in the Cartagena Protocol, the material, institutional, and ideational relations of force did allow for a change, and the global struggle over GEOs is far from over.

Notes

Introduction
1 The individuals who consented to interviews for this work were guaranteed anonymity. They are referred to by numbers in the text. The date and place of each interview is listed in the References.

Chapter 1. Theorizing International Environmental Diplomacy
1 "Economistic" (Gramsci 1971) or "structural" (Cox 1996) Marxists see historical change as primarily the result of economic and technological forces. Ideologies are simply reflections of underlying economic structures and interests and are not accorded a significant role in social change in their own right. By contrast, the Gramscian approach stresses the role of consciousness, ideas, and political action on the part of classes and class-fractions in global political transformations (Gill and Law 1988).

Chapter 2. The Biotech Bloc
1 The lack of research on these issues in North America can be demonstrated by comparing research figures for 2000. That year, the US Department of Agriculture allocated approximately $1.5 million for peer-selected risk assessment research (about ten projects). Meanwhile, the UK government allocated nearly ten times as much in scaled-up ecological and agronomic experiments to assess the presence of possible real differences in the ecological performance of GE versus non-GE crops at the farm scale (C. Stewart 2001).
2 This figure of 30 percent includes crops stacked with both herbicide tolerance and Bt traits (James 2005).

Chapter 4. Biosafety as a Field of International Politics
1 One industry representative noted that the UNEP guidelines were written by only a small group of "Piet [van der Meer] and Helen [Marquard]'s friends," and lost credibility in the wider international community as a result. The process of preparing the UNEP guidelines, in the view of this representative, was the "fumble" that led everyone down the road to the Biosafety Protocol of today (interview #3).

Chapter 5. Staking Out Positions
1 Eventually, Chile, Uruguay, and Singapore also left the G-77/China.
2 The regulatory systems of Japan, Australia, France, and Germany would eventually abandon the assumed equivalency of GEOs and non-GEOs, in order to allow the labelling of GEOs, for example. But in the mid-1990s, this was not yet the direction they were taking.
3 North American industry groups raised the same arguments in the fall of 1996, when pressure was being placed on Europe to accept the import of genetically modified soybeans. At that time, Monsanto argued that it would be impossible to segregate as-yet-unapproved Roundup Ready soybean varieties from conventional beans. However, interviews with former

Monsanto executives reveal that the grain traders and Monsanto had actually developed a plan for segregating a separate supply chain for Roundup Ready soybeans if need be, but this was not revealed publicly at the time: "We never made it public. It would have taken the pressure off the Europeans," said Denise Bertrand, Monsanto's former manager for Roundup Ready crops (Charles 2001, 164).

4 Were this process to have been managed by AAFC, for example, only food and farm industry groups would likely have been seen as stakeholders (interview #8), and only select ones among them (e.g., the organic industry was not taken seriously as part of AAFC's constituency, despite that it was growing at a rate of 10 to 20 percent per year [interview #11]).

5 The WTO treaties are constantly being renegotiated, so whether the Cartagena Protocol is the more recent treaty is not clear-cut. To calculate succession, would one adopt the 1947 date of the introduction of the GATT, the 1994 conclusion of the Uruguay round, or the completion of the next (Doha) round (Kerr 2002)?

Chapter 6. A Precautionary Protocol

1 Even though the Miami Group did get these separate procedures for LMO-FFPs in the final text, states may still require AIA for all LMOs under domestic regulatory frameworks.

2 The TEP is a more formal process than the TransAtlantic Business Dialogue, which was established to work towards a binding trade and investment agreement between Europe and North America (USTR 1998).

Chapter 7. The Politics of Precaution in the Wake of the Cartagena Protocol

1 A lower threshold of 0.5 percent is set for transitional GE products, a group of up to twenty-four GMOs that had received a favourable review from community scientific committees or the European Food Safety Authority by the time the new regulations were implemented in 2004 but which did not yet have full authorization for use in the EU (European Parliament 2003a; Greenpeace 2004).

2 Zedan introduced the GMO ERA side event at COP-MOP-2 in Montreal on 30 May 2005. A letter of support from Zedan can be found on the GMO ERA website at http://www.gmo-guidelines.info/.

3 Dr. Rod MacRae of the Toronto Food Policy Council asked this question at a public debate on rBGH in Peterborough, Canada, in 1996. Similar questions are raised in Mausberg and Press-Merkur 1995.

References

Interviews

Interview #1. 2002. Ottawa, Canada. April 9.
Interview #2. 2002. The Hague, the Netherlands. April 23.
Interview #3. 2002. The Hague, the Netherlands. April 22.
Interview #4. 2002. The Hague, the Netherlands. April 23.
Interview #5. 2002. Ottawa, Canada. March 8.
Interview #6. 2002. Ottawa, Canada. April 8.
Interview #7. 2002. Ottawa, Canada. March 22.
Interview #8. 2002. Ottawa, Canada. March 8.
Interview #9. 2002. The Hague, the Netherlands. April 24.
Interview #10. 2002. The Hague, the Netherlands. April 23.
Interview #11. 2002. Peterborough, Canada. March 4.
Interview #12. 2002. Ottawa, Canada. April 9.
Interview #13. 2002. Ottawa, Canada. March 22.
Interview #14. 2002. The Hague, the Netherlands. April 25
Interview #15. 2002. The Hague, the Netherlands. April 22.
Interview #16. 2002. The Hague, the Netherlands. April 25.
Interview #17. 2002. Ottawa, Canada. March 22.
Interview #18. 2002. The Hague, the Netherlands. April 24.
Interview #19. 2002. The Hague, the Netherlands. April 24.
Interview #20. 2002. De Bilt, the Netherlands. April 30.
Interview #21. 2002. De Bilt, the Netherlands. April 30.
Interview #22. 2002. De Bilt, the Netherlands. April 30.
Interview #23. 2005. Montreal, Canada. June 1.
Interview #24. 2005. Montreal, Canada. May 30.
Interview #25. 2002. The Hague, the Netherlands. April 25.
Interview #26. 2005. Montreal, Canada. June 2.

Texts

AAFC (Agriculture and Agri-Food Canada). 1996. *Agriculture and Agri-Food Canada's Action Plan.* Ottawa: Government of Canada.
–. 2004. Documentation requirements for living modified organisms for food or feed, or for processing (LMO/FFP's). Ottawa: Agriculture and Agri-Food Canada. http://www.agr.gc.ca/itpd-dpci/english/topics/bsp_trilateral.htm.
Abley, Mark. 2000. Biotech lobby got millions from Ottawa: Public cash used to alter image. *Montreal Gazette,* 20 February, A1.
Afonso, Margarida. 2002. The relationship with other international agreements: An EU perspective. In Bail, Falkner, and Marquard 2002, 423-37.

Agriculture Canada. 1993. *Biotechnology in Agriculture, Science for Better Living*. Ottawa: Agriculture Canada.

Akasaka, Kiyo. 2002. Japan. In Bail, Falkner, and Marquard 2002, 200-6.

Anderson, David. 2002. Environment ministers: Canada. In Bail, Falkner, and Marquard 2002, 237-43.

Andow, David A., and Angelika Hilbeck. 2004. Science-based risk assessment for non-target effects of transgenic crops. *Bioscience* 54(7): 637-49.

Andrée, Peter. 1996. Cultivating sustainability: Strategies for agriculture in the Kawarthas. Frost Centre Occasional Papers 1: Canadian Heritage and Development Studies, Trent University.

–. 2006. An analysis of efforts to improve GM food regulation in Canada. *Science and Public Policy* 33(5): 377-89.

Andrée, Peter, and Lucy Sharratt. 2004. *Genetically Modified Organisms and Precaution: Is the Canadian Government Implementing the Royal Society of Canada's Recommendations?* Ottawa: Polaris Institute.

Anonymous. 2002a. Codex committee moves forward on risk analysis standard. *BRIDGES Trade BioRes* 2(8): 2-4.

Anonymous. 2002b. Pharmacrops. *Washington Post*, 22 October, E01.

Anonymous. 2004. The conflict. Percy Schmeiser website. http://www.percyschmeiser.com/conflict.htm (accessed 19 November 2005).

Anonymous. 2005a. EU ministers vote to allow national bans. *BRIDGES Trade BioRes* 5(12): 6-7.

Anonymous. 2005b. EU court rules against Austrian regional ban on GMOs. *BRIDGES Trade BioRes* 5(18): 3.

Arundel, Anthony, Matthias Hocke, and Joyce Tait. 2000. How important is genetic engineering to European seed firms? *Nature Biotechnology* 18(6): 578.

Avery, Natalie, Martine Drake, and Tim Lang. 1993. *Cracking the Codex: An Analysis of Who Sets World Food Standards*. London: National Food Alliance.

Bail, Christoph, Jean Paul Decaestecker, and Matthias Jorgensen. 2002. European Union. In Bail, Falkner, and Marquard, 2002, 166-85.

Bail, Christoph, Robert Falkner, and Helen Marquard, eds. 2002. *The Cartagena Protocol on Biosafety: Reconciling Trade in Biotechnology with Environment and Development?* London: The Royal Institute of International Affairs.

Ballhorn, Richard. 2002. Miami Group: Canada. In Bail, Falkner, and Marquard 2002, 105-14.

Ban Terminator Campaign. 2005. Canada grants new controversial terminator patent to US company. Press release, 9 November. (on file with author)

Barnes, Tonya. 2001. Negotiations on the International Treaty on Plant Genetic Resources for Food and Agriculture: 30 October – 3 November 2001. *Earth Negotiations Bulletin* 9(213): 1-14.

Barrett, Katherine, and Carolyn Raffensperger. 2002. From principle to action: Applying the precautionary principle to agricultural biotechnology. *International Journal of Biotechnology* 4(1): 4-17.

Barrett, Spencer C.H., Joyce L. Beare-Rogers, Conrad G. Brunk, Timothy A. Caulfield, Brian E. Ellis, Marc G. Fortin, Antony J. Ham Pong, Jeffrey A. Hutchings, John J. Kennelly, Jeremy N. McNeil, Leonard Ritter, Karin M. Wittenberg, R. Campbell Wyndham, and Rickey Yoshio Yada. 2001. *Elements of Precaution: Recommendations for the Regulation of Food Biotechnology in Canada*. Ottawa: Royal Society of Canada.

BCH (Biosafety Clearing-House). 2006a. Search decisions on LMOs for food, for feed or for processing under Article 11. http://bch.biodiv.org/decisions/decisionsundera11.shtml (accessed 23 January 2006).

–. 2006b. Norway decision on LMO under Advance Informed Agreement. http://bch.biodiv.org/database/results.aspx?searchid=174051&page=1&documenttype=5 (accessed 23 January 2006).

Beachy, Roger N. 1991. The very structure of scientific research does not mitigate against developing products to help the environment, the poor, and the hungry. *Journal of Agricultural and Environmental Ethics* 4(1): 159-65.

Beck, Ulrich. 1992. *Risk Society*. London: Sage Publications.
–. 1995. *Ecological Politics in an Age of Risk*. Cambridge: Polity Press.
–. 2000. *What Is Globalization?* Cambridge: Cambridge University Press.
Belmonte, R.V. 2006. Mexico and Paraguay Block Agreement on Biosafety. Inter Press Service News Agency, 17 March.
Bereano, Phil. 2002. Codex matters. Council for Responsible Genetics website, 4 April. http://www.gene-watch.org/programs/biosafety/codex-phil.html.
–. 2006. Dispatch from Curitiba. *Gene Watch* 19(3): 4-5.
Berg, Paul, David Baltimore, Herbert W. Boyer, Stanley N. Cohen, Ronald W. Davis, David S. Hogness, Daniel Nathans, Richard Roblin, James D. Watson, Sherman Weissman, and Norton D. Zinder. 1974. Potential biohazards of recombinant DNA molecules. *Science* 185: 303. Quoted in Wright 1994, 179.
Bergleson, Joy, Colin B. Purrington, and Gale Wichmann. 1998. Promiscuity in transgenic plants. *Nature* (395): 25.
Bernstein, J., A. Cherian, L.J. Goree IV, D. McGraw, and S. Wise. 1994. First meeting of the conference of the parties to the convention on biological diversity: 28 November – 9 December 1994. *Earth Negotiations Bulletin* 9(28).
Bernstein, J., P. Chasek, and L.J. Goree IV. 1993. First session of the Intergovernmental committee on the Convention on Biological Diversity: 11-15 October 1993. *Earth Negotiation Bulletin* 9(6).
Bernstein, Steven, and Benjamin Cashore. 2002. Globalization, internationalization, and liberal environmentalism: Exploring non-domestic sources of influence on Canadian environmental policy. In *Canadian Environmental Policy*, 2nd ed., ed. Debora L. VanNijnatten and Robert Boardman, 221-32. Oxford: Oxford University Press.
Betsill, Michelle M., and Elisabeth Corell. 2001. NGO influence in international environmental negotiations: A framework for analysis. *Global Environmental Politics* 1(4): 65-85.
BIO (Biotechnology Industry Organization). 2002. "BIO Staff: L. Val Giddings, Ph.D." www.bio.org/aboutbio/vp/vgiddings.asp (accessed 27 November 2002; page now discontinued).
Biosafety (Independent Group of Scientific and Legal Experts on Biosafety). 1995. *Biosafety: Scientific findings and elements of a protocol*. Penang, Malaysia: Third World Network. http://www.twnside.org.sg/title/bios-cn.htm.
BIOTECanada. 1999. Public awareness and risk assessment in agricultural biotechnology. http://www.agbiotechnet.com/proceedings.
Birchard, K. 2000. European Commission to end de facto moratorium on GM products. *Lancet* (356): 320-22.
Blojkovac, Craig, and Paul Muldoon. 2000. Persistence versus persistents. *Alternatives* 26(4): 10-13.
Bodegard, Johan. 2002. Documentation. In Bail, Falkner, and Marquard 2002, 338-43.
Bowdens Media Monitoring 2000. Press Conference: Biosafety Protocol Conference. 25-26 January. Ottawa: Bowdens Media Monitoring. (on file with author)
Boyd, David R. 2003. *Unnatural Law: Rethinking Canadian Environmental Law and Policy*. Vancouver: UBC Press.
Braithwaite, John, and Peter Drahos. 2000. *Global Business Regulation*. Cambridge: Cambridge University Press.
Brown, Lester R., and John E. Young. 1990. Feeding the world in the nineties. In *State of the World 1990*, ed. L.R. Brown, 59-78. New York: W.W. Norton and Worldwatch Books.
Brunk, Conrad, and Brian Ellis. 2001. Letter to Ian C. Green, deputy minister, Health Canada. (retrieved under the Access to Information Act; on file with author)
BSCO (Biotechnology Strategies and Coordination Office), Canadian Food Inspection Agency. 1997. *Regulatory Impact Analysis Statement: Environmental Assessment of Biotechnology Field Releases*. Ottawa: Canadian Food Inspection Agency.
Burchell, Graham, Colin Gordon, and Peter Miller, eds. 1991. *The Foucault Effect: Studies in Governmentality*. Hemel Hempstead: Harvester Wheatsheaf.
Canada. 2001. Notice of Canada's signature and government position regarding implementation of the Cartagena Protocol on Biosafety. Ottawa: Environment Canada. http://www.ec.gc.ca/press/2001/010405-2_n_e.htm.

–. 2003. A framework for the application of precaution in science-based decision making about risk. Ottawa: Privy Council Office. http://www.pco-bcp.gc.ca/docs/Publications/precaution/precaution_e.pdf.

Carson, Rachel. 1962. *Silent Spring*. Boston: Houghton Mifflin.

CAST (Council for Agricultural Science and Technology). 1999. CAST biotechnology briefing speakers. http://www.cast-science.org/9902biot.htm (accessed 5 December 2002; site now discontinued).

CBD (Convention on Biological Diversity). 1994. Decision I/9: Medium-term programme of work of the Conference of the Parties. UNEP/CBD/COP/1/17.

–. 1995a. Report of the Open-Ended Ad Hoc Group of Experts on Biosafety. Jakarta: United Nations Environment Programme, second Meeting of the Conference of the Parties to the CBD. UNEP/CBD/COP/2/7.

–. 1995b. Decision II/5: Consideration of the need for and modalities of a protocol for the safe transfer, handling and use of living modified organisms. UNEP/CBD/COP/2/19.

–. 1997. Compilation of the views of governments on the contents of the future protocol. Montreal: CBD Open-Ended Ad Hoc Working Group on Biosafety. UNEP/CBD/BSWG/2/2.

–. 1999a. Report of the sixth Meeting of the Open-Ended Ad Hoc Working Group on Biosafety. UNEP/CBD/ExCOP/1/2.

–. 1999b. Informal consultations on Biosafety Protocol held in Vienna from 15 to 19 September: Chairman's summary. UNEP/CBD/ExCOP/1/INF/3.

–. 2000a. Draft Cartagena Protocol on Biosafety. UNEP/ExCOP/1/L.5. 28 January.

–. 2000b. Biosafety Capacity-Building: Completed, Ongoing and Planned Projects/Programmes. UNEP/CBD/ICCP/1/INF/1. 23 October 23.

–. 2002. Report of the Meeting of the Technical Experts on the Requirements of Paragraph 2 (a) of Article 18 of the Cartagena Protocol on Biosafety. Intergovernmental Committee for the Cartagena Protocol on Biosafety. UNEP/CBD/ICCP/3/7/Add.1. 22 March 22.

–. 2003. The Cartagena Protocol on Biosafety: A record of the negotiations. Montreal: Secretariat of the CBD.

–. 2005. Report of the second Meeting of the Conference of the Parties to the Convention on Biological Diversity serving as the Meeting of the Parties to the Cartagena Protocol on Biosafety. UNEP/CBD/BS/COP-MOP/2/15. 6 June.

–. 2006a. Cartagena Protocol on Biosafety: Status of ratification and entry into force. http://www.biodiv.org/biosafety/signinglist.aspx?sts=rtf&ord=dt (accessed 25 August 2006).

–. 2006b. Report of the Third Meeting of the Conference of the Parties to the Convention on Biological Diversity Serving as the Meeting of the Parties to the Cartagena Protocol on Biosafety. UNEP/CBD/BS/COP-MOP/3/15. 8 May.

CBI (Council for Biotechnology Information). 2000. Biotechnology: Good ideas are growing. Ottawa: Council for Biotechnology Information. (on file with author)

CEN Biotechnology Caucus. 1997. Draft CEN Biotechnology Caucus discussion paper on the elements of the Biosafety Protocol. Ottawa: Canadian Environment Network. (on file with author)

–. 2002. Canada must ratify the Biosafety Protocol now. Press release, 10 September. (on file with author)

CFIA (Canadian Food Inspection Agency). 2000. Food safety and you. Brochure #A62-52/2000. Nepean, ON: Canadian Food Inspection Agency.

Charles, Daniel. 2001. *Lords of the Harvest*. Cambridge, MA: Perseus.

Charnovitz, Steve. 2000. The supervision of health and biosafety regulation by world trade rules. *Tulane Environmental Law Journal* 13(2): 271-302.

Chasek, Pamela, ed. 1996. Summary of the first meeting of the Open-Ended Ad Hoc Working Group on Biosafety: 22-26 July 1996. *Earth Negotiations Bulletin* 9(48).

–. 1998. Report of the fourth session of the Ad Hoc Working Group on Biosafety. *Earth Negotiations Bulletin* 9(85).

–. 1999. Report of the sixth session of the Open-Ended Ad Hoc Working Group on Biosafety and the First Extraordinary Session of the CBD Conference of the Parties: 14-23 February 1999. *Earth Negotiations Bulletin* 9(117).

–. 2000. Report of the Resumed Session of the Extraordinary Meeting of the Conference of the Parties for the Adoption of the Protocol on Biosafety to the Convention on Biological Diversity: 24-28 January 2000. *Earth Negotiations Bulletin* 9(137).

–. 2004. Summary of the First Meeting of the Conference of the Parties to the Convention on Biological Diversity Serving as the Meeting of the Parties to the Cartagena Protocol on Biosafety: 23-27 February 2004. *Earth Negotiations Bulletin* 9(289).

–. 2005. Summary of the First Meeting of the Ad Hoc Group on Liability and Redress and the Second Meeting of the Parties to the Cartagena Protocol on Biosafety: 25 May-3 June 2005. *Earth Negotiations Bulletin* 9(320).

–. 2006. Summary of the Third Meeting of the Parties to the Cartagena Protocol on Biosafety: 13-17 March 2006. *Earth Negotiations Bulletin* 9(351).

Chatterjee, Pratap, and Matthias Finger. 1994. *The Earth Brokers*. London: Routledge.

Clapp, Stephen. 2002. Canada urged to reject Cartagena Biosafety Protocol. *Food Chemical News* 44(31).

Clark, E. Anne. 2000. What is sound science? Faculty member website, University of Guelph. http://www.plant.uoguelph.ca/research/homepages/eclark/vermont.htm.

Clarke, Tom. 2003. Banana lab opens in Uganda. www.nature.com. 22 August.

Clement, Matthew. 2004. Rice imperialism: The agribusiness threat to Third World rice production. *Monthly Review* 55(9): 15-22.

Codex Alimentarius Commission. 2003a. Principles for the risk analysis of foods derived from modern biotechnology. Report CAC/GL 44-2003. ftp://ftp.fao.org/es/esn/food/princ_gmfoods_en.pdf.

–. 2003b. Guidelines for the conduct of food safety assessments of foods derived from recombinant DNA plants. Report CAC/GL 45-2003. ftp://ftp.fao.org/es/esn/food/guide_plants_en.pdf.

–. 2003c. Guidelines for the conduct of food safety assessments of foods produced using recombinant-DNA microorganisms. Report CAC/GL 46-2003. ftp://ftp.fao.org/es/esn/food/guide_mos_en.pdf.

–. 2004a. Working principles for risk analysis for application in the framework of the Codex Alimentarius Commission. In *Procedural Manual*, 14th ed., 101-8. Rome: WHO/FAO. ftp://ftp.fao.org/docrep/fao/007/y5817e/y5817e00.pdf.

–. 2004b. Appendix: General decisions of the commission. *Procedural Manual*, 14th ed., 188-191. Rome: WHO/FAO. ftp://ftp.fao.org/docrep/fao/007/y5817e/y5817e00.pdf.

Cohen, Maurie J. 2001. Ecological modernization and its discontents. Paper presented at the International Society for Industrial Ecology, Leeuwenhorst, the Netherlands.

Collins, C.A., and C. Gutherie. 1999. Allele-specific genetic interactions between Prp8 and RNA active site residues suggest a function for Prp8 at the catalytic core of the spliceosome. *Genes Dev* 13(15): 1970-82. In Commoner 2002, 6.

Commoner, Barry. 2002. Unraveling the DNA myth: The spurious foundation of genetic engineering. *Harper's Magazine*, February, 1-15.

Consumers International. 2002. Precaution and risk: A consumer response. Food Policy Briefing Paper No. 3, April. (on file with author)

Corell, Elisabeth, and Michelle M. Betsill. 2001. A comparative look at NGO influence in international environmental negotiations: Desertification and climate change. *Global Environmental Politics* 1(4): 86-107.

Cors, Thomas A. 2000. Biosafety and international trade: Conflict or convergence? *International Journal of Biotechnology*. 1/2/3: 27-43.

Cosbey, Aaron, and Stas Burgiel. 2000. *The Cartagena Protocol on Biosafety: An analysis of results*. Calgary: International Institute for Sustainable Development.

Cottier, Thomas. 2002. Implications for trade law and policy: Towards convergence and integration. In Bail, Falkner, and Marquard 2002, 467-81.

Council of Canadians. 2000. National Poll and Cross-Country Protest Demonstrate Consumers Won't Be Fooled by GE Foods (Press release). Ottawa: Council of Canadians. 31 March.

Cox, Robert. 1996. *Approaches to World Order*. Cambridge: Cambridge University Press.

Crick, F.H.C. 1958. *On Protein Synthesis. Symposium of the Society for Experimental Biology XII.* New York: Academic Press.

CropLife International. 2006. Key decision on documentation requirements for GMOs taken in Brazil. Press release, 20 March. (on file with author)

Crouch, Martha. 1991. The very structure of scientific research mitigates against developing products to help the environment, the poor, and the hungry. *Journal of Agricultural and Environmental Ethics* 4(2): 151-58.

CSE (Centre for Science and the Environment). 2001. Seeds of Discontent? *Down To Earth* 10(8): 6.

CWB (Canadian Wheat Board). 2001. CWB biotechnology position statement. Winnipeg: Canadian Wheat Board. http://www.cwb.ca/public/en/hot/biotechnology/#bio-position.

Daniel, Ann. 1997. Briefing note for Canadian delegation Biosafety Protocol negotiations October 13-17, 1997: Preambles, objectives, general obligations and principles (draft) (937-3556). Ottawa: Department of Justice. (on file with author)

Darier, Eric, ed. 1999. *Discourses of the Environment.* Oxford: Blackwell.

Dawkins, Kristen. 2000a. Biotech, from Seattle to Montreal and beyond: The battle royale of the 21st century. *Seedling* 7 (March): 2-8.

–. 2000b. Who should pay for the costs of the StarLink scandal? Minneapolis, MN: Institute for Agriculture and Trade Policy. (on file with author)

DBT (Department of Biotechnology), Ministry of Science and Technology, Government of India. 1998. Revised guidelines for research in transgenic plants and guidelines for toxicity and allergenicity evaluations of transgenic seeds, plants and plant parts. New Delhi: Ministry of Science and Technology.

Dexter, Lewis A. 1970. *Elite and Specialized Interviewing.* Evanston, IL: Northwestern University Press.

Diamond, Nancy K. 1992. Bye, bye biodiversity: A Rio Earth Summit diary. *Diversity* 8(2): 10.

Dreyfus, Herbert, and Paul Rabinow, eds. 1983. *Michel Foucault: Beyond Structuralism and Hermeneutics.* Chicago: University of Chicago Press.

Duvick, Donald. 1995. Biotechnology is compatible with sustainable agriculture. *Journal of Agriculture and Environmental Ethics* 8(8): 112-25.

Dyer, Gwynne. 1999. Frankenstein foods. *Globe and Mail,* 20 February D1-5.

Economist. 1999. Food for thought. *Economist,* 19-25 June. 19-22.

Edge, T., and S. Forsyth. 1997. Biosafety Protocol negotiations October 13-17, 1997: Unintentional transboundary movements (including accidental and emergency cases). Ottawa: Biodiversity Convention Office (Draft – 953-1666). (on file with author)

Edge, Tom. 1998. Fax to Biosafety Protocol Advisory Group, Biodiversity Convention Office, Environment Canada. 21 April. (on file with author)

EEC (Council of the European Economic Community). 1990. Council Directive 90/220/EEC on the deliberate release into the environment of genetically modified organisms. *Official Journal of the European Communities* 8.5.90: L 117/15.

Egziabher, Tewolde B.G. 1999. Of power affirmed to men and of safety denied to life. *Third World Resurgence,* June (106). http://www.twnside.org.sg/title/tewolde1-cn.htm.

–. 2002. Ethiopia. In Bail, Falkner, and Marquard 2002, 115-23.

Eichenwald, K., G. Kolata, and M. Peterson. 2001. Biotechnology food: From the lab to a debacle. *New York Times,* 25 January, A1.

Elliot, Lorraine. 1998. *The Global Politics of the Environment.* New York: New York University Press.

Enright, Cathleen A. 2002. United States. In Bail, Falkner, and Marquard 2002, 95-104.

Environment Canada. 2002. Living modified organism regulations. CEPA Environmental Registry. Ottawa: Environment Canada.

European Commission. 2000. Communication from the Commission on the Precautionary Principle (COM (2000[1]). Brussels: Commission of the European Communities.

–. 2003. European Commission regrets US decision to file WTO case on GMOs as misguided and unnecessary. Press release IP/03/681. 13 May.

–. 2004. EU Commission pushes for GMO "green light." Press release 10/04, Delegation of the European Commission to the USA, 28 January.

–. 2005. Invocation of Article 16 under Directive 90/220/EEC and Article 23 under Directive 2001/18/EC (safeguard clause). 15 March. http://europa.eu.int/comm/environment/biotechnology/safeguard_clauses.htm.

–. 2006. Commission proposes practical improvements to the way the European GMO legislative framework is implemented. Press release IP/06/498, 12 April.

European Communities. 2004. European Communities: Measures affecting the approval and marketing of biotech products (DS291, DS292, DS293). First written submission, Geneva, 17 May.

European Council and Commission. 1999. 2194th Council meeting – Environment – Luxembourg 24/25 June. C/99/203.

European Parliament. 1997a. Regulation (EC) No. 258/97 of the European Parliament and of the Council of 27 January 1997 concerning novel foods and novel food ingredients. *Official Journal of the European Communities* L (043): 0001-6.

–. 1997b. Regulation (EC) No. 1813/97 concerning the compulsory indication on the labelling of certain foodstuffs produced from genetically modified organisms of particulars other than those provided for in Directive 79/112/EEC. *Official Journal of the European Communities* L (257):0007-8.

–. 1998. Council Regulation (EC) No. 1139/98 of 26 May 1998 concerning the compulsory indication of the labelling of certain foodstuffs produced from genetically modified organisms of particulars other than those provided for in Directive 79/112/EEC. *Official Journal of the European Communities* L (159): 0004-7.

–. 2001. Directive 2001/18/EC on the deliberate release into the environment of genetically modified organisms and repealing Council Directive 90/220/EEC. *Official Journal of the European Communities* L (106): 0001-38.

European Parliament and the Council of the European Union. 2003a. Regulation EC No. 1829/2003 of 22 September 2003 on the regulation of genetically modified food and feed. http://europa.eu.int/eur-lex/pri/en/oj/dat/2003/l_268/l_26820031018en00010023.pdf.

–. 2003b. Regulation of EC No. 1830/2003 of 22 September 2003 concerning the traceability and labelling of genetically modified organisms and the traceability of food and feed products produced from genetically modified organisms and amending Directive 2001/18/EC. http://europa.eu.int/eur-lex/pri/en/oj/dat/2003/l_268/l_26820031018en00240028.pdf.

Ewan, S.W.B., and A. Pustzai. 1999. Effects of diets containing genetically modified potatoes expressing Galanthus nivalis lectin on rat small intestine. *Lancet* (354): 1353-55.

FAO (Food and Agricultural Organization of the United Nations). 2001. International convention on plant genetic resources for food and agriculture approved by FAO conference. Press release 01/81 C5. 5 November.

FAO/WHO (Food and Agriculture Organization of the United Nations and World Health Organization Joint Consultation). 1991. Strategies for assessing the safety of foods produced through biotechnology. Geneva: FAO/WHO.

–. 1996. Biotechnology and food safety. Rome: FAO/WHO.

Falck-Zepeda, J.B., B. Traxler, and R.G. Nelson. 1999. Rent creation and distribution from the first three years of planting Bt cotton. ISAAA Briefs No. 14. Ithaca, NY: International Service for the Acquisition of Agri-biotech Applications.

Falkner, Robert. 2000. Regulating biotech trade: The Cartagena Protocol on Biosafety. *International Affairs* 76(2): 299-313.

–. 2002. Negotiating the Biosafety Protocol: The international process. In Bail, Falkner, and Marquard 2002, 3-22.

Falkner, Robert, and Aarti Gupta. 2004. Implementing the Biosafety Protocol: Key challenges. Chatham House Sustainable Development Programme Briefing Paper (SDP BP 04/04), November.

FCPMC (Food and Consumer Products Manufacturers of Canada). 1999. Food and Consumer Products Manufacturers of Canada supports development of voluntary labelling. Press release, 17 September. (on file with author)

Financial Express. 2005. EPO revokes neem patent rights. 10 March. http://www.financialexpress.com/fe_full_story.php?content_id=84816.

Fisher, E. 2002. Jamaica. In Bail, Falkner, and Marquard 2002, 124-28.

FOEE (Friends of the Earth Europe). 2003. Genetically modified food and feed. FOEE Biotechnology Programme and European GMO Campaign. http://www.foeeurope.org./GMOs/
european_legislation/genetically_modified.htm.

FOEI (Friends of the Earth International). 2000. The precautionary principle: Accept only
the genuine article. Montreal: FOEI. (on file with author)

–. 2001. Trade case study: Beef-hormone dispute. London: FOEI Trade, Environment and
Sustainability Program. http:www.foei.org/trade/activistguide/hormone.htm.

–. 2005. Tackling GMO contamination. Briefing paper, June. http://www.foei.org/
publications/pdfs/contamination3.pdf.

–. 2006a. Looking behind the US spin: WTO ruling does not prevent countries from restricting or banning GMOs. Briefing paper, February. http://www.foeeurope.org/
publications/2006/WTO_briefing.pdf.

–. 2006b. Brazil 2006: Global Standard on Identification of GMOs to be decided by international treaty. Briefing paper, February. http://www.foei.org/gmo/Briefing_Curitiba.pdf.

Forgacs, David. 2000. *The Antonio Gramsci Reader: Selected Writings 1916-1936.* New York:
New York University Press.

Foucault, Michel. 1978. *The History of Sexuality: An Introduction.* Vol. 1. New York: Random
House.

–. 1983. The subject and power. In Dreyfus and Rabinow 1983, 208-28.

–. 1991. On governmentality. In Burchell, Gordon, and Miller 1991, 87-104.

Fransen, Lindsey, Antonio LaViña, Favian Dayrit, Loraine Gatlabayan, Dwi Andreas Santosa,
and Soeryo Adiwibowo. 2005. *Integrating Socio-Economic Considerations into Biosafety Decisions: The Role of Public Participation.* Washington, DC: World Resources Institute.

Friedmann, Harriet. 1994. Distance and durability: Shaky foundations of the world food
system. In *The Global Restructuring of Agro-Food Systems,* ed. Phillip McMichael, 258-76.
Ithaca, NY: Cornell University Press.

FSANZ (Food Standards Australia New Zealand). 2003. Report on the review of labelling of
genetically modified foods. December. http://www.foodstandards.gov.au/_srcfiles/
GM_label_REVIEW%20REPORT%20_Final%203_.pdf.

Gale, Louise. 2002. Greenpeace International. In Bail, Falkner, and Marquard 2002, 251-62.

Galvez, Amanda. 2002. Mexico. In Bail, Falkner, and Marquard 2002, 207-11.

Garton, A., R. Falkner, and R. Tarasofsky. 2006. Documentation requirements under the
Cartagena Protocol on Biosafety: The decision by the 3rd Meeting of Parties on Article
18.2(a). Background note for expert meeting: "Will the Decision on the Biosafety Protocol at the 3rd Meeting of Parties Work?" Chatham House, London, 4 July. Royal Institute
of International Affairs.

GATT (General Agreement on Tariffs and Trade). 1984. Panel on Japanese Measures on
Imports of Leather. Report of the panel adopted on 15/16 May 1984 (L/5623 – 31S/94).
http://www.worldtradelaw.net/reports/gattpanels/japanleatherIII.pdf. Cited in Mackenzie et al. 2003, 238.

–. 1986. Text of the General Agreement (from 1947, with amendments to 1986). http://www.
wto.org/english/docs_e/legal_e/gatt47_e.pdf.

Gaugitsch, Helmut. 2002. Scientific aspects of the biosafety debate. In Bail, Falkner, and
Marquard 2002, 83-94.

Gear, Brian. 1999. Biotechnology in the WTO and other international fora. Briefing note
BNM 020489 for Minister's Round Table, December 17, 1999. 9 December. (on file with
author)

Germain, Randall D., and Michael Kenny. 1998. Engaging Gramsci: International relations
theory and the new Gramscians. *Review of International Studies* 24(2): 3-21.

Ghijsen, Herb. 1998. Plant variety protection in a developing and demanding world. *Biotechnology and Development Monitor* 36: 2-5.

GIC (Global Industry Coalition). 1999. Biodiversity jeopardized in Cartagena biosafety
negotiations. Press release, 18 February. (on file with author)

Gill, Stephen. 1991. *American Hegemony and the Trilateral Commission.* Cambridge: Cambridge University Press.

–. 1998. New constitutionalism, democratisation, and global political economy. *Pacifica Review* 10(1): 23-40.

Gill, Stephen, and David Law. 1988. *The Global Political Economy: Perspectives, Problems and Policies*. Baltimore: Johns Hopkins Press.

–. 1989. Global hegemony and the structural power of capital. *International Studies Quarterly* 33(4): 475-99.

Glover, Domenic. 2003. "GMOs and the politics of international trade." Democratising biotechnology: Genetically Modified Crops in Developing Countries Briefing Series. Briefing 5. Brighton, UK: Institute of Development Studies.

Glowka, Lyle, Françoise Burhenne-Guilmin, and Hugh Synge, with Jeffrey A. McNeely and Lothar Gündling. 1994. *A Guide to the Convention on Biological Diversity*. IUCN Biodiversity Programme. Environmental Policy and Law Paper 30.

GMO ERA. 2005. International Project on GMO Environmental Risk Assessment Methodologies. Booklet distributed at COP-MOP-2. http://www.gmo-guidelines.info/public/publications/download/bookletnov05en.pdf.

Gordon, Colin. 1991. Governmental rationality: An introduction. In Burchell, Gordon, and Miller 1991, 1-51.

Gottweis, Herbert. 1998. *Governing Molecules: The Discursive Politics of Genetic Engineering in Europe and the United States*. Cambridge, MA: MIT Press.

Graff, Laurence. 2002. The precautionary principle. In Bail, Falkner, and Marquard 2002, 410-22.

Gramsci, Antonio. 1971. *Selections from the Prison Notebooks*. New York: International Publishers.

Gray, J.S. 1990. Statistics and the precautionary principle. *Marine Pollution Bulletin* 21(4): 174-76.

Green, Ian C. 2001. Letter to William Leiss, president, Royal Society of Canada. (retrieved under the Access to Information Act; on file with author)

Greenpeace. 2004. The European Union's new labeling rules for genetically engineered food and feed: Implications for the market of GMO and non-GMO products. Greenpeace European Unit, Brussels, Belgium. http://weblog.greenpeace.org/ge/archives/eufoodfeedtoupdate2-1.pdf.

–. 2005a. Coca Cola and 106 food brands commit to supply non-GE products in China. Press release, 13 November.

–. 2005b. June: Background briefing. MON863. Greenpeace European Unit, Brussels, Belgium. http://eu.greenpeace.org/downloads/gmo/Mon863June05.pdf.

–. 2005c. Debate at Environment Council, December 2: Next steps on GMOs. Press release, 30 November.

Grocery Manufacturers Association. 2006. WTO biotech decision reinforces need for science-based regulations globally. Press release, 7 February.

Gupta, Aarti. 1999. Framing "Biosafety" in an international context: The Biosafety Protocol negotiations. Belfer Center for Science and International Affairs (BCSIA) Discussion Paper E-99-10. Cambridge, MA: Kennedy School of Government, Harvard University.

–. 2000a. Creating a global biosafety regime. *International Journal of Biotechnology* 2: 205-30.

–. 2000b. Governing Biosafety in India: The relevance of the Cartagena Protocol. Belfer Center for Science and International Affairs (BCSIA) Discussion Paper, Environment and Natural Resources Program, Kennedy School of Government, Harvard University.

–. 2000c. Governing trade in genetically modified organisms: The Cartagena Protocol on Biosafety. *Environment* 42(4): 23-33.

Haas, Peter M. 1989. Do regimes matter? Epistemic communities and Mediterranean pollution control. *International Organization* 43(3):377-403.

–. 1992. Obtaining international environmental protection through epistemic consensus. In *Global Environmental Change and International Relations*, ed. I. Rowlands and M. Greene, 38-59. London: Macmillan.

Haas, P.M., R.O. Keohane, and M.A. Levy, ed. 1993. *Institutions for the Earth: Sources of effective international environmental protection*. Cambridge MA: MIT Press.

Hajer, M. 1995. *The Politics of Environmental Discourse: Ecological Modernisation and the Policy Process*. New York: Oxford University Press.

Hall, Stuart. 1988. The toad in the garden: Thatcherism among the theorists. In *Marxism and the Interpretation of Culture*, ed. C. Nelson and L. Grossberg, 35-73. Chicago: University of Illinois Press.

Haraway, Donna. 2000. *How Like a Leaf.* London: Routledge.

Harstock, Nancy. 1990. Foucault on power. In *Feminism/Postmodernism*, ed. Linda J. Nicholson, 157-75. New York: Routledge, Chapman and Hall.

Health Canada. 1999. Health Canada rejects bovine growth hormone in Canada. Press release 1999-03, 14 January.

–. 2001. Action Plan of the Government of Canada in response to the Royal Society of Canada Expert Panel Report. Ottawa: Health Canada.

–. 2002. Frequently asked questions: Biotechnology and genetically modified foods. Ottawa: Health Canada. http:www.hc-sc.gc.ca/food-aliment/mh-dm/ofb-bba/nfi-ani/e_faq.html (accessed 2 June 2002).

–. 2004. Approved Products: Novel food decisions. http://www.hc-sc.gc.ca/fn-an/gmf-agm/appro/index_e.html.

Heffernan, William. 1999. Biotechnology and mature capitalism. In *World Food Security and Sustainability: The Impacts of Biotechnology and Industrial Consolidation*, ed. Donald P. Weeks, Jane B. Seglken, and Ralph W.F. Hardy, 121-36. Ithaca, NY: National Agricultural Biotechnology Council, Report 11.

Held, Tamilla, ed. 2002. Summary of the ninth session of the intergovernmental legally binding instrument for the application of the prior informed consent procedure for certain hazardous chemicals and pesticides in international trade: 30 September–4 October, 2002. *Earth Negotiations Bulletin* 15(75).

Herity, John. 1996. CANDEL summary report. Ottawa: Environment Canada. (on file with author)

–. 1997. The Biosafety Protocol: Status and outlook. Paper presented at "Biotechnology in Agriculture and Food: Living with the New Realities," Canada Grains Council conference, Winnipeg, Manitoba. http://www.canadgrainscouncil.ca/herity.htm (accessed 29 March 2002; page now discontinued).

–. 2002. Capacity-building and the Biosafety Clearing-House. In Bail, Falkner, and Marquard 2002, 344-50.

Hilbeck, A., and D.A. Andow, eds. 2004. *A Case Study of Bt Maize in Kenya.* Vol. 1 of *Environmental Risk Assessment of Genetically Modified Organisms.* Wallingford, UK: CABI Publishing.

Hilbeck, A., D.A. Andow, and E.M.G. Fontes, eds. 2006. *Methodologies for Assessing Bt Cotton in Brazil.* Vol. 2 of *Environmental Risk Assessment of Genetically Modified Organisms.* Wallingford, UK: CABI Publishing.

Hisano, Shuji. 2005. A critical observation on the mainstream discourse of biotechnology for the poor. *Tailoring Biotechnologies* 1(2):81-106.

Holdrege, Craig. 1996. *A Question of Genes: Understanding Life in Context.* Hudson, NY/Edinburgh, Scotland: Lindisfarne Press/Floris Books.

Hovi, Jon, Detlef F. Sprinz, and Arild Underdal. 2003. The Oslo-Potsdam solution to measuring regime effectiveness: Critique, response, and the road ahead. *Global Environmental Politics* 3(3): 74-96.

Humphreys, David. 1996. Hegemonic ideology and the international Tropical Timber Organisation. In *The Environment and International Relations*, ed. J. Vogler and M.F. Imber, 215-33. London: Routledge.

Humphreys, M., and N. Nishikawa. 1999. Update on voluntary labelling of foods obtained through biotechnology. Briefing note BNM 020489 for Minister's Roundtable, December 17, 1999. Office of Biotechnology, Canadian Food Inspection Agency, 6 December. (on file with author)

Hutchinson, Cameron. 2001. International environmental law attempts to be "mutually supportive" with international trade law: A compatibility analysis of the Cartagena Protocol to the Convention on Biological Diversity with the World Trade Organisation Agreement on the Application of Sanitary and Phytosanitary Measures. *Journal of International Wildlife Law and Policy* 4: 1-34.

ICCBD (Intergovernmental Committee on the Convention on Biological Diversity). 1994. Consideration of the need for, and modalities of, a protocol on biosafety: Note by the interim secretariat. Nairobi: United Nations Environment Programme, Convention on Biological Diversity. UNEP/CBD/IC/2/12.

ICTSD (International Centre for Trade and Sustainable Development). 2006. Constructive ambiguity saves LMO labelling discussions at Cartagena Protocol summit. *BRIDGES Weekly Trade News Digest* 10(10): 7.

IFIC (International Food Information Council). 2000. US consumer attitudes towards food biotechnology. http://www.ific.org/publications/.

IGTC (International Grain Trade Coalition). 2002. Presentation at the ICCP-3 meeting in The Hague, the Netherlands. (on file with author)

Irwin, Alan. 1995. *Citizen Science.* London: Routledge.

Isaac, G.E., M. Phillipson, and W.A. Kerr. 2001. *International Regulation of Trade in the Products of Biotechnology.* Saskatoon: Estey Centre for Law and Economics in International Trade.

Ivars, Birthe. 2002. Norway. In Bail, Falkner, and Marquard 2002, 193-99.

Jaffe, Gregory. 2004. Regulating transgenic crops: A comparative analysis of different regulatory processes. *Transgenic Research* 13: 5-19.

James, Clive. 2001. Global review of commercialized transgenic crops: 2001 feature: Bt cotton, International Service for Agricultural Acquisitions of Agri-Biotech Applications. http://www.isaaa.org.

–. 2002. Global status of commercialized transgenic crops: 2002. International Service for Agricultural Acquisitions of Agri-Biotech Applications. http://www.isaaa.org.

–. 2004. Preview: Global status of commercialized biotech/GM crops: 2004. ISAAA Briefs No. 32. Ithaca, NY: International Service for the Acquisition of Agri-biotech Applications. http://www.isaaa.org.

–. 2005. Global Status of Commercialized Biotech/GM Crops: 2005. ISAAA Briefs No. 34. Ithaca, NY: International Service for the Acquisition of Agri-biotech Applications. http://www.isaaa.org.

Janicke, M. 1985. *Preventative Environmental Policy as Ecological Modernization and Structural Policy.* Berlin: Berlin Science Center.

Jasanoff, Sheila. 1997. NGOs and the environment: From knowledge to action. *Third World Quarterly* 18(3): 579-94.

Jeffs, A. 2003. Wheat sales won't be risked: Impact on exports must be studied – agriculture minister. *Edmonton Journal,* 13 March, H1.

Jenkins, Peter I. 2002. Monsanto drops pursuit of federal approval for genetically engineered lawn grass. Washington, DC: International Center for Technology Assessment. http://www.icta.org/releases/prgrass31002.htm.

Jordan, Andrew, and Timothy O'Riordan. 1999. The precautionary principle in contemporary environmental policy and politics. In *Protecting Public Health and the Environment,* ed. Carolyn Raffensperger and Joel Tickner, 15-35. Washington, DC: Island Press.

Kay, Lily E. 2000. *Who Wrote the Book of Life? A History of the Genetic Code.* Stanford, CA: Stanford University Press.

Keller, Evelyn Fox. 2000. *The Century of the Gene.* Cambridge, MA: Harvard University Press.

Kerr, William A. 2002. Who should make the rules of trade? The complex issue of multilateral environmental agreements. *Estey Centre Journal of International Law and Trade Policy* 3(2): 163-75.

Khor, Martin. 1993. UNEP Expert Panel calls for biosafety protocol under Biodiversity Convention. Biodiversity Convention Briefings No. 3. Penang, Malaysia: Third World Network. http://www.panasia.org.sg/souths/twn/title/bio3-cn.htm.

King, Jonathan, and Doreen Stabinsky. 1998/99. Biotechnology under globalisation: The corporate expropriation of plant, animal and microbial species. *Race and Class* 40(2/3): 73-89.

Kleiss, Melanie E. 2003. NEPA and scientific uncertainty: Using the precautionary principle to bridge the gap. In *Jurisdynamics of Environmental Protection,* ed. Jim Chen, 117-42. Washington, DC: Environmental Law Institute.

Kloppenburg, Jack. 1988. *First the Seed: The Political Economy of Plant Biotechnology, 1492-2000.* Cambridge: Cambridge University Press.

Kneen, Brewster. 1999. *Farmageddon: Food and the Culture of Biotechnology.* Gabriola Island, BC: New Society.

Koester, Viet. 2002. The Biosafety Working Group (BSWG) process: A personal account from the chair. In Bail, Falkner, and Marquard 2002, 44-61.

Krimsky, Sheldon. 1982. *Genetic Alchemy: The Social History of the Recombinant DNA Controversy.* Cambridge, MA: MIT Press.

–. 1991. *Biotechnics and Society: The Rise of Industrial Genetics.* Westport, CT: Praeger.

Krimsky, Sheldon, and Roger Wrubel. 1996. *Agricultural Biotechnology and the Environment: Science, Policy and Social Issues.* Chicago: University of Illinois Press.

Kuiper, Harry. 2000. Profiling techniques to identify differences between foods derived from biotechnology and their counterparts. Geneva: Joint FAO/WHO Expert Consultation on Foods Derived from Biotechnology, 29 May.

Kuyek, Devlin. 2002. *The Real Board of Directors: The Construction of Biotechnology Policy in Canada, 1980-2002.* Sorrento, BC: The Ram's Horn.

Kwaja, Rajen Habib. 2002. Socio-economic considerations. In Bail, Falkner, and Marquard 2002, 361-65.

Landsberg, Michelle. 2000. Canadians the winners in "Seattle, the sequel." *Toronto Star,* 6 February, A2.

Lane, Michael, and Richard Schweiger. 1995. The report of the UNEP Expert Panel IV, two years later: Analysis and critique. Paper presented at the first meeting of the Expert Panel on Bio-Safety, established by the Conference of the Parties of the Convention on Biological Diversity, Madrid, Spain. (on file with author)

LaViña, Antonio G.M. 2002. A mandate for a biosafety protocol: The Jakarta negotiations. In Bail, Falkner, and Marquard 2002, 34-43.

Lehmann, V. 1998. Patent on seed sterility threatens seed saving. *Biotechnology and Development Monitor* 35: 6-8.

Lehmann, V., and W.A. Pengue. 2000. Herbicide tolerant soybean: Just another step in a technology treadmill? *Biotechnology and Development Monitor* 43: 11-14.

Levidow, Les. 1999. Blocking biotechnology as pollution: Political cultures in the UK risk controversy. Paper presented at the Alternate Futures and Popular Protest conference, Manchester Metropolitan University, Milton Keynes, UK.

Levidow, Les, Susan Carr, Rene von Shomber, and David Wield. 1996. Regulating agricultural biotechnology in Europe: Harmonisation difficulties, opportunities, dilemmas. *Science and Public Policy* 23(3): 135-57.

Levy, David L., and David Egan. 2003. A neo-Gramscian approach to corporate political strategy: Conflict and accommodation in the climate change negotiations. *Journal of Management Studies* 40(4): 803-30.

Levy, David L., and Peter J. Newell. 2000. Oceans apart? Business responses to global environmental issues in Europe and North America. *Environment* 42(9): 9-20.

–. 2002. Business strategy and international environmental governance: Toward a neo-Gramscian synthesis. *Global Environmental Politics* 2(4): 84-101.

Levy, Marc A., Oran Young, and Michael Zürn. 1995. The study of international regimes. *European Journal of International Relations* 1(3): 267-330.

Lewis, Glennis. 1997a. *An Analysis of Legal Text Submitted for the Biosafety Protocol.* Calgary: Lewis Consulting. (on file with author)

–. 1997b. Biosafety Protocol worksheet – draft: Canadian policy considerations and preferred legal text for proposed articles. Ottawa: Lewis Consulting. (on file with author)

–. 1998. Discussion document for the Biosafety Advisory Group meeting January 15 and 16, 1998. Calgary: Lewis Consulting. (on file with author)

Lijie, Cai. 2002. China. In Bail, Falkner, and Marquard, 2002, 160-65.

Lin, Lim Li, and Lim Li Ching. 2005. Analysis of key decision at Biosafety Protocol meeting (MOP3). *South-North Development Monitor,* 21 March.

–. 2006. No Agreement on Article 18.2(a) Experts' Meeting. Briefings for the Biosafety Protocol Liability and Redress WG (1) and MOP2. No. 4. Penang, Malaysia: Third World Network. Ref Doc TWN/Biosafety/2005/E.

Ling, Chee Yoke. 1996. Concerted moves to undermine a strong biosafety agreement. Penang, Malaysia: Third World Network. http://www.panasia.org.sg/souths/twn/title/chee-cn.htm.

Lipshutz, Ronnie D. 1996. *Global Civil Society and Global Environmental Governance.* Albany, NY: State University of New York Press.

Litfin, Karen. 1994. *Ozone Discourses: Science and Politics in Global Environmental Cooperation.* New York: Columbia University Press.

–. 1995. Framing science: Precautionary discourse and the ozone treaty. *Millennium* 24(2): 251-77.

Losey, J.E., L.S. Rayno, and M.E. Carter. 1999. Transgenic pollen harms monarch larvae. *Nature* 399:6733.

Louwaars, N.P. (1998). *Sui generis* rights: From opposing to complementary approaches. *Biotechnology and Development Monitor* 36: 13-16.

Luke, Timothy W. 1995. On environmentality: Geo-power and eco-knowledge in the discourses of contemporary environmentalism. *Cultural Critique* 31(Fall): 57-81.

–. 1999. Environmentality as green governmentality. In Darier 1999, 121-51.

Mackenzie, Ruth. 2004. The Cartagena Protocol after the First Meeting of the Parties. *RECIEL* 13(3): 270-78.

Mackenzie, Ruth, Françoise Burhenne-Guilmin, Antonio G.M. LaViña, and Jacob D. Werksman, with Alfonso Ascencio, Julian Kinderlerer, Katharina Kummer, and Richard Tapper. 2003. An explanatory guide to the Cartagena Protocol on Biosafety. IUCN Environmental Policy and Law Paper No. 46. Gland, Switzerland, and Cambridge, UK: IUCN Environmental Law Centre.

Mackenzie, R., and S. Francescon. 2000. The regulation of genetically modified foods in the European Union: An overview. *NYU Environmental Law Journal* 8: 530-55.

MacRae, Rod. 2001. Mixed messages: Canada's domestic regulatory system for GEOs contradicts basic principles underlying the Cartagena Protocol on Biosafety. Toronto: Canadian Institute for Environmental Law and Policy.

Marquard, Helen. 2002. Scope. In Bail, Falkner, and Marquard 2002, 289-99.

Mausberg, Burkhard, and Maureen Press-Merkur. 1995. *The Citizen's Guide to Biotechnology.* Toronto: Canadian Institute for Environmental Law and Policy.

Mayr, Juan. 2002. Environment ministers: Columbia. In Bail, Falkner, and Marquard 2002, 218-29.

Meacher, Michael. 2002. Environment ministers: United Kingdom. In Bail, Falkner and Marquard 2002, 230-36.

Meadows, D.H., D.L. Meadows, J. Randers, and W.W. III Behrens. 1972. *The Limits to Growth.* New York: Universe Books.

Miami Group. 2000. Miami Group Proposal for Moving Forward on Cartagena, rev. 2 text. 21 January. (on file with author)

Mikkelsen, T.R., B. Anderson, and R.B. Jorgensen. 1996. The risk of crop trans-gene spread. *Nature* 380: 31.

Miller, Henry. 1997. *Policy Controversy in Biotechnology: An Insider's View.* Austin: Academic Press and R.G. Landes.

Mitchell, Ronald B. 2002. A quantitative approach to evaluating international environmental regimes. *Global Environmental Politics* 2(4): 58-83.

Mitsch, F.J., and J.S. Mitchell. 1999. Ag Biotech: Thanks, but no thanks? Deutsche Bank, 12 July.

Monbiot, George. 2003. Let's do a Monsanto. *Guardian Unlimited,* 10 June.

Monsanto. 2000. Technology use agreement terms and conditions: Roundup Ready canola. Monsanto Canada. (on file with author)

Mooney, Pat R. 1996. Civil and uncivil societies. In *The Commodification of Life,* ed. Anthony K. Welch, 1-9. Vancouver: The Pomelo Project, Simon Fraser University.

Moore, Elizabeth. 2000. Food safety, labelling, and the role of science: Regulating genetically-engineered food crops in Canada and the United States. Paper presented at ECPR Joint Sessions Workshop on the Politics of Food, Copenhagen. (on file with author)

Muller, Bernarditas C. 2002. Philippines. In Bail, Falkner, and Marquard 2002, 138-45.

Munson, Abby. 1993. Genetically manipulated organisms: International policy-making and implications. *International Affairs* 69(3): 497-517.

Murphy, Craig. 1994. *International Organization and Industrial Change.* Cambridge: Cambridge University Press.

Myhr, A.I., and T. Traavik. 1999. The precautionary principle applied to deliberate release of genetically modified organisms (GMOs). *Microbial Ecology in Health and Disease* 1999 (11): 65-74.

NABC (National Agricultural Biotechnology Council). 2000. The biobased economy of the twenty-first century: Agriculture expanding into health, energy, chemicals and materials. Ithaca, NY: National Agricultural Biotechnology Council.

Nechay, Gabor. 2002. Central and Eastern Europe. In Bail, Falkner, and Marquard 2002, 212-17.

Nevill, John. 2002. Seychelles. In Bail, Falkner, and Marquard 2002, 146-54.

Newell, Peter J. 2005. Business and international environmental governance: The state of the art. In *The Business of Global Environmental Governance,* ed. David L. Levy and Peter J. Newell, 21-45. Cambridge, MA: MIT Press.

Nijar, Gurdial Singh. 2002. Third World Network. In Bail, Falkner, and Marquard 2002, 263-72.

NIN (National Institute of Nutrition). 1998. Board of trustees. www.nin.ca/board.htm (accessed 28 March 1998; page now discontinued).

Nissen, J.L. 1997. Achieving a balance between trade and the environment: The need to amend the WTO/GATT to include multilateral environmental agreements. *Law and Policy in International Business* 28(3): 901-28. In Stabinsky 2000, 280.

Nobs, Beat. 2002. Switzerland. In Bail, Falkner, and Marquard 2002, 186-92.

Nogeira, Arthur H.V. 2002. Brazil. In Bail, Falkner, and Marquard 2002, 129-37.

Norway (Kingdom of). 1993. *Gene Technology Act.* Act No. 38 of 2 April 1993.

NRC (National Research Council). 2000. *Genetically-Modified Pest-Protected Plants: Science and Regulation.* Washington, DC: National Academy Press.

Oberthür, Sebastian. 2001. Linkages between the Montreal and the Kyoto protocols: Enhancing synergies between protecting the ozone layer and the global climate. *International Environmental Agreements: Politics, Law and Economics* 1(3): 357-77.

O'Brien, Mary. 2000. *Making Better Environmental Decisions.* Boston: MIT Press.

OECD (Organisation for Economic Co-operation and Development). 1986. Recombinant DNA Safety Considerations. Paris: OECD Secretariat.

–. 1993. Safety considerations for biotechnology scale-up of crop plants. Paris: OECD Secretariat.

OFB (Office of Food Biotechnology), Canadian Food Inspection Agency. 2000. Novel foods information – food biotechnology: Glyphosate tolerant soybean 40-3-2. http://www.hc-sc.gc/food-aliment/mh-dm/ofb-bba/nfi-ani/.pdf/e_ofb-096-100-d-rev.pdf.

Osava, Mario. 2006. Biosafety protocol alive, but restricted. Inter Press Service News Agency, 19 March.

OSTP (Office of Science and Technology Policy), United States Executive Office of the President. 1986. Coordinated framework for regulation of biotechnology. *Federal Register* 51: 23302-50.

–. 1992. "Exercise of federal oversight within the scope of statutory authority: Planned introductions of biotechnology products into the environment." *Federal Register* 57: 6753-62.

Padgett, T., and A. Downie. 2003. Lula's next big fight. *Time* (Europe edition), 24 November. http://www.time.com/time/europe/magazine/article/0,13005,901031124543734, 00.html.

Palardy, Nancy. 1997. Biotechnology: Influence within the United Nations Biosafety Protocol. Master's thesis, York University, North York, Ontario.

PANNA (Pesticide Action Network of North America). 1998. NGOs call for rapid implementation of prior informed consent. Press release, 10 September.

Pardo-Quintillan, S. 1999. Free trade, public health protection and consumer information in the European and WTO context. *Journal of World Trade* 33(6): 147-97.

PBO (Plant Biosafety Office), Canadian Food Inspection Agency. 1995. Decision document DD95-01: Determination of environmental safety of Agrevo Canada Inc.'s glufosinate ammonium-tolerant canola. Ottawa: Canadian Food Inspection Agency, Plant Health and Production Division.

–. 2000. Regulatory Directive 2000-07: Guidelines for the environmental release of plants with novel traits within confined field trials in Canada. Ottawa: Canadian Food Inspection Agency, Plant Health and Protection Division.

–. 2003. Status of regulated plants with novel traits in Canada, October 30. Canadian Food Inspection Agency: Plant Products Division. http://www.inspection.gc.ca/english/plaveg/bio/pntvcne.shtml (accessed 20 July 2004).

Pew Initiative on Food and Biotechnology. 2005. U.S. vs. EU: An examination of the trade issues surrounding genetically modified food. http://pewagbiotech.org/resources/issuebriefs/useu.pdf.

PHAC (Public Health Agency of Canada). 1996. Laboratory Biosafety Guidelines, 2nd Edition. Office of Laboratory Security. http://www.phac-aspc.gc.ca/publicat/lbg-ldmbl-96/lbg1_e.html.

Porter, Garreth, and Jane Welsh Brown. 1996. *Global Environmental Politics*. Boulder, CO: Westview.

Purdue, Derrick. 1995. Hegemonic trips: World trade, intellectual property and biodiversity. *Environmental Politics* 4(1): 88-107.

–. 2000. *Anti-GenetiX: The Emergence of the Anti-GM Movement.* Aldershot, UK: Ashgate.

Quist, D., and I. Chapela. 2001. Transgenic DNA introgressed into traditional maize landraces in Oaxaca, Mexico. *Nature* 414: 541-43.

Radin, John. 1999. The technology protection system: Revolutionary or evolutionary? *Biotechnology and Development Monitor* 37: 24.

Raffensperger, Carolyn, and Joel Tickner, eds. 1999. *Protecting the Public Health and the Environment: Implementing the Precautionary Principle.* Washington, DC: Island Press.

RAFI (Rural Advancement Foundation International). 1998. US patent on new genetic technology will prevent farmers from saving seed. *RAFI GenoTypes,* 11 March. (on file with author)

–. 1999. Traitor tech: The terminator's wider implications. *RAFI Occasional Paper Series* 6(1). January.

Rajan, M.G. 1997. *Global Environmental Politics: India and the North-South Politics of Global Environmental Issues.* New Delhi: Oxford University Press.

Ramey, T.S., M.J. Wimmer, and R.M. Rocker. 1999. GMOs are dead. Deutsche Bank, 21 May.

Rajnova, K. 2003. Leader of the pack. worldpress.org, 25 September.

Reifshneider, Laura M. 2002. Industry: Global Industry Coalition. In Bail, Falkner, and Marquard, 2002, 273-77.

Reuters. 2006. WTO confirms ruling against EU GMO moratorium. Reuters, 11 May.

RFSTC (Research Foundation for Science, Technology and Ecology). 2005. Landmark victory in world's first case against biopiracy! European Patent Office upholds decision to revoke neem patent. Press release, 8 March.

Richardson, Mary, Joan Sherman, and Michael Gismondi. 1993. *Winning Back the Words.* Toronto: Garamond Press.

Rifkin, Jeremy. 1983. *Algeny.* New York: Viking.

Rissler, J., and M. Mellon. 1996. *The Ecological Risks of Engineered Crops.* Cambridge, MA: MIT Press.

Rose, Nicholas, and Peter Miller. 1992. Political power beyond the state: Problematics of government. *British Journal of Sociology* 43(2): 173-205.

Rupert, Mark. 1995. *Producing Hegemony: The Politics of Mass Production and American Global Power.* Cambridge: Cambridge University Press.

Rutherford, Paul. 1999. The entry of life into history. In Darier 1999, 37-62.

Sachs, Wolfgang. 1999. *Planet Dialectics.* Halifax: Fernwood.

Safrin, Sabrina. 2002. The relationship with other agreements: Much ado about a savings clause. In Bail, Falkner, and Marquard 2002, 438-54.

Samper, Christián. 2002. The Extraordinary Meeting of the Conference of the Parties (ExCOP). In Bail, Falkner, and Marquard 2002, 44-61.

Schmidheiny, S. 1992. *Changing Course.* Cambridge, MA: MIT Press.

Schoonejans, Eric. 2002. Advance informed agreement procedures. In Bail, Falkner, and Marquard 2002, 299-320.

Schubbert, R., D. Renz, B. Schmitz, and W. Doerfler. 1997. Foreign (M13) DNA ingested by mice reaches peripheral leukocytes, spleen and liver via intestinal wall mucosa and can be covalently linked to mouse DNA. *Proc Natl Acad Sci USA* 94: 961-66.

Schumacher, E.F. 1973. *Small Is Beautiful*. London: Blond and Briggs.

Schurman, Rachel. 2004. Fighting "Frankenfoods": Industry opportunity structures and the efficacy of the anti-biotech movement in Western Europe. *Social Problems* 51(2): 243-68.

Schurman, Rachel, and William Munro. 2003. Sustaining rage: Cultural capital, strategic location, and motivating sensibilities in the US anti-genetic engineering movement. Submitted to *Rural Sociology*.

Schweiger, Thomas G. 1999. How the GE Moratorium was won. Personal communication by email, 3 July. (on file with author)

–. 2001. Europe: Hostile lands for GMOs. In Tokar 2001, 351-61.

SCP (Strategic Communications and Planning). 2000. *Biosafety: Public Environment Analysis*. Montreal: Strategic Communication and Planning. (on file with author)

Shakley, Simon, and Brian Wynne. 1996. Representing uncertainty in global climate change science and policy: Boundary-ordering devices and authority. *Science, Technology and Human Values* 21(3): 275-302.

Shand, Hope. 2001. Gene giants: Understanding the "life industry." In Tokar 2001, 222-37.

Shapiro, Robert. 1999. How genetic engineering will save our planet. *The Futurist*, 28-29 April.

Sharratt, Lucy. 2001. No to bovine growth hormone: Ten years of resistance in Canada. In Tokar 2001, 385-96.

Shiva, Vandana. 1993. *Monocultures of the Mind*. London: Zed Books.

–. 1998. Protecting our biological and intellectual heritage in the age of biopiracy. New Delhi: Foundation for Science, Technology and Ecology. http:www.indiaserver.cm/betas/vshiva/piracy.htm (accessed 14 June 1999; page now discontinued).

Shrader-Frechette, K.S. 1991. *Risk and Rationality: Philosophical Foundations of Populist Reforms*. Berkeley: University of California Press.

Sinsheimer, Robert L. 1970. Genetic engineering: The modification of man. *Impacts of Science on Society* 20: 279-90.

Smith, Richard. 1998. Report of the Commissioner of the Environment and Sustainable Development. Ottawa: Office of the Auditor General. http://www.oag-bvg.gc.ca/domino/reports.nsf/html/c802ce.html.

–. 2000. Report of the Commissioner of the Environment and Sustainable Development. Ottawa: Office of the Auditor General. http://www.oag-bvg.gc.ca/domino/reports.nsf/html/comenu_e.html.

Stabinsky, Doreen. 2000. Bringing social analysis into a multilateral environmental agreement: Social impact assessment and the Biosafety Protocol. *Journal of Environment and Development* 9(3): 260-83.

Stanbury, W.T. 1992. Reforming the federal regulatory process in Canada, 1971-1992. Annex to the Report of the Sub-Committee on Regulations and Competitiveness of the Standing Committee on Finance. Ottawa. (on file with author)

Standing Committee (on Environment and Sustainable Development). 1996. Biotechnology regulatory policy: A matter of public confidence. Ottawa, November. (on file with author)

Stauber, J.C., and S. Rampton. 1995. *Toxic Sludge Is Good for You*. Monroe, ME: Common Courage Press.

Steffenhagen, Bretta. 2001. The influence of biotech industry on German and European negotiation positions regarding the 2000 "Cartagena Protocol on Biosafety." Master's thesis, Otto-Suhr-Institut fur Politikwissenschaft, Freie Universität Berlin. (on file with author)

Stewart, C. Neal. 2001. GM crop data: Agronomy and ecology in tandem. *Nature Biotechnology* 19: 3.

Stewart, Lyle. 2000. The bio-battle of words. *Montreal Gazette*, 29 January, C4.

–. 2002a. GM food meeting left bad taste. *Montreal Gazette*, 1 March, B1.

–. 2002b. Good PR is growing. *This Magazine*, May/June, 5-10.

Stockholm Convention. 2001. Stockholm Convention on Persistent Organic Pollutants. http://www.pops.int/documents/convtext/convtext_en.pdf.

–. 2006. Stockholm Convention on Persistent Organic Pollutants – Participants. http://www.pops.int/documents/signature/signstatus.htm.

Subraminian, A. 1990. TRIPS and the paradigm of the GATT: A tropical, temperate view. *World Economy* 13(4): 509-21.

Suppan, Steve. 2005. US vs. EC Biotech Products case: WTO dispute backgrounder. Minneapolis, MN: Institute for Agriculture and Trade Policy.

–. 2006. The WTO's EC-biotech products ruling and the Cartagena Protocol. Minneapolis, MN: Institute for Agriculture and Trade Policy.

Supreme Court of Canada. 2001. *114957 Canada Ltée (Spraytech, Société d'arrosage) v. Hudson (Town)*, [2001] 2 S.C.R. 241, 2001 SCC 40.

–. 2002. *Harvard College v. Canada (Commissioner of Patents)*, [2002] 4 S.C.R. 45, 2002 SCC 76.

–. 2004. *Monsanto Canada Inc. v. Schmeiser*, [2004] 1 S.C.R. 902, 2004 SCC 34.

Swaminathan, M.S. 1998. Farmers' rights and plant genetic resources. *Biotechnology and Development Monitor* 36:6-9.

Swenarchuk, Michelle. 2003. *The Harvard Mouse and All That: Life Patents in Canada*. Toronto: Canadian Environmental Law Association. Publication #454.

TABD (TransAtlantic Business Dialogue). 2002. TransAtlantic Business Dialogue. http://www.tabd.com (accessed 10 October 2002).

Tapper, Richard. 2002. Environment, Business and Development Group. In Bail, Falkner, and Marquard 2002, 268-77.

Thomas, Jim. 2001. Princes, aliens, superheroes and snowballs: The playful world of the UK genetic resistance. In Tokar 2001, 337-50.

Tiedje, James, Robert K. Colwell, Yaffa L. Grossman, Robert E. Hodson, Richard E. Lenski, Richard N. Mack, Philip J. Regal. 1989. The planned introduction of genetically engineered organisms: Ecological considerations and recommendations. *Ecology* 70 (April): 297-315.

Tokar, Brian, ed. 2001. *Redesigning Life?* Zed Books: London.

Torgerson, Helge. 1996. Ecological impacts of traditional crop plants: A basis for the assessment of transgenic plants? Wein, Austria: Umweltbundesamt Monographien Band 75.

Traxler, A., A. Helssenberger, G. Frank, C. Lethmayer, and H. Gauglitsch. 2001. Ecological monitoring of genetically modified organisms. Wien, Austria: Umweltbundesamt Monographien Band 147.

UN (United Nations). 1969. Vienna Convention on the Law of Treaties. *Treaty Series* 1155:331.

UNCED (United Nations Conference on Environment and Development). 1992a. Sustainable Development Agenda 21, Chapter 16. New York: United Nations. http://www.un.org/esa/susdev/agenda21.chapter16.htm.

–. 1992b. Rio Declaration. Rio de Janeiro: United Nations. http://www.un.org/documents/ga/conf151/aconf15126-1annex1.htm.

UNEP (United Nations Environment Programme). 1992. Convention on Biological Diversity. http://www.biodiv.org/convention/articles.asp.

–. 1993. Report of Panel IV: Consideration of the need for and modalities of a protocol setting out appropriate procedures including, in particular, advance informed agreement in the field of the safe transfer, handling and use of any living modified organism resulting from biotechnology that may have adverse effect on the conservation and sustainable use of biological diversity. UNEP/Bio. Div./Panel/Inf.4. Nairobi, 28 April.

–. 1995. International technical guidelines for safety in biotechnology. Nairobi, Kenya: UNEP.

–. 1998. Rotterdam Convention for the PIC Procedure for Certain Hazardous Chemicals and Pesticides in International Trade. Rotterdam: UNEP. http://www.pic.int.

–. 2002. UNEP/GEF building capacity for the implementation of the Cartagena Protocol on Biosafety. Nairobi, Kenya: UNEP.

–. 2003. Synthesis report of the subregional workshops on: Risk assessment and management systems and public awareness, education and participation (November 2002 to May 2003). UNEP/GEF Project on Development of National Biosafety Frameworks (CD-Rom distributed at COP-MOP-2).

–. 2005. Information note on UNEP-GEF biosafety activities. UNEP/GEF Biosafety Projects (CD-Rom distributed at COP-MOP-2).

United States. 2004. European communities: Measures affecting the approval and marketing of biotech products (WT/DS291, 292, 293). First submission of the United States, 21 April.

USDA (United States Department of Agriculture). 2003a. US and cooperating countries file WTO case against EU moratorium on biotech foods and crops. Press release 0156-03. 13 May.

–. 2003b. United States requests dispute panel in WTO challenge to EU biotech moratorium. Press release 03-54. 7 August.

–. 2005a. France biotechnology exploring coexistence. USDA Foreign Agricultural Service GAIN Report FR5084, 1 December. http://www.fas.usda.gov/gainfiles/200512/146131663.pdf.

–. 2005b. Peru biotechnology annual 2005. USDA Foreign Agricultural Service, GAIN Report PE5012, 7 July. Quoted in Garton, Falkner, and Tarasofsky 2006, 5.

US FDA (US Food and Drug Administration). 1992. Statement of policy: Foods derived from new plant varieties. *Federal Register* 57(104): 22984-23001.

–. 1995. FDA's policy for foods developed by biotechnology. http://vm.cfsan.fda.gov/~lrd/biopolcy.html.

US Supreme Court. 1980. *Diamond v. Chakrabarty*. 447 U.S. 303. http://caselaw.lp.findlaw.com/scripts/getcase.pl?court=US&vol=447&invol=303.

USTR (United States Trade Representative). 1998. United States and European Union conclude joint action plan for the Transatlantic Economic Partnership. Office of the United States Trade Representative, 9 November. http://www.ustr.gov/regions/eu-med/westeur/98-99.pdf.

Villar, Juan Lopez. 2002. U.S. steps up pressure to force acceptance of genetically modified organisms worldwide. Friends of the Earth International Briefing Paper, April. (on file with author)

Vint, Robert. 2002. Force-feeding the world. UK*abc*. http://www.UKabc.org/forcefeeding.htm.

Vogler, John. 1996. The environment in international relations: Legacies and contentions. In *The Environment and International Relations,* ed. John Vogler and Mark F. Imber, 1-21. London: Routledge.

von Hayek, F.A. 1976. *The Road to Serfdom*. London: Routledge.

von Shomberg, Rene. 1998. An appraisal of the working in practice of Directive 90/220/EEC on the deliberate release of genetically modified organisms. Scientific and Technical Office of Assessment (STOA) of the European Parliament, 2 January. (on file with author)

Wallström, Margot. 2002. European Commission. In Bail, Falkner, and Marquard 2002, 244-50.

Waltz, Kenneth. 1979. *Theory of International Politics*. Reading, MA: Addison-Wesley.

Wapner, Paul. 1996. *Environmental Activists and World Civic Politics*. Albany: State University of New York Press.

WCED (World Commission on Environment and Development). 1987. *Our Common Future*. New York: Oxford University Press.

White, Kenneth. 2000. Economic profile of the biotechnology sector. Paper prepared for the Canadian Biotechnology Advisory Committee Project Steering Committee on Intellectual Property and the Patenting of Higher Life Forms, Ottawa.

WHO (World Health Organization). 1975. Certification scheme on the quality of pharmaceutical products moving in international commerce. Twenty-eighth World Health Assembly, Geneva. Official Records of the WHO No. 226. http://www.who.int/medicines/organization/qsm/activities/drugregul/certification/certifscheme.shtml.

Wiener, Jonathan B., and Michael D. Rogers. 2002. Comparing precaution in the United States and Europe. *Journal of Risk Research* 5(4): 317-49.

Winfield, Mark. 1997. *Report on the Second Meeting of Biosafety Protocol Working Group.* Toronto: Canadian Institute for Environmental Law and Policy. (on file with author)

–. 1998. *Summary of Key Issues in the Negotiations on the Protocol on Biosafety under the Convention on Biological Diversity.* Toronto: Canadian Institute of Environmental Law and Policy. (on file with author)

–. 1999. Notes on June 10 Biosafety Protocol Advisory Committee Meeting. Toronto: Canadian Institute for Environmental Law and Policy. (on file with author)

–. 2000. *Reflections on the Biosafety Protocol Negotiations in Montreal.* Toronto: Canadian Institute for Environmental Law and Policy. (on file with author)

Winson, Anthony. 1993. *The Intimate Commodity: Food and the Development of the Agro-Industrial Complex in Canada.* Toronto: Garamond Press.

WRI (World Resources Institute). 1992. Global biodiversity strategy: Guidelines for action to save, study and use earth's biotic wealth sustainably and equitably. World Resources Institute.

Wright, Susan. 1994. *Molecular Politics: Developing American and British Regulatory Policy for Genetic Engineering, 1972-1982.* Chicago: University of Chicago Press.

WTO (World Trade Organization). 1994a. Agreement on Sanitary and Phytosanitary Measures. Geneva: World Trade Organization. http://www.wto.org/english/docs_c/legal e/ legal_e.htm.

–. 1994b. Agreement on Technical Barriers to Trade. Geneva: World Trade Organization. http://www.wto.org/english/docs_e/legal_e/17-tbt.pdf.

–. 1994c. General Agreement on Tariffs and Trade. Geneva: World Trade Organization. http://www.wto.org/english/docs_e/legal_e/gatt.pdf.

–. 1998a. EC – measures concerning meat and meat products (hormones). Report of the Appellate Body (AB-1997-4). Geneva: World Trade Organization.

–. 1998b. United States – import prohibition of certain shrimp and shrimp products. Report of the Appellate Body (AB-1998-4). Geneva: World Trade Organization.

–. 1999. Japan – measures affecting agricultural products. Report of the Appellate Body (AB-1998-8). Geneva: World Trade Organization.

–. 2006. European Communities – measures affecting the approval and marketing of biotech products. Interim reports of the panel (WT/DS291/INTERIM, WT/DS292/INTERIM, WT/DS293/INTERIM). Geneva: World Trade Organization.

Wynne, B. 1989. Sheepfarming after Chernobyl. *Environment* 31(2): 10-39.

–. 1994. Scientific knowledge and the global environment. In *Social Theory and the Global Environment,* ed. Michael Redclift and Ted Benton, 169-89. London: Routledge.

Young, Oran R., ed. 1997. *Global Governance: Drawing Insights from the Environmental Experience.* Cambridge, MA: MIT Press.

–. 2001. Inferences and indices: Evaluating the effectiveness of international environmental regimes. *Global Environmental Politics* 1(1): 99-121.

–. 2002. *The Institutional Dimensions of Environmental Change: Fit, Interplay and Scale.* Cambridge, MA: MIT Press.

Young, T. 2006. Commentary: Cartagena Protocol on Biosafety MOP-3. *BRIDGES Trade BioRes* 6(6): 7.

Zarrilli, Simonetta. 2005. International trade in GMOs and GM products: National and multilateral legal frameworks. Policy issues in international trade in commodities. Study Series 29. New York and Geneva: United Nations Conference on Trade and Development.

Zedan, Hamdallah. 2002. The road to the biosafety protocol. In Bail, Falkner and Marquard 2002, 23-33.

Index

Note: Figures are indicated by *f* following the page number.